Molecular Biology and its Application to Medical Mycology

NATO ASI Series

Advanced Science Institutes Series

A series presenting the results of activities sponsored by the NATO Science Committee, which aims at the dissemination of advanced scientific and technological knowledge, with a view to strengthening links between scientific communities.

The Series is published by an international board of publishers in conjunction with the NATO Scientific Affairs Division

A Life Sciences B Physics	Plenum Publishing Corporation London and New York
C Mathematical and Physical Sciences D Behavioural and Social Sciences E Applied Sciences	Kluwer Academic Publishers Dordrecht, Boston and London
F Computer and Systems Sciences G Ecological Sciences H Cell Biology I Global Environmental Change	Springer-Verlag Berlin Heidelberg New York London Paris Tokyo Hong Kong Barcelona Budapest

NATO-PCO DATABASE

The electronic index to the NATO ASI Series provides full bibliographical references (with keywords and/or abstracts) to more than 30000 contributions from international scientists published in all sections of the NATO ASI Series. Access to the NATO-PCO DATABASE compiled by the NATO Publication Coordination Office is possible in two ways:

- via online FILE 128 (NATO-PCO DATABASE) hosted by ESRIN, Via Galileo Galilei, I-00044 Frascati, Italy.
- via CD-ROM "NATO-PCO DATABASE" with user-friendly retrieval software in English, French and German (© WTV GmbH and DATAWARE Technologies Inc. 1989).

The CD-ROM can be ordered through any member of the Board of Publishers or through NATO-PCO, Overijse, Belgium.

Series H: Cell Biology, Vol. 69

Molecular Biology and its Application to Medical Mycology

Edited by

Bruno Maresca

International Institute of Genetics and Biophysics
Via Marconi 12, 80125 Naples, Italy

George S. Kobayashi

Washington University School of Medicine
Division of Infectious Diseases
St. Louis, Missouri 63110, USA

Hideyo Yamaguchi

Research Center for Medical Mycology
Teikyo University School of Medicine
359 Otsuka, Hachioji-shi, Tokyo 192-03, Japan

Springer-Verlag
Berlin Heidelberg New York London Paris Tokyo
Hong Kong Barcelona Budapest
Published in cooperation with NATO Scientific Affairs Division

Proceedings of the NATO Advanced Research Workshop on Molecular Biology and its Application to Medical Mycology, held at Taormina, Italy, January 6–8, 1992

Book design by Donatella Capone, Secretary IIGB Press, Naples, Italy

ISBN-13:978-3-642-84627-4 e-ISBN-13:978-3-642-84625-0
DOI: 10.1007/978-3-642-84625-0

Library of Congress Cataloging-in-Publication Data
Molecular biology and its application to medical mycology / edited by Bruno Maresca, and George S Kobayashi, and Hideyo Yamaguchi
(NATO ASI series Series H, Cell biology , vol 69)
"Proceedings of the NATO Advanced Research Workshop on Molecular Biology and Its Application to Medical Mycology, held at Taormina, Italy, January 6-8, 1992" – T p verso Includes bibliographical references and index

1 Medical mycology–Congresses 2 Molecular biology–Congresses 3 Mycoses–Molecular aspects–Congresses I Maresca, B (Bruno), 1950- II Kobayashi, George S , 1927- III Yamaguchi, Hideyo, 1934- IV North Atlantic Treaty Organization Scientific Affairs Division V NATO Advanced Research Workshop on Molecular Biology and Its Application to Medical Mycology (1992 Taormina, Italy) VI Series
[DNLM 1 Fungi–genetics–congresses 2 Molecular Biology–congresses 3 Morphogenesis–congresses 4 Mycology–congresses 5 Yeasts–genetics–congresses QW 180 M718 1992] QR245 M645 1993 616' 015–dc20

Softcover reprint of the hardcover 1st edition 1993

Typesetting Camera ready by author
Book design Donatella Capone, Secretary IIGB Press, Naples, Italy
31/3145 - 5 4 3 2 1 0 - Printed on acid-free paper

Preface

With a spectacular view of a grumbling and lava spewing mount Etna in the background, the NATO sponsored workshop "Molecular Biology and its Application to Medical Mycology" provided a stimulating forum for a group of researchers who work with fungi pathogenic for humans to interact and discuss problems with molecular biologists who use non-pathogenic fungi primarily as laboratory models for investigation. The workshop was held January 6-8, 1992 in the Congress Center of the San Domenico Palace Hotel, a former monastery dating back to 1430 located in Taormina, situated high above the coastline of the Mediterranean sea overlooking Isola Bella on the island of Sicily. In this setting the two groups were encouraged to participate in active discussions and exchange of ideas. Thirty invited papers and one poster session were presented in the three day workshop that was divided into six sessions covering: (1) the molecular biology of yeasts; (2) the molecular biology of filamentous fungi; (3) fungal morphogenesis; and (4) the use and application of the newer molecular biology tools in the study of medically important fungi. Each session was designed to include studies using pathogenic and non-pathogenic fungi as laboratory models.

The opening presentation by Sergio Moreno from Paul Nurse's group at Oxford University set the standard for the formal presentations. In his paper "Cell cycle regulation in fission yeast", he showed that the cdc2^{+} *gene of* Schizosaccharomyces pombe *encodes a 34kDa phosphoprotein with protein serine/threonine kinase activity. He described current molecular biology studies on its role in mitosis during cell cycle of this yeast. This talk was followed by Claudina Roudrigues-Pousada from Oeiras, Portugal who described her studies on acquisition of thermotolerance in* Saccharomyces cerevisiae *using 1,10-phenantroline an iron chelating compound that induces arrest of growth at "start", during which time they acquire thermotolerance without producing a full set of heat shock proteins. Bridging the gap between studies on non-pathogenic fungi and pathogens was the paper given by Gerald Fink of the Whitehead Institute in Boston, Massachusetts. He described his thought provoking studies with nutritionally deprived diploid strains of* S. cerevisiae *that undergo changes in cell shape and pattern of cell division that results in filamentous growth. These observations implied some very important implications for animal pathogenesis in that filamentation in yeasts of the genus* Candida *is often correlated with the process of pathogenicity. The difficulty of using classical genetic technics to investigate yeast belonging to the genus* Candida *was emphasized by Masamichi Takagi from the University of Tokyo with his strategy of reverse genetics by using recombinant DNA technology. Using technics of molecular cloning, Richard Calderone of Georgetown University School of Medicine described the isolation of a gene fragment from* C. albicans *which shows homology with human complement receptor (CR2) gene. Closing the first session was Jacob Reiser from the Swiss Federal Institute of Technology of Zurich, Switzerland who described his molecular biological studies with the genus* Trichosporon, *which include yeast-like fungi that have been implicated as opportunistic pathogens of humans.*

The contributions presented in the two sessions on filamentous fungi focused mainly on studies using Aspergillus nidulans *and* Neurospora crassa. *These studies clearly illustrated the advantages these fungi offer over those that cause disease in man and animals particularly in the area of molecular genetics. Based on advances made in this area, Yuzuru Iimura described his studies on the development of molecular genetics systems for several industrial species including* A. niger *and* A. oryzae. *Studies from Geoffrey Turner's laboratory on expression of genes for the biosynthesis of penicillin further support the contention that rapid basic and molecular biological advances are made when the problem involves an industrially important metabolite particularly when it is produced by a non-pathogen. The exception, however is in the area of phytopathology where Flora Banuett described her elegant studies on* Ustilago maydis. *Her work and those of others who work with plant pathogens clearly point out the deficiencies and paucity of information that exists when it comes to similar studies on the molecular biology involving fungi that cause disease of humans and animals.*

In the second session, molecular biological studies with yeast, it became evident that great inroads are being made in similar areas with medically important fungi. Several investigators have been motivated to study Candida albicans *and* Cryptococcus neoformans *because of important clinical questions that remain unanswered concerning host-parasite interaction and the paucity of information on its virulence, and pathogenesis. Jim Cutler described his studies on the characterization of* C. albicans-*specific DNA fragment that may be useful diagnostically. Interest in and studies with* C. neoformans *are rapidly increasing and a great deal of progress has been made in our knowledge of the genetics and biochemistry of this opportunistic pathogen. Jeffrey Edman reviewed work on the nature of DNA-mediated transformation and the usefulness of this fungus in studies of host-parasite interaction.*

William Timberlake discussed transcriptional regulators of morphogenetic genes in Aspergillus *and emphasized that genetic and molecular biological studies with this group have been facilitated because these fungi have well defined sexual phases, stable mutants are available and easy to generate, efficient transformation systems exist, and the biochemistry of these organisms has been extensively studied. While research using non-pathogenic filamentous fungi and yeast as experimental models have contributed a greatly to our knowledge of molecular genetics and biology, none of these organisms cause disease. A great deal of basic information was presented on* Histoplasma capsulatum *and* Wangiella dermatitidis *in the session on morphogenesis.*

Bruno Maresca of IIGB, Naples, emphasized that transition of mycelia to yeast is necessary for pathogenesis and showed that when H. capsulatum *goes from an ambient environment in nature to 37°C of the host heat shock genes are turned on. Furthermore he provided data to show that the phospholipid composition differs in strain of* H. capsulatum *that have different susceptibilities to temperature and virulence and the difference unsaturated fatty acids in these strains could explain why each strain has a different capacity to adapt to the new environment during host invasion. In contrast to the dimorphic nature of* H. capsulatum, *the dematiaceous fungus* W. dermatitidis *is polymorphic. it can grow as a budding yeast, a filamentous mold, or as a multicellular mass. The findings that chitin content increases and deposition is delocalized during yeast-to-multicellular form transition in* W. dermatitidis *and that* S. cerevisiae *possessed several chitin synthases led Paul Szaniszlo from the University of Texas, Austin, to determine whether* W. dermatitidis *possesses multiple chitin synthases. These studies were described in the session on morphogenesis.*

It became apparent in the session on prospectives for medical mycology that the newer technologies of molecular biology are being applied to the study of fungi pathogenic for humans and animals. The studies of Antonio Cassone from Istituto Superiore di Sanità, Rome, and those of Somay Murayama of Teikyo University of Tokyo underscored the current needs for tools that can rapidly and with specificity diagnose the life threatening fungal infections. In addition to the need for these tests, it was apparent from the presentation of Gerald Medoff that the current available chemotherapeutic agents used to treat these infections are inadequate and that newer strategies need to be considered for development of more effective antibiotics. Hugo Vanden Bossche, Steven Kelly reviewed the current approach and targets that are being emphasized and Hideyo Yamaguchi described newer fungal targets that need to be exploited.

The workshop was concluded with a round table discussion on the topic of integrating the technics of molecular biology to the needs of medical mycology. Audience participation was active, with constructive dialogue between those with expertise in molecular biology and researchers who study medically important fungi. During this session, it became evident that many molecular biological technics described in standard handbooks are not directly applicable to the study of pathogenic fungi and that a need exists to focus on the uniqueness of fungal pathogens of humans and animals. This NATO sponsored workshop provided the proper forum to address these problems by bringing together groups of scientist who study fungi with disparate interests to discuss the shortcomings and needs of clinically relevant problems involving pathogenic fungi. In summary, the participants were unanimous in their opinion that the workshop achieved its goals and recommended that future meetings be held every other year in order to update interested parties on research progress and advance in newer technics that have been developed.

Bruno Maresca, George Kobayashi and Hideyo Yamaguchi

First row from left: Bill Timberlake, Judith Medoff, Sergio Moreno, Bruno Maresca, Gerry Fink.
Second row from left: Robert Brambl, Paul Szaniszlo, Ray Dixon, Richard Calderone, Jerry Medoff.
Third row from left: IIGB group of Naples, Hugo Vanden Bossche, George Kobayashi, Hideyo Yamaguchi.

The NATO Workshop was also supported by

Consiglio Nazionale delle Ricerche, Italy

Regione Siciliana, Italy

Istituto Italiano per gli Studi Filosofici, Naples, Italy

Pfizer Pharmaceuticals Inc., Tokyo, Japan

Tsumura & Co., Tokyo, Japan

Bayer Yakuhin, Ltd., Osaka, Japan

Nichinichi Foods Co., Ltd., Mie-ken, Japan

Sandoz Pharmaceuticals, Ltd., Tokyo, Japan

Boehring-Pfizer, U.S.A.

Monsanto, U.S.A.

Abbott Labs, U.S.A.

Contents

Molecular Biology of Yeast

Molecular Biology of Filamentous Fungi

Contributors

The numbers in brackets are the opening page numbers of the contributors' articles.

Robert Brambl, University of Minnesota, Dept. of Plant Biology, 220 Biological Sciences, 1445 Gartner Ave., St. Paul, 55108-1095 MN USA, **(105)**

Richard Calderone, Dept. of Microbiology Georgetown University School of Medicine, Washington D.C. 20007 USA, **(23)**

Antonio Cassone, Istituto Superiore di Sanità, Divisione di Micologia Medica, Viale Regina Elena, 299, Roma 00161 Italy, **(159)**

Jim E. Cutler, Department of Microbiology, Montana State University, Bozeman 59717, Montana USA, **(47)**

Jeffrey C. Edman, Hormone Res. Institute, University of California School of Medicine San Francisco 94143-0534 CA USA, **(53)**

Laura Frontali, Università di Roma "La Sapienza" Dip. di Biol Cell e Sviluppo-Sez. di Scienze Fisiol., Piazzale Aldo Moro, Roma 00195 Italy, **(63)**

John Galgiani, University of Arizona Medical Center, Sec. of Infectious Diseases, Tucson 85723, Arizona USA, **(221)**

Yuzuru Iimura, National Research of Brewing, 2-6-30 Takinogawa, Kita-ku Tokyo 114 Japan, **(87)**

Steven L. Kelly, The University of Sheffield Dept. of Mol. Biol. & Biotechnology, P.O. Box 594 Firth Court Western Bank, Sheffield S10 2UH U.K., **(199)**

Giuseppe Macino, Dept. of Biopathology , Policlinico, University of Rome "La Sapienza", Viale Regina Elena, 324, Rome 00161 Italy, **(117)**

Bruno Maresca, International Institute of Genetics and Biophysics, Via Marconi, 12, Naples 80125 Italy, **(243)**

Gerald Medoff, Department of Infectious Deseases, Washington University, Box 8051, St. Louis 63110, MO USA, **(215)**

Sergio Moreno, Dept. of Biochemistry University of Oxford, South Parks Road, Oxford OX1 3QU U.K., **(3)**

N. Ronald Morris, Dept. of Pharmacology, UMDNJ-RWJMS, Piscataway 08854-5635, NJ USA, **(97)**

Somay Y. Murayama, Dept. of Microbiology, Teikyo University School of Medicine, 2-11-1 Kaga Itabashi-ku Tokyo 173 Japan, **(149)**

Yoshinori Ohsumi, Dept. of Biology, College of Arts and Sciences, The University of Tokyo, Komaba, Meguro-ku Tokyo 153 Japan, **(39)**

Luciano Polonelli, Istituto di Microbiologia Facoltà di Medicina e Chirurgia, Via Gramsci, 14, Parma 40300 Italy, **(141)**

John Pulitzer, International Institute of Genetics and Biophysics, Via Marconi, 12, Naples 80125 Italy, **(75)**

Jakob Reiser, Institut für Biotechnologie ETH-Hönggerberg, Zürich CH-8093 Switzerland, **(31)**

Claudina Rodrigues-Pousada, Instituto Gulbenkian de Ciencia, Rua da Quinta Grande Apartado 14, Oeiras Codex 2781 Portugal, **(7)**

Paul E. Steele, Dept. of Pathology School of Medicine, University of Cincinnatti, 231 Bethesda Ave., Cincinnatti 45267-0529, OH USA, **(227)**

Paul J. Szaniszlo, Dept. of Microbiology, University of Texas Austin, Austin 78712, TX USA, **(229)**

Masamichi Takagi, Dept. of Agricultural Chemistry, The University of Tokyo, Yayoi, Bunkyo-ku Tokyo 113 Japan, **(13)**

Geoffrey Turner, The University of Sheffield , Dept. of Molecular Biology and Biotech., Sheffield S10 2TN U.K., **(125)**

Hugo Vanden Bossche, Dept. of Comparative Biochem., Janssen Res. Foundation, Turnhoutseweg 30, Beerse B-2340 Belgium, **(179)**

Hideyo Yamaguchi, Res. Center for Medical Mycology, Teikyo University School of Medicine, 359 Otsuka, Tokyo 192-03 Hachioji Japan, **(169)**

Molecular Biology of Yeasts

Cell cycle regulation in fission yeast

S. Moreno and P. Nurse
ICRF Cell Cycle Group, Microbiology Unit
Department of Biochemistry
Oxford OX1 3QU
U. K.

Introduction

The cell cycle has undergone major advances in the last four years. A better understanding of how cells divide has been possible thanks to the fusion of two independent areas of research, one exploiting fungal genetics and molecular biology in *Schizosaccharomyces pombe*, *Saccharomyces cerevisiae* and *Aspergillus nidulans*, and the second based on biochemistry in *Xenopus* and mammalian cell lines. Here we review how the genes involved in the control of the cell cycle in fission yeast were discovered and how their gene products interact with each other.

Cell cycle mutants

In *S. pombe* like any other eukaryotic organism the cell cycle consists of S phase when the DNA becomes replicated, followed by the segregation of sister chromatids during mitosis. In between these two phases there are two gaps named G1 and G2. Two kinds of mutants have been identified in *S. pombe* that define functions important for controlling the cell cycle:
1. Recesive temperature sensitive *cdc* (cell division cycle) mutants defective in progression through the cell cycle but which carry on growing and, therefore become elongated at the restrictive temperature.
2. The *wee* mutants, are dominant mutants that divide at approximately half the size of the wild type and define functions that are rate limiting for the initiation of mitosis.
Mutations in around 30 different genes can cause a *cdc* phenotype and mutants arrest at different points in the cell cycle. However originally, only mutations in two genes were found to be able to generate the wee phenotype: *wee1+* and *wee2+*. Interestingly, *wee2+* was found to be allelic to a *cdc* gene, *cdc2+*. Furthermore, the cdc2 function was shown to be required twice in the cell cycle at the transitions G1/S and G2/M (Nurse and Bissett, 1981).
The *cdc2+* gene was cloned by complementation of the temperature sensitive strain and shown to encode a 34kd protein kinase. This gene is functionally and structurally homologous to the *S. cerevisiae CDC28* gene, and is present in all eukaryotic cells. The cdc2 protein forms a homodimer with another protein called cyclin B (the product of the gene *cdc13+* in *S. pombe*); this complex has been shown to be the same as MPF (M-phase Promoting Factor) in

NATO ASI Series, Vol. H 69
Molecular Biology and its Application to Medical Mycology
Edited by B. Maresca, G. S. Kobayashi and H. Yamaguchi

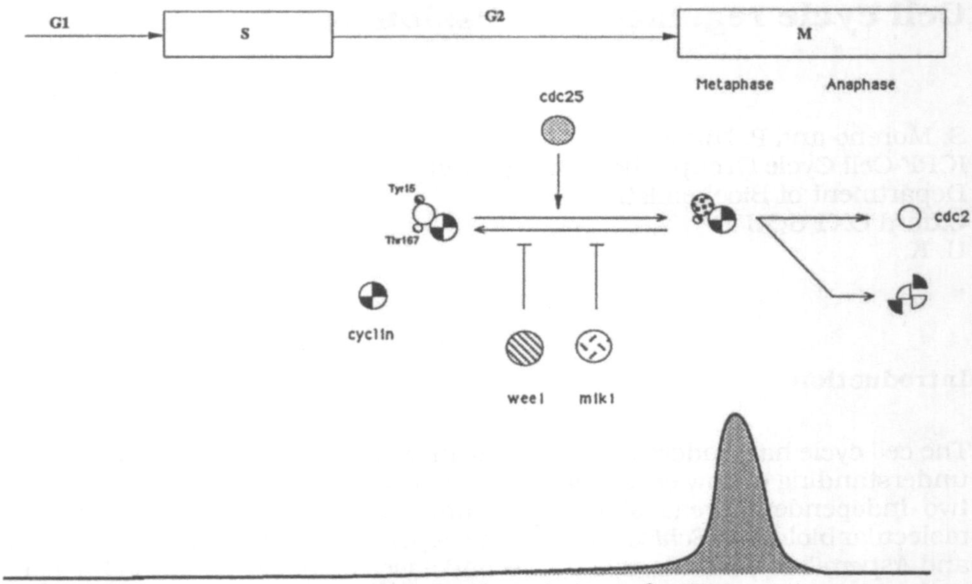

Figure 1. Regulation of cdc2 activity at the initiation of mitosis.

Xenopus and the growth associated histone H1 kinase (Arion *et al.*, 1988; Dunphy *et al.*, 1988; Gautier et al., 1988; Labbe *et al.*, 1988). The cdc2 protein kinase activity is low during interphase and high at the beginning of mitosis (Draetta and Beach, 1988; Booher *et al.*, 1989; Moreno *et al.*, 1989). This periodic activation of cdc2 in each cell cycle promotes entry into mitosis. A number of proteins, such as lamins and histone H1, have been shown to be substrates of cdc2 at mitosis (review in Draetta, 1990; Moreno and Nurse, 1990).

How is the activity of the cdc2/cyclin complex regulated through the cell cycle?

Association between cdc2 and cyclin B is necessary for the activity of the complex, and inactivation correlates with cyclin destruction at anaphase (Booher *et al.*, 1989; Moreno *et al.*, 1989). However, the precise timing of cdc2 activation is regulated by the phosphorylation state of the cdc2 protein. In *S. pombe* cdc2 is phosphorylated on two residues: Thr167 and Tyr15 (Gould and Nurse, 1989; Gould *et al.*, 1991). Phosphorylation of Thr167 is required for activity and might play a role in stabilising the association between cdc2 and cyclinB (Gould *et al.*, 1991). On the other hand, phosphorylation of the Tyr15 inhibits cdc2 activity even when is bound to cyclin B. Dephosphorylation of this residue promotes activation of the cdc2/cyclin B complex.
The protein phosphatase that catalyses this dephosphorylation event is the product of the *cdc25+* gene. Recently, it has been shown that cdc25 shares a limited homology to protein tyrosine phosphatases and is able to dephosphorylate cdc2 on the Tyr15 *in vitro* (Moreno and Nurse, 1991; Strausfeld *et al.*, 1991; Gautier *et al.*, 1991). The products of the genes *wee1+* and *mik1+* counteract cdc25 activity and together maintain cdc2 in the

tyrosine phosphorylated, inactive state (Russell and Nurse, 1987; Lundgren *et al.*, 1991). Single mutants of these genes are viable; however, in double mutants cdc2 is not phosphorylated on Tyr15 and cells try to enter into mitosis from any point in the cell cycle. This phenotype is similar to the cells containing a mutant form of cdc2 in which the Tyr15 has been mutated to Phe15 or cells overexpressing cdc25 to very high levels (Gould and Nurse, 1989). Taken together, these experiments indicate that tyrosine phosphorylation of cdc2 is important for coupling initiation of mitosis to completion of the previous DNA replication (Enoch and Nurse, 1991).

References

Arion, D, Meijer, L, Brizuela, L, Beach, D (1988) cdc2 is a component of the M phase-specific histone H1 kinase: evidence for identity with MPF. Cell 55: 371-378.
Booher, R N, Alfa, C E, Hyams, J S, Beach, D H (1989) The fission yeast cdc2/cdc13/suc1 protein kinase: regulation of catalytic activity and nuclear localization. Cell 58: 485-497.
Draetta, G (1990) Cell cycle control in eukaryotes: molecular mechanisms of cdc2 activation. Trends Biochem Sci 15: 378-83.
Draetta, G , Beach, D (1988) Activation of cdc2 protein kinase during mitosis in human cells: cell cycle-dependent phosphorylation and subunit rearrangement. Cell 54: 17-26.
Dunphy, W G, Brizuela, L, Beach, D, Newport, J (1988) The Xenopus homolog of cdc2 is a component of MPF, a cytoplasmic regulator of mitosis. Cell 54: 423-431.
Dunphy, W G, Kumagai A (1991) The cdc25 protein contains an intrinsic phosphatase activity. Cell: 189-196.
Enoch, T, Nurse, P (1991) Coupling M phase and S phase: Controls maintaining the dependence of mitosis on chromosome replication. Cell 65: 921-923.
Gautier, J, Norbury, C, Lohka, M, Nurse, P, Maller, J (1988) Purified maturation-promoting factor contains the product of a Xenopus homolog of the fission yeast cell cycle control gene cdc2+. Cell 54: 433-439.
Gautier, J, Solomon, M J, Booher, R N, Bazan, J F, Kirschner, M W (1991) cdc25 is a specific tyrisine phosphatase that directly activates p34cdc2. Cell 67: 197-211.
Gould, K L, Moreno, S, Owen, D J, Sazer, S, Nurse, P (1991) Phosphorylation at Thr 167 is required for Schizosaccharomyces pombe p34cdc2 function. EMBO J. 10: 3297-3309.
Gould, K L , Nurse, P (1989) Tyrosine phosphorylation of the fission yeast cdc2 protein kinase regulates entry into mitosis. Nature 342: 39-45.
Labbe, J C, Lee, M G, Nurse, P, Picard, A, Doree, M (1988) Activation at M-phase of a protein kinase encoded by a starfish homologue of the cell cycle control gene cdc2+. Nature 335: 251-254.
Lundgren, K, Walworth, N, Booher, R, Dembski, M, Kirschner, M, Beach, D (1991) mik1 and wee1 cooperate in the inhibitory tyrosine phosphorylation of cdc2. Cell 64: 1111-22.
Moreno, S, Hayles, J, Nurse, P (1989) Regulation of p34cdc2 protein kinase during mitosis. Cell 58: 361-372.
Moreno, S , Nurse, P (1990) Substrates for p34cdc2: in vivo veritas? Cell 61: 549-551.
Moreno, S, Nurse, P (1991) Clues to action of cdc25 protein. Nature 351: 194.
Nurse, P, Bissett, Y (1981) Gene required in Gl for commitment to cell cycle and in G2 for control of mitosis in fission yeast. Nature 292: 558-560.
Russell, P, Nurse, P (1987) Negative regulation of mitosis by wee1+ a gene encoding a protein kinase homolog. Cell 49: 559-567.
Strausfeld, U, Labbé, J-C, Fesquet, D, Cavadore, J C, Picard, A, Sadhu, K, Russell, P, Dorée, M (1991) Dephosphorylation and activation of a p34cdc2/cyclin B complex in vitro by human CDC25 protein. Nature 351: 242-245.

Yeast genes overcoming growth arrest induced by 1,10-phenantroline

P. Bossier, L. Fernandes and C. Rodrigues-Pousada
Instituto Gulbenkian de Ciência
Laboratório Genética Molecular
Apartado 14
2781 Oeiras Codex
Portugal

Introduction

In the yeast *Saccharomyces cerevisiae* various physiological effects can be observed upon the addition of 1,10-phenantroline to a culture. 1,10-phenantroline is an ion chelator with a very high affinity for Fe^{2+}, Zn^{2+} and Cu^{2+}. The stability constants of the respective complexes are 21, 17 and 18 (Callander and Barford, 1983). Because of this characteristic 1,10-phenantroline can be used to deprive microorganisms from these essential oligo-elements. In the presence of 100 to 250 µg/ml a total block mRNA synthesis is provoked within a period of maximum 10 min. This feature has often been exploited in the study of mRNA stability (Santiago *et al.*, 1986). In the presence of 5 to 20 µg/ml the effects are less drastic. Veinot-Drebot *et al.* (1989) have noticed an inhibition of pre-rRNA synthesis with very little effect upon poly(A)$^+$RNA synthesis. Concomitantly the culture arrests at "start" and the cells are accumulating in the unbudded state. During this process cells acquire thermotolerance (Barnes *et al.*, 1990) apparently without the production of the full set of heat shock proteins.
So far only one gene (HSP104) has been positively identified as being involved in the acquisition of thermotolerance in yeast (Sanchez and Lindquist, 1990). We are studying the latter phenomenon using the drug 1,10-phenantroline. We set out to clone yeast genes which at high copy number are able to overcome growth arrest induced by 1,10-phenantroline. It was hoped that this approach will yield genes which are involved in regulation of the transition into stationary phase and the acquisition of thermotolerance.

Results and discussion

A yeast library into the *LEU2*-based high copy number vector pMA3a was used to transform the yeast strain W303,α. The resulting transformants were screened for the ability to grow in the presence of 1,10-phenantroline. So far this search has yielded two characterized genes.
The gene in the first positive clone, responsible for the above-described phenotype, was found to be identical to the previously described *YAP*1 gene (Moye-Rowley *et al.*, 1989). The *YAP*1 gene codes for a *trans*-acting factor, the yAP-1 protein (MW: 72.5 kDa), which binds to the ARE element with the

NATO ASI Series, Vol. H 69
Molecular Biology and its Application to Medical Mycology
Edited by B. Maresca, G. S. Kobayashi and H. Yamaguchi

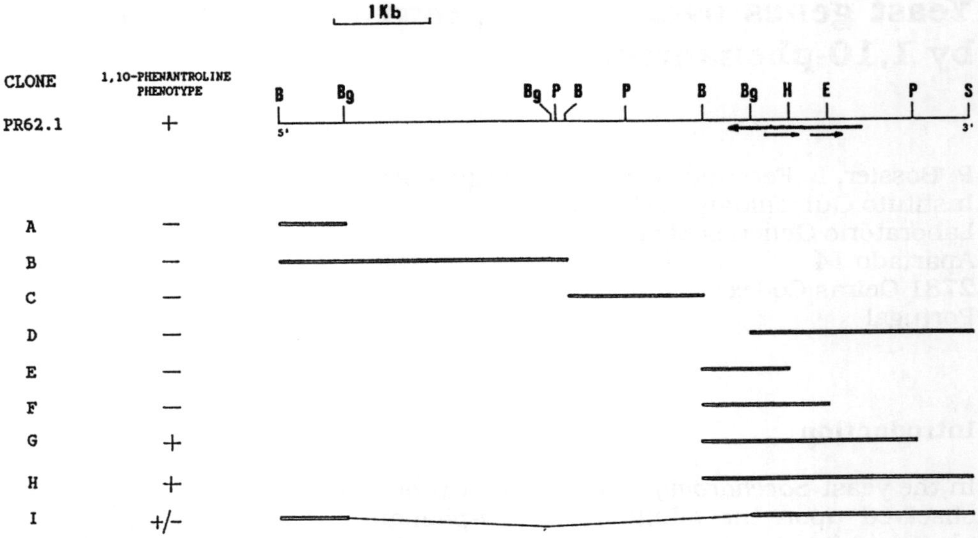

Figure 1. Restriction map of clone PR62.1 and the ability of various subclones of it back into pMA3a to confer growth in the presence of 1,10-phenantroline. The arrows indicate ORF's found in subclone PR62.1H as determined by sequencing. B: *BamHI*; Bg: *BglII*; E: *EcoRI*; H: *HindIII*; P: *PstI*; S: *Sau3AI*.

sequence C/GTGACTC/A. The DNA binding domain in yAP-1 consists of a stretch of amino acids rich in basic residues, followed immediately, in the direction of the C-terminal, by a Leu-Zipper. The Leu-Zipper is responsible for the dimerization of the protein and consists of leucine residues regularly spaced every seventh amino acid.

The gene in the second positive clone, designated PR62.1, was localized by subcloning fragments of it back into pMA3a (Fig. 1). Subclone PR62.1H was found, upon sequencing, to contain two small and one larger ORF. The possibility that the smaller ORF's would contain the gene responsible for the above-described phenotype was excluded through the inability of the yeast strain W303,α transformed with either PR62.1D, PR62.1E or PR62.1F to grow in the presence of 10 μg/ml 1,10-phenantroline (See Fig. 1).

The larger ORF codes for a protein with 416 residues and a potential MW of 45 kDa. The deduced amino acid sequence was used to search the protein data base for homologies. Surprisingly, the latter protein turned out to bear, in two domains, a high degree of homology to the yAP-1 protein (Fig. 2). The homology found in domain A (N-terminal in both proteins; Fig. 2A) corresponds to the DNA binding domain of yAP-1. The homology is very high in the stretch of amino acid rich in basic residues and is lower in the area of the Leu-Zipper, although the Leu residues making up the Leu-Zipper are conserved. The homology found in domain B (C-terminal in both proteins; Fig.2B) is of unknown function. Future research might establish whether or not domain B constitutes a nuclear targeting signal. The facts that both genes can be cloned in the same way and that they bear a high degree of homology in a DNA binding domain strongly suggest that the gene identified in the second

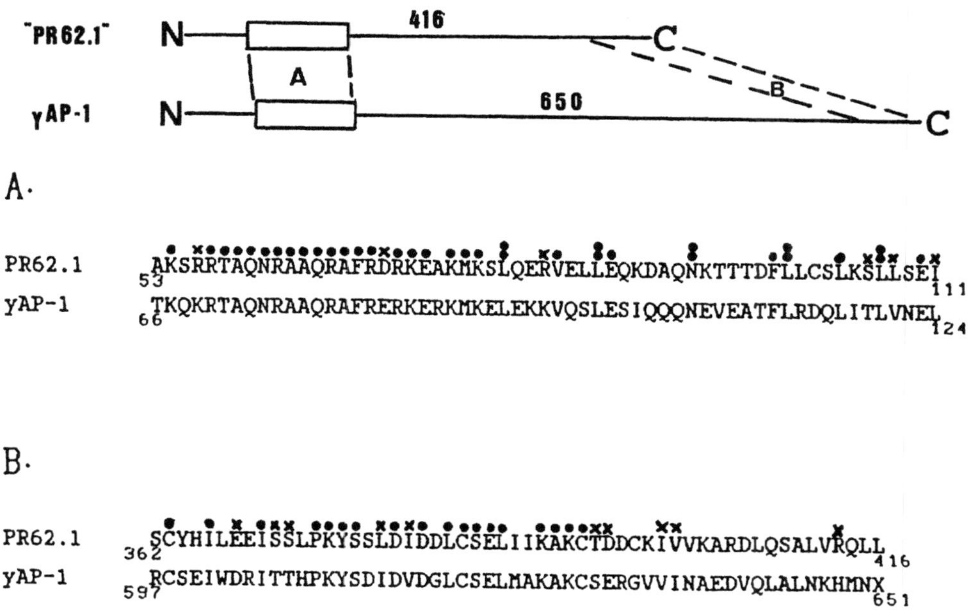

Figure 2. Homology between yAP-1 and the gene product identified in the clone PR62.1. The box indicates the position of the potential DNA binding domain in the two proteins. Closed circle: identical amino acid, crosses: conservative substitutions, double circles: the Leu residues of the Leu-Zipper.

clone is a *trans*-acting factor as well. It is interesting to speculate that both proteins would bind to the same *cis*-acting element. This speculation finds some support in the results of Moye-Rowley *et al.* (1989). They predicted, using DNA affinity blots, the existence of a second ARE-binding protein with an apparent molecular weight of 45 kDa, which corresponds exactly to the predicted molecular weight of protein coded for by the gene identified in the

	Concentration of cycloheximide (µg/ml)				
W303,α transformed with	0	0.5	1	2	5
pMA3a	++	-	-	-	-
pMA3a(*YAP*1)	++	++	++	++	+
pMA3a("PR62.1")	++	+	+	+/-	-

Table 1: Growth in cycloheximide containing plates after 4 days incubation at 30°C. Strain W303, α transformed with pMA3a is used as control. Plasmid pMA3a(*YAP*1) and pMA3a("PR62.1") are plasmids overexpressing respectively the *YAP*1 gene and the gene responsible for growth in the presence of 1,10-phenantroline identified in clone PR62.1. ++: good growth, as in absence of the chemical; +: reduced growth; +/-: growth visible retarded; - no growth.

Concentration of Cd (mM)

W303, α transformed with	0	0.05	0.1	0.2	0.4	0.8
pMA3a	++	-	-	-	-	-
pMA3a(*YAP*1)	++	+	+/-	+/-	-	-
pMA3a("PR62.1")	++	++	+	+	+/-	-

Table 2: Growth in cadmium containing plates after 4 days of incubation at 30°C. For further explanation see legend Table 1.

second clone.
Besides the ability of both genes to allow for growth in the presence of 1,10-phenantroline (10µg/ml), they also allow for growth at elevated concentrations of cycloheximide and cadmium (Tabel 1 and 2). These findings are remarkable in this sense that 1,10-phenantroline is primarily known as an inhibitor of transcription, while cycloheximide is known as a translational inhibitor. Future research will have to establish how this pleiotrophic phenotype is brought about.

PHE **CYCLO** **Cd**

YAP1

PR62.1

10 µg/ml **1 µg/ml** **0.2 mM**

Figure 3. Commensalistic nature of *YAP*1 and clone PR62.1 in 1,10-phenantroline containing plates. The yeast strain W303, α transformed either with pMA3a(*YAP*1) or pMA3a("PR62.1") was spotted in the middle of agar plates containing the chemicals at the indicated concentrations. At the same time W303, α transformed with pMA3a was streaked from the edge of the plate inwards. On plates containing the chemicals at the indicated concentrations W303, α transformed with pMA3a does not grow (See also Table 1 and 2). Only on 1,10-phenantroline containing plates W303, α transformed with pMA3a can benefit from the presence of the same strain overproducing the gene products of *YAP*1 and PR62.1. Phe: 1,10-phenantroline; cyclo:cycloheximide; Cd: cadmium.

During the course of this work, it was noticed that both genes, when present at high copy number in the cell, allow for the proliferation of control cells in their vicinity on 1,10-phenantroline containing plates (Fig. 3). This effect is absent in cycloheximide or cadmium containing plates. At this stage in the research we can only speculate on the mechanisms leading towards this commensalistic nature of both genes. It is possible that both gene products activate the transcription of a gene which codes for a secreted protein mobilizing oligo-elements in the medium.

Perspectives

The fact that both genes give rise to a very similar phenotype and that they are highly homologous in a potential DNA binding domain, supports the idea that they are both *trans*-acting factors, although for the gene in clone PR62.1 this will still have to be proven unambiguously. The pleiotrophic phenotype caused by both genes when overexpressed, rises both questions about their regulation and about the nature of their target genes. And finally, bearing in mind how these genes were cloned, we wonder how, if at all, they are involved in the regulation of the transition into stationary phase and in the regulation of the acquisition of thermotolerance.

Acknowledgements

The yeast DNA library was a generous gift from Dr. M.F. Tuite (University of Kent at Canterbury, England). Homology search was performed by Dr. J. Sgouros (MIPS, Martinsried, Germany). P. Bossier is the beneficiary of a post-doc grant from the Calouste Gulbenkian Foundation.

References

Barnes CA, Johnston GC, Singer RA, (1990) Thermotolerance is independent of induction of the full spectrum of heat shock proteins and of the cell cycle blockage in the yeast *Saccharomyces cerevisiae*, J Bacteriol 172: 4352-4358.

Callander IJ and Barford JP (1983) Precipitation, chelation and the availability of metals as nutrients in anaerobic digestion. I. Methodology. Biotech Bioeng 25: 1947-1957.

Moye-Rowley WS, Harshman KD, Parker CS (1989) Yeast *YAP1* encodes a novel form of the jun family of transcriptional activator proteins. Genes Devel 3: 283-292.

Sanchez Y and Lindquist S (1990) HSP104 required for induced thermotolerance. Science 248: 1112-1115.

Santiago TC, Purvis IJ, Bettany, AJE and Brown AJP (1986) The relationship between mRNA stability and length in *Saccharomyces cerevisiae*. Nucl Acids Res 14: 8347-8360.

Veinot-Drebot LM, Singer RA and Johnston GC (1989) rRNA transcription initiation is decreased by inhibitors of the yeast cell cylce control step "start". J Biol Chem 264: 19258-19534.

Reverse genetics in a non-conventional yeast, *Candida maltosa*

M.Takagi
Department of Agricultural Chemistry
The University of Tokyo
Bunkyo-ku,
Tokyo 113
Japan

Introduction

Most of eukaryotic microorganisms which are of industrial and clinical interest cannot be investigated by classical genetic analysis. Only the way we can promote genetic study with these microorganisms is the so-called reverse genetics by using recombinant DNA technology. The process of this approach consists of

i) a finding of a characteristic and interesting property (P) in a certain microorganism (M),
ii) isolation and structural analysis of a gene (G) related to (P),
iii) construction of a host-vector system (H-V) and gene disruption system(G-D) for (M),
iv) analysis of phenotype of a genetically engineered strain of (M) with a multicopied (G) or a disrupted (G) and
v) providing direct evidence that (G) has a specific function to determine (P) in (M).

We have been interested in two systems as follows, and here I will describe system-I only.

System-I

(M): *Candida maltosa*
(P): *n*-Alkane-hydroxylation by enzymes such as cytochrome P-450 and inducibility of biogenesis of organelles like ER and peroxisome and of proteins localized in these organelles; resistant to cycloheximide which is an antibiotic to inhibit protein biosynthesis in eukaryotes.
(G): Genes for cytochrome P-450, NADPH-cytochrome P-450-reductase and genes involved in induction phenomenon; genes for ribosomal protein L41.
(H-V) and (G-D): Plasmid vectors and integration vectors are available; a gene with two alleles can be disrupted.

System-II

(M): *Rhizopus* spp.
(P): Secretion of various enzymes in large amounts; rapid tip-growth of a long cell.
(G): Genes for acid protease and RNase; genes for chitin synthase and chitinase.
(H-V) and (G-D): Vector construction is in progress; a gene disruption system is unavailable.

NATO ASI Series, Vol. H 69
Molecular Biology and its Application to Medical Mycology
Edited by B. Maresca, G. S. Kobayashi and H. Yamaguchi

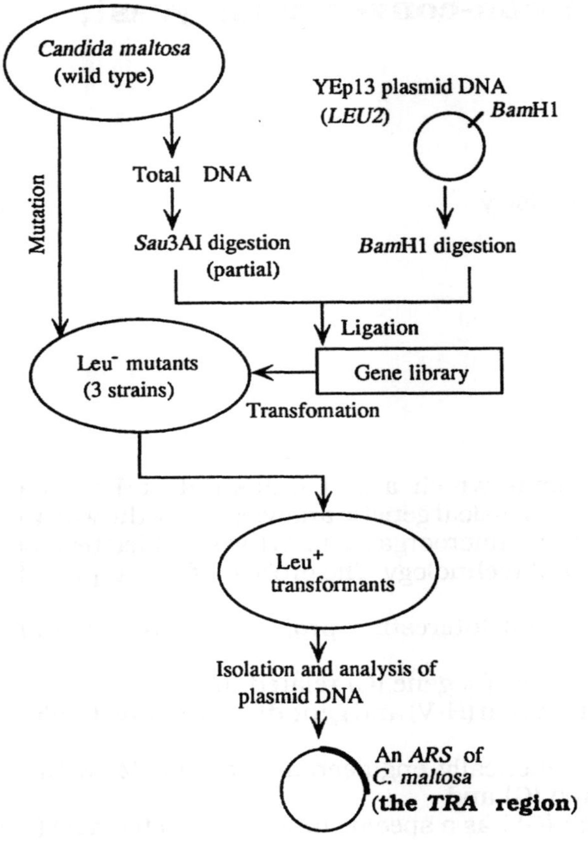

Figure 1. Isolation of an *ARS* from the Genome of *Candida maltosa*.

Construction of a host-vector system in *C. maltosa* by using an *ARS* site isolated from its genome

To construct a host-vector system, the isolation of an *ARS* site from its genome which replicates autonomously in *C.maltosa* was done as shown Fig. 1. Vectors(pTRA1 and pTRA11) were constructed that contained this *ARS* site. Their copy numbers in *C.maltosa* were between 10 and 20 (Takagi *et al*,, 1986a). It was found that a DNA fragment of about 200bp exhibited *ARS* activity in both *C.maltosa* and *S.cerevisiae*, and its nucleotide sequence analysis revealed that it contained five 11bp-sequences which are homologous to the consensus sequence of *ARS* site of *S.cerevisiae*. (Kawai *et al*, 1987).
We developed an improved host-vector system using *C.maltosa* CH1(*his⁻*)as a host. The vectors contain the *ARS* site and a DNA fragment isolated from the genome of *C.maltosa*. Since this DNA fragment could complemented both *C.maltosa* CH1 and *S.cerevisiae* (*his5⁻*), we termed the gene contained in this DNA fragment *C-HIS5* (Hikiji *et al*, 1989).
As *C.maltosa* is diploid or aneuploid it is expected that most of genes have two alleles. So, two genetic markers in a strain are required to investigate the function of a gene by the gene-disruption technique. So, in the first step, we

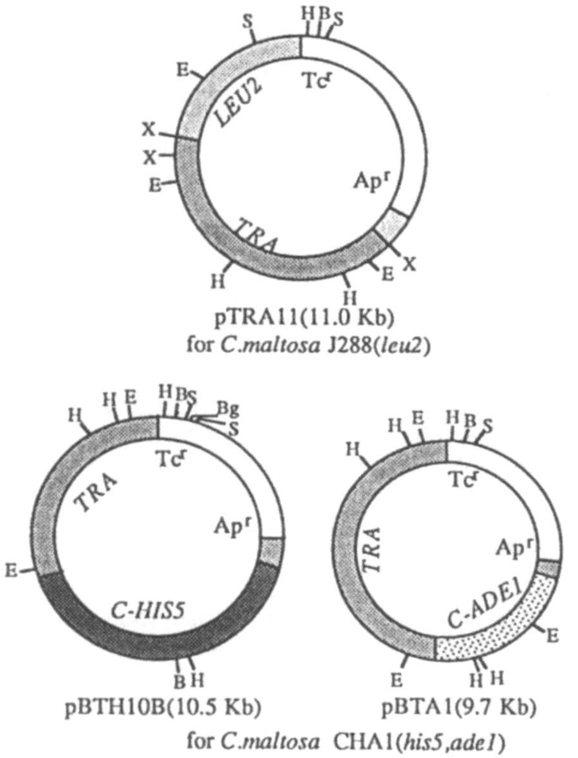

Figure 2. Cloning Vectors for *C. maltosa*

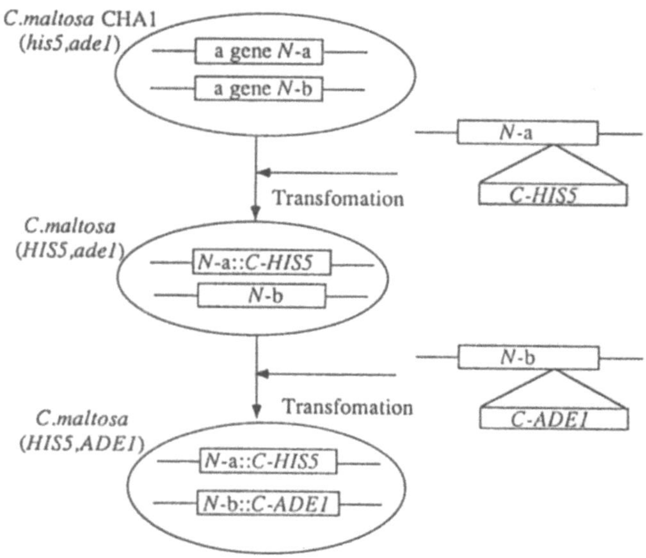

Figure 3. The Strategy for a two-step gene disruption in *C. maltosa* with two alleles for a gene *N*.

developed a new host-vector system using a host *C.maltosa* CHA1(*his5,ade1*) with two genetic markers. A DNA fragment which complements Ade was isolated from the *C.maltosa* genome. Since the DNA fragment also complemented adenine auxotrophy of *S.cerevisiae* (*ade1*), we termed a gene contained in this DNA fragment *C-ADE1* (Kawai *et al*, 1991). The structures of three plasmids for *C.maltosa* that we are routinely using in our laboratory are shown in Fig.2. Isolation of CHA1 and cloning of *C-HIS5* and *ADE1* enable us to disrupt two alleles of each gene,as shown in Fig.3.

Purification of cytochrome P-450 *alk* from *n*-alkane-grown cells of *Candida maltosa*,and cloning and nucleotide sequencing of the encoding gene

n-Alkane-inducible cytochrome P-450(P-450) catalyzes the terminal hydroxylation of alkanes and fatty acids in the alkane assimilation pathway. We purified the major alkane-inducible form of P-450,as summarized in Table 1, and determined partial amino acid sequences of the protein. Then, we cloned the gene encoding this major cytochrome P-450 by the procedure described in Fig.4 (Takagi *et al*, 1989).

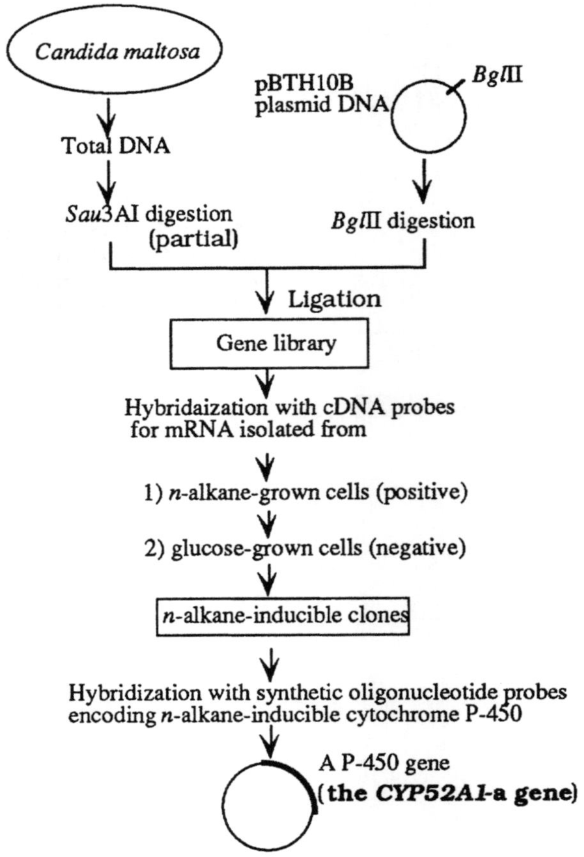

Figure 4.Isolation of a gene encoding cytochrome P-450 of *C. maltosa*.

Evidence that more than one gene encodes *n*-alkane-inducible cytochrome p450s in *C. maltosa*, revealed by the two-step gene disruption

To clarify the physiological function of P-450*alk*, we disrupted the P-450*alk* gene using the two-step gene-disruption system as shown in Fig.5 and confirmed successful disruption of two alleles of this gene by Southern-blot analysis. However,this disruptant still assimilated alkanes and contained alkane-inducible forms of P-450. This indicates the presence of at least one more gene encoding an alkane-inducible form of P-450,other than the disrupted two allelic genes (Ohkuma *et al*, 1991a). In fact, Southern-blot analysis under low-stringency conditions using the cloned gene as a probe revealed the presence of multiple P-450*alk*-related genes in the genome of *C.maltosa*.
Twelve P-450*alk*-related genes were isolated so far from a genomic library constructed from this strain, and these were classified on the basis of sequence similarities into four pairs of allelic genes and four non-allelic genes. At least two pairs of these alleles were found to be tandemly arranged in the genome. Nucleotide sequences of eight genes and one allele are determined and compared to each other. These results provide evidence for a P450*alk* [*CYP52*] multigene family in *C.maltosa* (Ohkuma *et al*, 1991b). We are now analyzing the enzymatic properties of each of these gene products expressed in *S. cerevisiae*.

Analysis of *cis*-acting elements regulated by carbon sources in the promoter region and *trans*-acting factors

We analyzed *cis*-acting elements regulated by carbon sources in the promoter region of the *CYP52A1*-a by using a vector shown in Fig.6. So far, we defined two upstream activation sequences(UAS,responsible for induction by alkane) and one upstream repression sequence (URS, responsible for repression by glucose).
We also isolated a gene named *ALI1* which is essential for assimilation of alkane by *C.maltosa*. The gene product is localized in the nucleus. We discuss the possibility that it is involved in the process of induction of genes related to assimilation of alkane or its derivatives, for example, as a transcription factor (Hwang *et al*, 1991).

Table 1.
Purification of cytochrome P-450alk from *C. maltosa* grown on *n*-alkane

Fraction	Specific content (nmol/mg protein)	(Ratio)	Yield(%)
Post-mitochondrial	0.034	(1)	100
Solubilized microsomal	0.19	(5.6)	68
AH-sepharose 4B treated	2.7	(80)	24
DEAE-sephacel treated	4.4	(130)	11
Bio-gel HTP treated	9.3	(270)	1.3

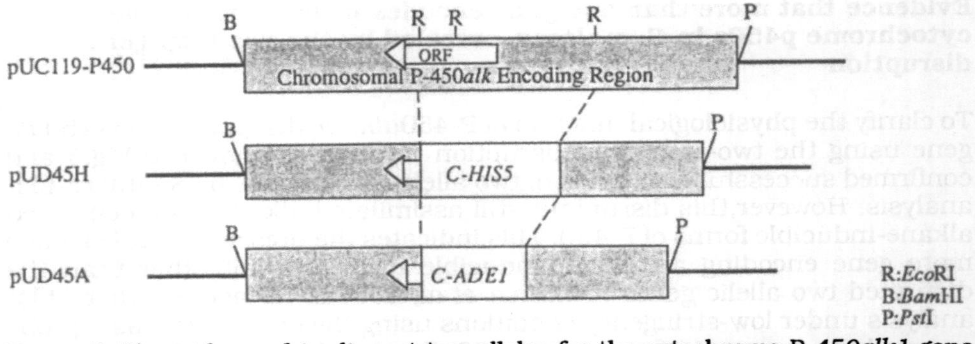

Figure 5. Plasmids used to disrupt two alleles for the cytochrome P-*450alk1* gene (*CYP52A1*).

Induction of cycloheximide resistance in *C. maltosa* by modifying the ribosomes

Cycloheximide is an antibiotic which inhibits eukaryotic protein synthesis. Although many yeasts including *Saccharomyces cerevisiae* are sensitive to cycloheximide, some yeast strains are resistant to this drug. Among the resistant strains, *C. maltosa* IAM12247 has an inducible resistance mechanism; that is, when *C. maltosa* is grown in the presence of cycloheximide, the growth is repressed for a distinct period of time depending on the concentration of cycloheximide, then the growth recovers. The cell-free translation experiments using polyuridylic acid as mRNA indicated that the protein-synthesizing machinery was altered to be resistant to cycloheximide. A reconstituted cell-free system was constructed consisting of ribosomes and soluble fraction(S-100), and it was concluded that ribosomal modification was induced during the cultivation of the cells in the presence of cycloheximide (Takagi *et al*, 1985).

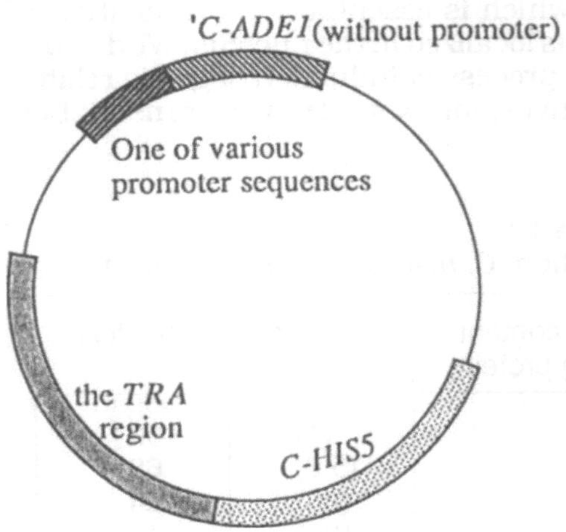

Figure 6. A plasmid vector to assay promoter activity.

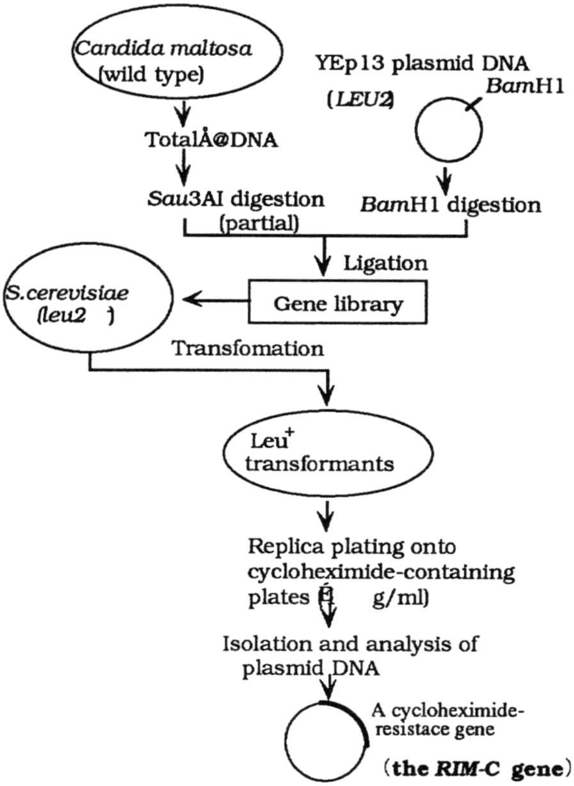

Figure 7. Isolation of a cycloheximide resistance gene of *C. maltosa.*

Cloning in *S. cerevisiae* of a cycloheximide resistance gene from the *C. maltosa* genome which modifies ribosomes

It was suggested that there is a gene(designated *RIM-C* for ribosome modification by cycloheximide) which functions in the presence of cycloheximide in such a way as to modify ribosomes so that protein synthesis in the cells is no longer inhibited by cycloheximide. So,we attempted to clone this presumptive gene by using a host-vector system of *S. cerevisiae* as shown in Fig.7 (Takagi et el. 1986b). We proved that this gene provide a convenient dominant vector marker for recombinant DNA technology in a yeast without genetic makers (Hino *et al,* 1992).

Drastic alteration of cycloheximide sensitivity by substitution of one amino acid in the L41 ribosomal protein in yeasts

We sequenced the *RIM-C* gene and found that the gene encodes a protein homologous to the L41 ribosomal protein of *S. cerevisiae*, whose amino acid sequence had already been reported. Two genes for L41 protein, named L41a and L41b, independently present in the genome of *S. cerevisiae*, were isolated. L41-related genes were also isolated from a few other yeast species.Each of

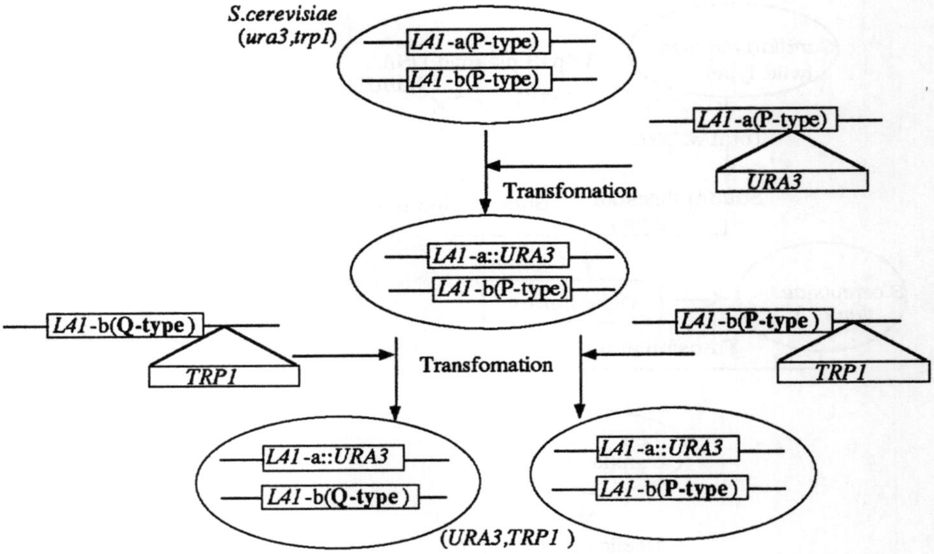

Figure 8. Construction of two strains of *S. cerevisiae* which are different only one amino acid in *L41*-b.

these genes has an intron at the same site of the open reading frame. Comparison of their deduced amino acid sequences and their ability to confer cycloheximide resistance to *S.cerevisiae*, when introduced in a high-copy number plasmid, suggested that the 56the amino acid residue of the L41 protein determines the sensitivity of the ribosome to cycloheximide; the amino acid is glutamine in the resistant ribosome, whereas that in the

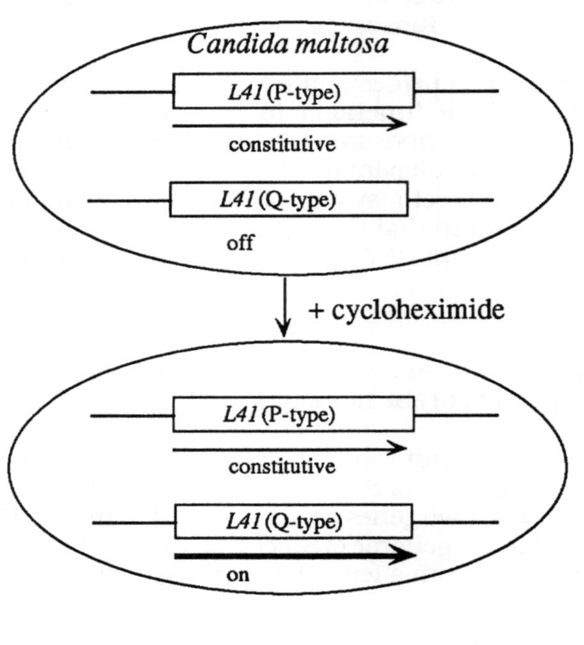

Figure 9. Cycloheximidine switches on the *L41*(Q-type) gene in *C. maltosa*.

Table 2.
One amino acid in ribosomal protein L41 determines sensitivity of protein synthesizing machinery to cycloheximide

Species	Amino acid at the 56th or corresponding position (codon)		Cycloheximide sensitivity
C. maltosa (*L41*:Q-type)	Glutamine	(CAA)	resistant
C. tropicalis (*L41*)	Glutamine	(CAA)	resistant
Kluyveromyces fragilis (*L41*)	Glutamine	(CAA)	resistant
C. maltosa (*L41*:P-type)	Proline	(CAA)	sensitive
S.cerevisiae (*L41*-a)	Proline	(CCT)	sensitive
S.cerevisiae (*L41*-b)	Proline	(CCT)	sensitive
Pichia guilliermondii (*L41*)	Proline	(CCC)	sensitive
tomato	Proline	(CCT)	sensitive
rat	Proline	(CCA)	sensitive
human	Proline	(CCA)	sensitive

sensitive ribosome is proline(Table 2). This was confirmed by constructing a cycloheximide-resistant strain of *S.cerevisiae* having a disrupted *L41*a gene and an *L41*b gene with a substitution of the glutamine codon for the proline codon, as shown in Fig.8 (Kawai *et al*, 1992).

We found that *C.maltosa* has at least two L41 genes, one is proline-type(P-type) as shown in Table 2. We also found that cycloheximide switches on the Q-type gene as shown in Fig.9. This fact may explain the inducibility of cycloheximide resistance by cycloheximide in *C. maltosa*.

References

Hikiji T, Ohkuma M, Takagi M, Yano K (1989) An improved host-vector system for *Candida maltosa* using a gene isolated from its genome that complements the *his5* mutation of *Saccharomyces cerevisiae*. Curr. Genet. 16: 261-266.

Hino A, Wongkhalaung C, Kawai S, Murao S, Yano K, Takano H, Takagi M (1992)Construction of a transformation system for a freeze-tolerant yeast *Kluyveromyces thermotolerans* Agric.Biol.Chem. 56: 228-232.

Hwang CW, Yano K, Takagi M (1991) Sequences of two tandem genes regulated by carbon sources, one being essential for n-alkane assimilation in *Candida maltosa*. Gene 106: 61-69.

Kawai S, Hikiji T, Murao S, Takagi M, Yano K (1991) Isolation and sequencing of a gene, *C-ADE1*, and its use for a host-vector system in *Candida maltosa* with two genetic markers. Agric.Biol.Chem.55:59-65.

Kawai S, Hwang CW, Sugimoto M, Takagi M, Yano K (1987) Subcloning and nucleotide sequencing of an *ARS* site of *Candida maltosa* which also functions in *Saccharomyces cerevisiae*. Agric. Biol. Chem. 51: 1587-1591.

Kawai S, Murao S, Mochizuki M, Shibuya I, Yano K, Takagi M (1992) Drastic alteration of cycloheximide sensitivity by substitution of one amino acid in L41 ribosomal protein of yeasts. J.Bacteriol. 174: 254-262.

Ohkuma M, Hikiji K, Tanimoto T, Schunck WH, Muller HG, Yano K, Takagi M (1991a) Evidence that more than one gene encodes n-alkane inducible cytochrome P-450s in *Candida maltosa*, revealed by the two-step gene disruption. Agric Biol. Chem. 55: 1757-1764.

Ohkuma M, Tanimoto T, Yano K, Takagi M (1991b) *Cyp52*(cytochrome P450alk) multigene family in *Candida maltosa*: Molecular cloning and nucleotide sequence of the two tandemly arranged genes. DNA and Cell Biology 10: 271-282.

Takagi M, Kawai S, Chang MC, Shibuya I, Yano K (1986a) Construction of a host-vector system in *Candida maltosa* by using an *ARS* site isolated from its genome. J.Bacteriol. 167: 551-555.

Takagi M, Kawai S, Shibuya I, Miyazaki M,Yano K (1986b) Cloning in *Saccharomyces cerevisiae* of a cycloheximide resistance gene from the *Candida maltosa* genome which modifies ribosomes. J.Bacteriol. 168: 417-419.

Takagi M, Kawai S, Takaya Y, Tanaka N, Sunairi M, Miyazaki M, Yano K (1985) Induction of cycloheximide resistance in *Candida maltosa* by modifying the ribosomes. J.Gen. Appl. Microbiol. 31: 267-275.

Takagi M, Ohkuma M, Kobayashi N, Watanabe M, Yano K (1989) Purification of cytochrome P-450alk from *n*-alkane-grown cells of *Candida maltosa*, and cloning and nucleotide sequencing of the encoding gene. Agric. Biol. Chem. 53: 2217-222.

The complement C3D-binding receptor (CR2) of *Candida albicans*: Cloning and sequence analysis of a gene fragment homologous with a human CR2 cDNA clone

Richard A. Calderone, Susan Southard, Cynthia Hale
Department of Microbiology
Georgetown University School of Medicine
3900 Reservoir Rd., NW.
Washington, D.C. 20007
USA

Abstract

We have previously purified and characterized a complement receptor-like protein (CR2) from *Candida albicans*, which recognizes the C3 complement conversion product C3d. The *Candida* CR2 is a 60 kD mannoprotein while the native protein is approximately 240 kD. The CR2 is found on the cell surface of germinating blastoconidia, hyphae and pseudohyphae but is not observed on the surface of blastoconidia. CR2 is expressed during infection in mice with the organism as determined by immunoelectron microscopy and lymphocyte responses to the purified CR2. Recently, we have expanded our studies to include the molecular characterization of the CR2 gene. Southern blot analysis of EcoR1-digested and Hind III-digested *C. albicans* genomic DNA was pursued using a cDNA clone of the human B-cell CR2 gene as a probe. Homology to four Eco R1-generated fragments of 11, 8, 3.2, and 1.3 kb and two Hind III-generated fragments of 9.5 and 0.7 kb was demonstrated. The hybridization patterns also suggested that the *C. albicans* CR2 gene was most likely a single copy gene. The 3.2 kb Eco R1 fragment has been isolated by plaque hybridization from a partial genomic library constructed in λgt11 using the same human CR2 to detect recombinants. Cross-hybridization analysis of the fragment localized homology to a 900 kb region. Sequence analysis of the 900 kb fragment has been completed.

Introduction

The outer cell surface of *C. albicans* is composed of an amorphous or fibrillar material whose composition is primarily mannoprotein. There would appear to be a variety of mannoprotein species, differing in the relative proportions of oligosaccharide and protein (Calderone and Braun, 1991; Shepherd, 1987). The mannoproteins of *C. albicans* are differentially expressed according to the growth form of the organism (yeast or filamentous) as well as the stage of growth (log or stationary phase) (Calderone and Braun, 1991). In some instances, mannoproteins which appear to be hyphal-specific, as determined

NATO ASI Series, Vol. H 69
Molecular Biology and its Application to Medical Mycology
Edited by B. Maresca, G. S. Kobayashi and H. Yamaguchi

by fluorescent antibody techniques, upon further analyses, are also observed from yeast cells but appear buried within the cell wall or at the level of the plasma membrane (Ollert *et al*, 1990; Li and Cutler, 1991).
Functionally, the mannoproteins fall into at least three different categories. The first category is represented by the aspartyl proteinase which has been shown to confer pathogenic properties to *C. albicans* (Ruchel, 1981). A second category would include the adhesin-receptor mannoproteins. A lectin-like mannoprotein which recognizes either fucose or amino sugars of human epithelial cells has been described by Douglas (1987a; 1987b; 1991) (Critchley and Douglas, 1987a; 1987b). This mannoprotein promotes the adherence of the yeast form of the organism to epithelial cells. Other examples in this category would include the complement-binding proteins which recognize the C3 complement cleavage components C3d (complement receptor 2 or CR2) and iC3b (complement receptor 3 or CR3 (Calderone and Braun, 1991; Calderone *et al.* 1988; Edwards *et al*, 1986; Eigentler *et al*, 1989; Gilmore *et al*, 1988; Heidenreich and Dierich, 1985; Linehan *et al*, 1988; Saxena and Calderone, 1990). Finally, by nature of their association with other cell wall structural components such as glucan, mannoprotein(s) may also fulfill a structural requirement for the cell, although little is known about this type of mannoprotein.
The intent of this review is to focus upon the complement receptors (CR) of *C. albicans*. While the CR2 of this organism has been purified and characterized to a great extent, the CR3 has not. However, functionally, there seems to be several good reasons to postulate that the CR3 possesses adhesin activity for the organism while the function of the CR2 is not known. More recently, a fragment of the gene encoding the *Candida* CR2 has been cloned and sequenced as well as mapped. These data will be described below.

The CR2 of *C. albicans*

Heidenreich and Dierich (1985) first described the rosetting of antibody-sensitized sheep RBC (EA) conjugated with C3d (EAC3d) by hyphal forms of *C. albicans*. Rosetting was also observed with EAiC3b but not with other intermediates of the C3 conversion pathway such as C3b (EAC3b). Of those *Candida* species tested, only *C. albicans* and *C. stellatiodea* demonstrated CR2- and CR3-like activity. Hyphal forms of the organism more readily exhibited receptor activity, an observation which has also been reported (Edwards *et al*, 1986; Calderone *et al*, 1988). Heidenreich and Dierich speculated that a mannan or mannoprotein conferred receptor function since inhibition of rosetting could be observed when mannose was included in the assay medium, although the degree of inhibition was only approximately 30%.
Edwards *et al* 1986 made observations similar to Heidenreich and Dierich in regard to the specificity of the complement intermediate as well as *Candida* species specificity (Edwards *et al* 1986). However, they extended the observations on the CR2 of *C. albicans* by showing that an antibody to the mammalian CR2 exhibited cross-reactivity with the *Candida* receptor. Cross-reactivity was demonstrated by an inhibition of rosetting of EAC3d when hyphae were preincubated with the anti-CR2 antibody. This was the first demonstration of homology between the mammalian and *Candida* CR2. As

discussed below, this observation was exploited in cloning the gene encoding the *Candida* CR2 by using a mammalian CR2 gene fragment (human B-cells) to screen a partial *Candida* genomic DNA library.

Purification of the Candida CR2: Purification of the *C. albicans* CR2 from DTT or whole cell extracts of hyphae was accomplished by affinity chromatography using ligand (C3d) (Calderone *et al*, 1988) or monoclonal (MAb) (Linehan *et al*, 1988) and polyclonal, monospecific antibodies (PAb) to the CR2 (Saxena and Calderone, 1990). *Candida* proteins of 60-62 and 68 kD were eluted from a C3d-thiol Sepharose column. MAbs to the *Candida* CR2 were isolated using an assay which measured their ability to block rosetting of EAC3d by hyphae. One of these MAbs (CA-A) reacted with a cell-surface epitope from hyphae but not yeast cells, blocked rosetting of EAC3d by hyphae and could be used in an affinity system to partially purify the CR2 of *Candida*. Four proteins of 60, 68, 70 and 74 kD, each glycosylated with mannose, could be eluted from the CA-A-conjugated column. Further purification by HPLC resolved the 60 kD protein in quantities sufficient to inhibit rosetting of EAC3d by hyphal forms of the organism.

Using a hyperimmune rabbit PAb against the *Candida* CR2 and isoelectric focusing, mannoproteins of 55 and 60 kD were purified from culture supernatants of hyphal-grown cells. These proteins were also active in blocking rosetting of EAC3d and hyphae, and, thus, demonstrated receptor activity. When partially deglycosylated using endoglycosidase F and then analyzed by Western blotting, the 55 and 60 kD doublet were converted to a faster-moving single species of approximately 45 kD. Thus, multiple forms of the CR2, which probably differ in their degree of glycosylation, are observed in various preparations of the organism.

As an additional approach to the identification of the *Candida* CR2, direct blotting of the C3d ligand to the *Candida* CR2 has been achieved (Fukayama *et al*, 1992). Electrophoretically-separated *Candida* proteins, transferred to nitrocellulose, were blotted with human or rabbit C3d and subsequently stained with anti-C3d antibody conjugated with alkaline phosphatase. Additionally, anti-CR2 antibody to the human CR2 was used in a Western blot with these same proteins. Both the C3d ligand and the antibody to the human CR2 reacted with a mannoprotein of 60 kD, similar in molecular mass to a protein run in parallel which reacted with the CA-A MAb. The approaches mentioned in the preceeding section which have been utilized to purify the CR2 of *C. albicans* are illustrated in Figure 1. Regardless of the method used, i.e., ligand or MAb affinity chromatography or direct binding assays using C3d or MAb against the mammalian CR2, the *Candida* CR2 appears to be 60 kD mannoprotein.

Cloning of the Candida CR2: Because of the homology of the *Candida* CR2 with the mammalian CR2 (as described above), we used a cDNA clone of the mammalian CR2 gene (from human B-cells) (Moore *et al*, 1987) in Southern blot analyses of EcoR1-and Hind III-digested *C. albicans* genomic DNA. The human CR2 gene clone (λE41) contained three EcoR1-derived fragments, including a 1.8kb fragment which hybridized to four EcoR1-generated fragments of 11, 8, 3.2 and 1.3 kb and two Hind III-generated fragments of 9.5 and 0.7 kb (Figure 2a). The 3.2 kb EcoR1 fragment was isolated by plaque hybridization from a EcoR1 partial genomic DNA library of *C. albicans*

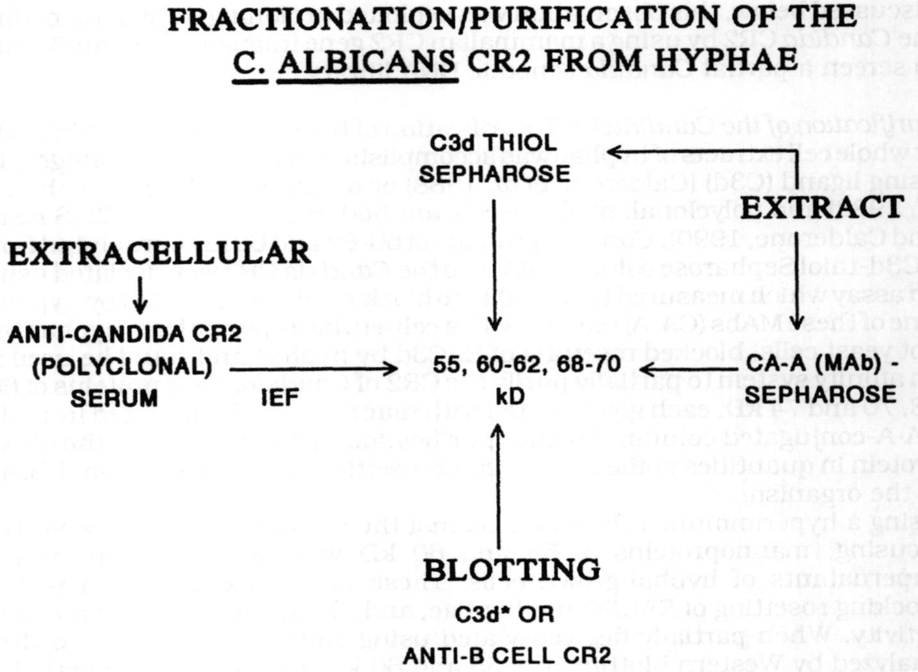

Figure 1. Purification of the *Candida albicans* CR2 using ligand and antibody affinity chromatography.

constructed in λgt11 using the same human CR2 sequence to detect recombinants.
When the 3.2 kb EcoR1 fragment was digested with Hind III and subjected to Southern blot analysis, we were able to demonstrate hybridization of the mammalian probe to a 900 bp region (CA-900) (Figure 2). Cross-hybridization of the CA-900 used as a probe and the mammalian 1.8 kb gene fragment was also demonstrated (data not shown).
The 900bp fragment (CA-900) was sequenced using the procedure of Sanger *et al* (1977). We observed an overall homology of 43% and 45% with the know sequences of the murine (Fingeroth, 1990) and human CR2 (Moore *et al*, 1987). A putative open-reading frame beginning at position 350 and continuing through position 962 was observed.

The CR3 of *C. albicans*

As with the CR2 studies described above, the first demonstration of a CR3-like activity was reported (Heidenreich and Dierich, 1985). The activity was limited to *C. albicans* and *C. stellatiodea*. When binding of iC3b was measured by rosetting of complement-conjugated erythrocytes (EAiC3b), receptor activity could be demonstrated only with filamentous forms of the organism (Edwards *et al*, 1986). However, binding of radiolabeled iC3b to blastoconidia

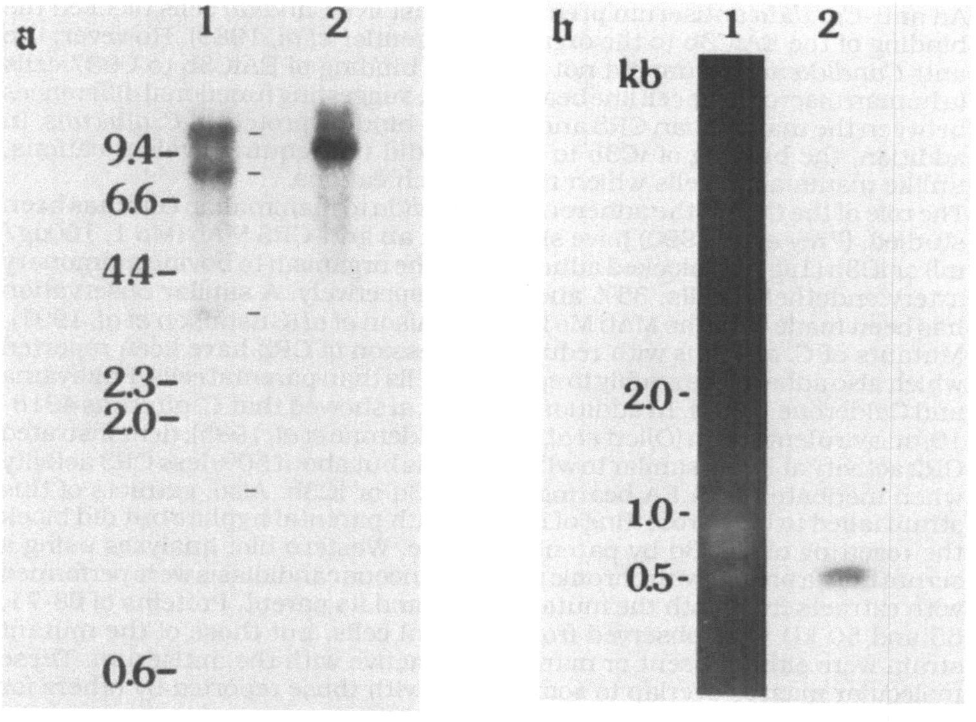

Figure 2.Homology of the human B-cell CR2 gene with *C. albicans*. (a) Genomic DNA of *C. albicans* digested with EcoR1 (lane 1) or Hind III (lane 2) and probed with the human CR2 gene. (b) A 3.2 kb EcoR1 *Candida* clone was digested with Hind III (lane 1) and blotted with the human CR2 gene (lane 2).

has been reported (Gilmore *et al*, 1988). Thus, both growth forms of the organism possess a CR3 activity, although the relative number of receptor molecules is probably considerably less with blastoconidial forms of the organism, similar to the observation by Bouchara *et al*, who reported a much higher level of laminin receptors on filamentous forms of the organism (Bouchara *et al*, 1990).

A number of MAbs against either the α- or β-chain of the mammalian cell CR3 have been tested against *C. albicans* (Edwards *et al*, 1986; Eigentler *et al*, 1989; Gilmore *et al*, 1988; Hostetter *et al*, 1990; Marcantonio and Hynes, 1988; Mayer *et al*, 1990). Only MAbs against the α-chain of the CR3 appear reactive with the organism, although Marcantonio and Hynes reported that a β-chain MAb recognized a protein of *C. albicans* in a Western blot assay (Marcantonio and Hynes, 1988).

The anti-CR3 MAb OKM-1 was used by Eigentler *et al* to isolate the *C. albicans* CR3 (Eigentler *et al*, 1989). Solubilized proteins of the organism were prepared from radiolabeled hyphae and OKM-1 was used to immunoprecipitate reactive proteins. A protein of 130 kD was observed in SDS-PAGE profiles along with minor bands of 50 and 100 kD. These molecular masses are different from that reported for the mammalian CR3 (α-chain, 165 kD and β-chain, 95 kD, Stoolman, 1989; Ross, 1986).

An anti-*Candida* antiserum prepared against live *Candida* cells blocked the binding of the EAiC3b to the organism (Eigentler *et al*, 1989). However, the anti-*Candida* antiserum did not block the binding of EAiC3b to U937 cells (a human macrophage cell line bearing CR3), suggesting functional differences between the mammalian CR3 and the iC3b-binding protein of *C. albicans*. In addition, the binding of iC3b to *Candida* did not require divalent cations, unlike mammalian cells which require such cations.
The role of the CR3 in the adherence of *Candida* to mammalian cells has been studied. (Frey *et al*, 1990) have shown that an anti-CR3 MAb (Mo 1, 100μg/ml) or iC3b (1μg/ml) blocked adherence of the organism to bovine pulmonary artery endothelial cells, 35% and 74%, respectively. A similar observation has been made with the MAb Mo1 by Gustafson *et al* (Gustafson *et al*, 1991). Mutants of *C. albicans* with reduced expression of CR3 have been reported which also adhere less readily to epithelial cells than parental cells (Fukayama and Calderone, 1991). In addition, Ollert *et al* showed that *C. albicans* 4918-10, an avirulent strain (Ollert *et al*, 1990; Calderone *et al*, 1985), demonstrated CR2 activity at levels similar to wild type cells but about 50% less CR3 activity when incubated with EA bearing either C3d or iC3b. Also, extracts of this strain failed to block rosetting of EAiC3b with parental hyphae but did block the rosetting of EAC3d by parental hyphae. Western blot analyses using a serum from a patient with chronic mucocutaneous candidiasis were performed with extracts from both the mutant strain and its parent. Proteins of 68-71, 55 and 50 kD were observed from parental cells, but those of the mutant strain were either absent or marginally reactive with the antiserum. These molecular masses overlap to some extent with those reported by others for the *Candida* CR3 (Eigentler *et al*, 1989).
Other investigators have demonstrated a role for the CR3 in the adherence of *C. albicans* to host cells by using peptides, which are common to ligands recognized by the mammalian CR3, to block the adherence of the organism to endothelium (Klotz, 1990; Klotz and Smith, 1990). The arginine-glycine-aspartic acid peptide (RGD-sequence) common to ligands such as fibronectin has been shown to represent one of the binding sites for the CR3 of mammalian cells (Stoolman, 1989). Previously, *C. albicans* has been shown to recognize ligands such as fibronectin (Skerl *et al*, 1983) and fibrinogen (Tronchin *et al*, 1989). This observation is quite provocative and may suggest new approaches in the prophylaxis against candidiasis. Thus, the observations described above in regard to the inhibiton of adherence by ligands or MAb with specificities for the CR3 as well as the experiments with an adherence-negative, CR3-reduced mutant, suggest that the *Candida* CR3 may have an adhesion function.
The CR3-adhesin of *C. albicans* would appear to be different from the adhesin involved in the recognition of epithelial cells as described by other investigators (see above). In the latter instance, the adhesin of epithelial cells, while a mannoprotein, apparently recognizes a carbohydrate ligand containing fucose or glucosamine in comparison to the CR3 which, as stated above, binds to proteins containing RGD amino acid sequences. Thus, at least two adhesins have been desribed for *C. albicans*, each with an apparent specificity for either epithelial or endothelial cells.

Acknowledgements

This research was supported in part by Public Health Services grants (NIAID25738 and HL21370).

References

Bouchara J, Tronchin G, Annaix V, Robert R, Senet JM (1990) Presence of laminin receptors on *Candida albicans*. Infect. Immun. 58: 48-54
Calderone RA, Braun P (1991) Adherence and receptor relationships of *Candida albicans*. Microbiol. Rev. 55: 1-20.
Calderone RA, Linehan L, Wadsworth E, Sandberg AL (1988) Identification of C3d receptors on *Candida albicans*. Infect. Immun. 56: 252-258.
Calderone RA, Cihlar RL, Lee D, Hoberg K, Scheld WM (1985) Yeast adhesion in the pathogenesis of endocarditis due to *Candida albicans* studies with adherence-negative mutants. J. Infect. Dis. 152: 710-715.
Critchley IA, Douglas LJ (1987) Role of glycosides as epithelial cell receptors for *Candida albicans*. J. Gen. Microbiol. 133: 637-643.
Critchley IA, Douglas LJ (1987) Isolation and partial characterization of an adhesin from *Candida albicans*. J. Gen. Microbiol. 133: 629-636.
Douglas LJ (1987) Adhesion to epithelial surfaces. In: O'Leary W (ed) Crit Rev in Microbiol, vol xv. CRC Press, Inc, Boca Raton FL, p 27.
Douglas LJ (1991) Adhesion of *Candida albicans* to host surfaces. In: Tumbay E (ed) Proceedings of FEMS symposium on *Candida* and *candidamycosis*, Plenum Press, New York, p 43-48.
Edwards JE, Gaither TA, O'Shea JJ, Rotrosen D, Lawley TJ, Wright SA, Frank FF, Green I (1986) Expression of specific binding sites on *Candida* with functional and antigenic characteristics of human complement receptors. J. Immunol. 137: 3577-3583.
Eigentler A, Schulz TF, Larcher C, Breitwieser EM, Myones B, Petzer AL, Dierich MP (1989) iC3b-binding protein on *Candida albicans*: temperature-dependent expression and relationship to human complement receptor type 3. Infect. Immun. 57: 616-622.
Fingeroth J (1991) Comparative structure and evolution of murine CR2. The homology of the human C3d/EBV Receptor (CD 21). J. Immunol. 144: 3458-3467.
Frey CL, Barone JM, Drutz D (1990) The role of the *Candida albicans* receptor in fungal adherence to endothelial cells. Abstracts of the American Society for Microbiology, F101.
Fukayama M, Calderone RA (1991) Adherence of cell surface mutants of *Candida albicans* to buccal epithelial cells and analyses of the cell surface proteins of the mutants. Infect. Immun. 59: 1341-1345.
Fukayama M, Wadsworth EA, Calderone RA (1992) Expression of the C3d-binding protein (CR2) from *Candida albicans* during experimental candidiasis as measured by lymphoproliferation. Infect. Immun. 60: in press.
Gilmore BJ, Resinas EM, Lorenz JS, Hostetter MK (1988) An iC3b receptor on *Candida albicans*: structure, function and correlates for pathogenicity. J. Infect. Dis. 157: 38-46.
Gustafson KS, Vercelloti GM, Lorenz JS, Hostetter MK (1991) Molecular mimicry in *Candida albicans*. Role of an integrin in adhesion of the yeast to human endothelium. J. Clin. Invest. 87: 1896-1902.
Heidenreich F, Dierich MP (1985) *Candida albicans* and *Candida stellatoidea*, in contrast to other *Candida* species, bind iC3b and C3d but not C3b. Infect. Immun. 50: 598-600.
Hoberg K, Cihlar RL, Calderone RA (1986) Characterization of cerulenin-resistant mutants of *Candida albicans*. Infect. Immun. 51: 102-109.

Hostetter MK, Lorenz JS, Preus L, Kendrick KE (1990) The iC3b receptor of *Candida albicans*: subcellular localization and modulation of receptor expression by glucose. J. Infect. Dis. 161: 761-768.
Li RK, Cutler JE (1991) A cell surface/plasma membrane antigen of *Candida albicans*. J. Gen. Microbiol. 137: 455-464.
Klotz SA (1990) Adherence of *Candida albicans* to components of the subendothelial extracellular matrix. FEMS Microbiol. Lett. 68: 249-254.
Klotz SA, Smith RL (1990) *Candida albicans* adherence to subendothelial ECM is inhibited by arginine-glycine-aspartic acid peptides. Clin Res, 38: 13A, AFCR Immunol. Rheumatol. Conf. New Orleans, LA.
Linehan L, Wadsworth E, Calderone RA (1988) *Candida albicans* C3d receptor, isolated by using a monoclonal antibody. Infect. Immun. 56: 1981-1986.
Marcantonio E, Hynes RO (1988) Antibodies to the conserved cytoplasmic domain of the integrin 1 subunit react with proteins in vertebrates, invertebrates and fungi. J. Cell. Biol. 106: 1765-1772.
Mayer CL, Diamond RD, Edwards JE (1990) Recognition of binding sites on *Candida albicans* by monoclonal antibodies to human leukocyte antigens. Infect. Immun. 58: 3765-3769.
Moore MM, Cooper NR, Tack BF (1987) Nemerow GR Molecular cloning of the cDNA encoding the Epstein-Barr virus-C3d receptor (complement receptor type 2) of human B-lymphocytes. Proc. Natl. Acad. Sci. USA, 84: 9194-9198.
Ollert MK, Wadsworth E, Calderone RA (1990) Reduced expression of the functionally active complement receptor for iC3b but not for C3d on an avirulent mutant of *Candida albicans*. Infect. Immun. 58: 909-913.
Ross GD (1986) Opsonization and membrane complement receptors. In: Ross GD (ed) Immunobiology of the complement system. Academic Press, Inc, New York p.87.
Ruchel R (1981) Properties of a purified proteinase from the yeast *Candida albicans*. Biochim. Biophys. Acta, 659: 99-113.
Sanger F, Nicklen S, Coulson AR (1977) DNA sequencing with chain-terminating inhibitors. Proc. Natl. Acad. Sci. USA, 74: 5463-5467.
Saxena A, Calderone RA (1990) Purification and characterization of the extracellular C3d-binding protein of *Candida albicans*. Infect. Immun. 58: 309-314.
Shepherd MG (1987) Cell envelope of *Candida albicans*. In: O'Leary WO (ed) CRC Critical Rev in Microbiol, Boca Raton, FL, 15: 7-27.
Skerl KG, Calderone RA, Segal E, Sreevalsan T, Scheld WM (1983) *In vitro* binding of *Candida albicans* yeast cells to human fibronectin. Canad. J. Microbiol. 30: 221-227.
Stoolman LM (1989) Adhesion molecules controlling lymphocyte migration. Cell, 56: 907-910.
Tronchin G, Bouchara JP, Robert R (1989) Dynamic changes of the cell wall surface of *Candida albicans* associated with germination and adherence. Eur. J. Cell. Biol. 50: 285-290.
Whelan WL, Delga JM, Wadsworth E, Walsh TJ, Kwon-Chung KJ, Calderone RA, Lipke P (1990) Isolation and characterization of cell surface mutants of *Candida albicans*. Infect. Immun. 58: 1552-1557.

Biology, physiology, biochemistry and molecular genetics of *Trichosporon* yeasts

J. Reiser, V. Glumoff*, M. Kälin, U. Ochsner and A. Fiechter
Institut für Biotechnologie
ETH-Hönggerberg
CH-8093 Zürich
Switzerland

Taxonomy of *Trichosporon* yeasts

The genus *Trichosporon* includes yeasts which are characterized by budding cells of various shapes, a more or less developed pseudomycelium or a true mycelium and arthrospores. *Trichosporon* yeasts may form asexual endospores but sexual reproduction has not been demonstrated so far (Do Carmo-Sousa, 1970). Biochemical characteristics such as hydrolysis of urea, utilization of mono-, di-, tri-, or polysaccharides etc., as well as studies concerning DNA base composition and DNA relatedness led Guého *et al.* (1984) to propose that *Trichosporon* yeasts should be classified into two separate groups. The first group, which appears to be related to the Ascomycetes, includes thirteen species with a G+C content lower than 50% (34.7%-48.8%) and lacks urease. The second group appears to be related to the Basidiomycetes and contains fifteen species with a G+C content higher than 50% (57%-64%) including *T. cutaneum*, *T. beigelii* and *T. pullulans*, and has the ability to hydrolyze urea. The basidiomycetous nature of some of the *Trichosporon* yeasts is demonstrated by the lamellar structure of the cell walls (Kreger-Van Rij and Veenhuis, 1971) and the presence of xylose (Weijman, 1979). Hara *et al.* (1989) have grouped 44 different strains of *T. beigelii* and related organisms based on differences in the ubiquinones and assimilation of melibiose and raffinose. The findings suggest that *T. beigelii* is a heterogeneous group of yeasts consisting of at least four different types of organisms.

T. cutaneum has been isolated from a number of sources including soil, industrial waste water, wood pulp, sludge and clinical specimens. *T. beigelii* has been found to be the causative agent of white piedra, which is a relatively inconsequential infection of the hair. However, this organism does not appear to be part of the skin flora in healthy subjects (see e.g. McBride *et al.*, 1988). Certain strains have also been found as opportunistic pathogens causing deep-seated and disseminated infections in immunocompromised patients (reviewed in Hoy *et al.*, 1986). Kemker *et al.* (1991) have analysed fifteen clinical and environmental strains of *T. beigelii* for similarities by using morphological features, biochemical profiles based on carbon compound

---

* Present address: Department of Medical Biochemistry, University of Turku, 20520 Turku, Finland.

NATO ASI Series, Vol. H 69
Molecular Biology and its Application to Medical Mycology
Edited by B. Maresca, G. S. Kobayashi and H. Yamaguchi

assimilation and uric acid utilization, isoenzyme electrophoresis, and restriction fragment length polymorphisms in rRNA genes. The findings suggested that strains that cause invasive disease are distinct from the superficial and nonclinical isolates and that isolates from the skin and mucosae represent a number of different organisms, including some environmental forms. The study also revealed that *T. beigelii* is a complex of genetically distinct organisms and that more than one type is found in clinical samples.

Like *T. cutaneum*, *T. pullulans* has a wide substrate spectrum. The maximum temperature of growth for *T. pullulans* is around 23-27°C (Do Carmo-Sousa, 1970) but growth of *T. pullulans* was observed even at 5°C (Berg *et al.*, 1987). Recently, a thermotolerant *T. adeninovorans* strain was described (Middelhoven *et al.*, 1984; Gienow *et al.*, 1990). This yeast is able to grow at 45°C and its cell wall structure was found to be Ascomycete-like. *T. adeninovorans* is able to use a variety of carbon sources including mono- and disaccharides, starch and *n*-alkanes. This organism has recently been reclassified and now belongs to the genus *Arxula* (K. Lietz, pers. commun.).

Physiology and biochemistry of *Trichosporon* yeasts

Trichosporon yeasts have the potential to use a very large variety of carbon sources (Laaser *et al.*, 1989). *T. cutaneum*, *T. beigelii* and *T. pullulans* have been shown to grow on various monosaccharides including pentoses and hexoses, on disaccharides, such as cellobiose, maltose, lactose, sucrose, melibiose and trehalose (Mörtberg and Neujahr, 1986) and on polysaccharides including xylans (Hrmovâ *et al.*, 1984), ball-milled filter paper (Stevens and Payne, 1977), starch (De Mot *et al.*, 1989), and on pectic acid, carboxymethylcellulose and locust-bean gum (Zimmermann and Emeis, 1989). Other carbon sources include ethylamine (Veenhuis *et al.*, 1986), uric acid (Middelhoven *et al.*, 1983), propionate, D-alanine, D-methionine and oleic acid (M. Veenhuis, pers. commun.), all of which lead to the formation of peroxisomes (Veenhuis *et al.*, 1985; 1986). D-glucarate, galactarate and L-tartarate (Schneider *et al.*, 1989), and cyclohexanecarboxylic acid (Hasegawa *et al.*, 1982) have also been found to serve as carbon sources for certain strains of *Trichosporon*. Furthermore, *T. cutaneum* and *T. beigelii* are capable of using various aromatic compounds as sole carbon and energy sources. The biochemistry and physiology of phenol, cresol, salicylate, benzoate and anthranilate degradation and the metabolism of aromatic amino acids have been the subjects of extensive research (reviewed in Dagley, 1985; Neujahr, 1990). These properties indicate the extraordinary potential of *T. cutaneum* for the efficient conversion of various carbon sources into biomass. *T. cutaneum* has a purely oxidative metabolism and does not form ethanol even under oxygen limitation (Käppeli and Fiechter, 1982) and it shows a high biomass yield of around 50 to 55% on glucose (Fiechter *et al.*, 1987). Hess (1988) has worked out the conditions for high density cultivation of *T. cutaneum*. Biomass concentrations as high as 200 g (dry weight) per liter could be harvested in a continuous cultivation process using a cell recycling system. The highest productivity, 22 g per liter per h, was obtained at a cell density of 120 g per liter. Thus, the broad substrate range in conjunction with the bioreactor technology developed for *T. cutaneum* are attractive features in view of the synthesis of valuable foreign proteins.

Trichosporon yeasts have the capacitiy to accumulate lipids and thus they belong to the oleaginous yeasts (Ratledge, 1988; West *et al.*, 1990). The process of lipid accumulation occurs when an oleaginous yeast is grown in a medium with a high carbon/nitrogen ratio (usually about 30:1), so that the excess carbon is assimilated without conversion to protein or nucleic acids. *T. pullulans* has been found to accumulate more than 65% of its biomass as lipid. The order of abundance of the fatty acyl groups of the lipids was oleate> palmitate>linoleate>stearate (Ratledge, 1988).

T. cutaneum cells can be grown in the presence of 0.05% phenol as a carbon source and the growth rate was found to be equal on both phenol and glucose (Spånning and Neujahr, 1987). The first three enzymes involved in phenol metabolism including phenol hydroxylase (Neujahr and Gaal, 1973), catechol 1,2-oxygenase (Varga and Neujahr, 1970) and *cis,cis*-muconate cyclase (Gaal and Neujahr, 1980) have been isolated and characterized in detail. The levels of the enzymes involved in phenol degradation were found to be some 50-400 times higher in phenol-grown cells than in glucose-grown cells (Gaal and Neujahr, 1981) and phenol hydroxylase comprises 2-5% of the total cell protein in fully induced cells (Neujahr and Gaal, 1973). In addition to phenol, resorcinol, catechol, cresols and fluorophenols can induce phenol hydroxylase (Gaal and Neujahr, 1981). A full length phenol hydroxylase cDNA has recently been sequenced and overexpressed in an enzymatically active form in *E. coli* (M. Kälin, unpublished results). This should facilitate the purification of the enzyme and allow the design of enzyme variants by site-directed mutagenesis.

Similar to *Kluyveromyces lactis*, *T. cutaneum* contains an inducible β-galactosidase (Mörtberg and Neujahr, 1986; West *et al.*, 1990). The *T. cutaneum* ATCC 46490 strain studied by Mörtberg and Neujahr revealed a cell-wall-bound extracellular enzyme, whereas the protein of the ATCC 20509 strain studied by West *et al.* (1990) was intracellular. A four to seven-fold induction of β-galactosidase was detected in cultures grown on lactose, lactulose or galactose and the enzyme hydrolysed lactose, lactulose and nitrophenyl-β-D-galactoside (West *et al.*, 1990). A second intracellular lactose hydrolase, a β-glycosidase was described by West *et al.* (1990). It had a wider substrate spectrum and hydrolysed lactose, nitrophenyl-β-D-galactosides, 4-nitrophenyl-β-D-glucoside, cellobiose, laminaribiose, laminaritriose and sophorose efficiently. This enzyme was induced by lactose, lactulose or galactose and also by cellobiose. Hrmová *et al.* (1984) had earlier detected an intracellular β-glucosidase capable of hydrolyzing cellobiose. The best inducer of this β-glucosidase was thiocellobiose.

Using ball-milled filter paper as a growth substrate, *T. cutaneum* and *T. pullulans* were found to produce appreciable amounts of cellulase activity (Stevens and Payne, 1977). The main products of cellulose degradation were cellobiose and glucose. Xylanase activity was also present in the same culture filtrates.

A xylan and xylose inducible endo-1,4-β-D-xylanase has been purified from the culture supernatant of the *T. cutaneum* DSM 70698 strain grown on xylan (Stüttgen and Sahm, 1982). This enzyme comprised about 27% of the secreted proteins at levels around 50 mg per liter. When the cells were grown on glucose or cellobiose, no β-xylanase activity could be detected. Hrmová *et al.* (1984) described a different strain of *T. cutaneum* which produced a xylan-

inducible β-xylanase. The enzyme could not be induced by xylose but the activity was increased 200-fold by xylan, about 700-fold by methyl-β-xyloside but only slightly by cellobiose. Zimmermann and Emeis (1988) have isolated and characterized a number of polysaccharidases from the culture supernatant of *T. beigelii*, strain CBS 5790. An amylase was present when the cells were grown on starch, and pectinases were found in starch, carboxymethyl cellulose (CMC), xylan or locust-bean gum grown cells. The production of these enzymes was strongly repressed by glucose.
The *T. pullulans* ATCC 10677 strain produced α-amylase and glucoamylase activities in a medium containing corn steep liquor and corn starch or soluble starch (De Mot and Verachtert, 1986). For both amylases, the maximum concentrations were found in stationary phase cultures. In addition, pullulanase activity was found in the glucoamylase fraction and cyclodextrinase activity was present in the α-amylase fraction. The electrophoretic analysis revealed that the α-amylase activity was due to a single protein. The glucoamylase, however, occured in multiple forms. The four glucoamylases and the α-amylase were found to be glycosylated.
Büttner *et al.* (1987) have isolated an inducible amylase which hydrolyzed starch by cleaving single glucose units, thus showing that it is a glucoamylase. Maltose was found to be a better inducer than soluble starch. To relieve carbon catabolite repression, mutants of *T. adeninovorans* resistant to 2-deoxy-D-glucose were selected (Büttner *et al.*, 1989). Relative to wild type cells, a one hundred fold derepression of glucoamylase activity was detected in such mutant cells grown on glucose.

Molecular genetics of *Trichosporon* yeasts

Further development of *Trichosporon* yeasts for production purposes is dependent on efficient transformation systems that will permit the maintenance and expression of both homologous and heterologous genes. Ideally, such a system should permit the maintenance of these DNAs without selective pressure, an important consideration for industrial scale-up procedures. With this goal in mind we have recently succeeded in setting up a genetic transformation procedure based on dominant selection markers (Glumoff *et al.*, 1989). Transformation of the *T. cutaneum* DSM 70698, ATCC 46490 and ATCC 58094 strains was obtained with plasmids carrying either the *E. coli* hygromycin B phosphotransferase-encoding gene (Gritz and Davies, 1983) or the *Streptoalloteichus hindustanus* phleomycin-resistance gene (Drocourt *et al.*, 1990), as dominant selection markers. Expression of both resistance-conferring genes was controlled by the glyceraldehyde-3-phosphate dehydrogenase (*gpdA*) gene promoter and the *trpC* gene terminator, from *Aspergillus nidulans* (Punt *et al.*, 1987; Mattern and Punt, 1988). The transforming DNA was found to consist of multiple tandem plasmid copies of high molecular weight. This polymeric structure, in nonselective media, was mitotically unstable, indicating that it existed in an episomal state.
A transformation system for *T. cutaneum* based on auxotrophic markers, and techniques for the induction, isolation and characterization of mutants have recently been described by Ochsner *et al.* (1991). A number of auxotrophic mutants were isolated and characterized by using biosynthetic precursors and/or inhibitors. A mutant unable to grow in the presence of ornithine

could be complemented successfully in protoplast transformation experiments using the cloned *A. nidulans* ornithine carbamoyl transferase (*argB*) gene as a selection marker with an efficiency of 5 to 100 transformants per µg of DNA. In these transformants, the heterologous *argB* gene was present in multiple tandem copies and the transforming DNA was found to remain stable after more than 50 generations in non-selective media. The same mutants could be complemented by a *T. cutaneum* cosmid library and a complementing cosmid was subsequently isolated from this library by a sib-selection strategy. This cosmid transformed *T. cutaneum* protoplasts with an efficiency of 50 to 200 colonies per µg of DNA. Southern blot analyses are consistent with the view that the transforming sequences became stably integrated into the host genome at the homologous site. The cosmid was subsequently subcloned and the sequence of a complementing 4.5 kb DNA fragment determined (Glumoff, 1992). The DNA sequence revealed the presence of a gene encoding the large subunit of the mitochondrial carbamoyl phosphate synthetase (CPSA). The *T. cutaneum* carbamoyl phosphate synthetase (*argA*) gene has the capacity to encode a protein of 1170 amino acids and its coding region is interrupted by a short intron. The derived amino acid sequence is 62% identical and 77% similar to the sequence of the corresponding protein of *Saccharomyces cerevisiae*. The codon usage in the *argA* gene was found to be similar to the one in filamentous fungi.

For the design of expression vectors, the strongly expressed phenol hydroxylase gene has been isolated and the 5'-non coding region of the gene is currently being tested for heterologous gene expression (M. Kälin, unpublished results).

A *Trichosporon*-based production system for heterologous proteins is attractive because of several reasons. First, *Trichosporon* yeasts show a high biomass yield and can be cultivated at very high cell densities and their bioreactor performance has been studied extensively. Second, these organisms appear to be tremendously versatile concerning the use of a variety of carbon and nitrogen sources. Thus, it may be feasible to use industrial waste products such as whey or molasses for the production of foreign proteins. Third, certain *Trichosporon* strains have the potential to amplify foreign gene sequences, yielding up to 200 copies of a foreign gene per cell and our current results are compatible with the view that the amplified sequences are mitotically stable.

In an attempt to develop the genetics of *T. adeninovorans*, Samsonova *et al.* (1989) have isolated several auxotrophic strains. The mutations were induced by UV-irradiation or by using N-methyl-N'-nitro-N-nitrosoguanidine. Adenine, methionine and cysteine auxotrophs were the predominant types, whereas mutants with a blocked biosynthesis of histidine, isoleucine/valine and aromatic amino acids were absent in the samples. The diverse and biochemically defined mutants of *T. adeninovorans* provide suitable chromosomal markers for future genetic studies in this yeast (Büttner *et al.*, 1990. A transformation system based on lysine auxotrophic mutants in conjunction with cloned *S. cerevisiae* lysine biosynthetic genes has recently been worked out for *T. adeninovorans* (K. Lietz, pers. commun.).

Acknowledgements

Our work on *Trichosporon* yeasts is being supported by the Swiss National Science Foundation, grant No. 31-30053.90.

References

Berg GR, Inniss WE, Heikkila JJ (1987) Stress proteins and thermotolerance in psychotrophic yeasts from arctic environments. Can J Microb 33: 383-389.

Büttner R, Bode R, Birnbaum D (1987) Purification and characterization of extracellular glucoamylase from the yeast *Trichosporon adeninovorans*. J Basic Microb 27: 299-308.

Büttner R, Scheit A, Bode R, Birnbaum D (1989) Isolation and characterization of mutants of *Trichosporon adeninovorans* resistant to 2-deoxy-D-glucose. J Basic Microb 29: 67-72.

Büttner R, Bode R, Samsonova IA, Birnbaum D (1990) Mapping of the glucoamylase gene of *Trichosporon adeninovorans* by mitotic haploidization using hybrids from protoplast fusion. J Basic Microbiol 30: 227-231.

Dagley S (1985) Microbial metabolism of aromatic compounds. In: Moo-Young M, Bull AT, Dalton H (eds) Comprehensive Biotechnology, vol 1. Pergamon Press Ltd, Oxford, pp. 483-505.

De Mot H, Verachtert H (1986) Secretion of α-amylase and multiple forms of glucoamylase by the yeast *Trichosporon pullulans*. Can J Microb 32: 47-51.

Do Carmo-Sousa L (1970) *Trichosporon* Behrend. In: Lodder J (ed.) The yeasts A taxonomic study. Second edition. North-Holland Publ. Co., Amsterdam, pp. 1309-1352.

Drocourt D, Calmels T, Reynes J-P, Baron M, Tiraby G (1990) Cassettes of the *Streptoalloteichus hindustanus ble* gene for transformation of lower and higher eukaryotes to phleomycin resistance. Nucl Acids Res 18: 4009.

Fiechter A, Käppeli O, Meussdoerffer F (1987) Batch and continuous culture. In: Rose AH, Harrison JS (eds) The Yeasts, vol 2. Academic Press, London, pp. 99-129.

Gaal A, Neujahr HY (1980) *cis,cis*-muconate cyclase from *Trichosporon cutaneum*. Bioch J 191: 37-43.

Gaal A, Neujahr HY (1981) Induction of phenol-metabolizing enzymes in *Trichosporon cutaneum*. Arch Microb 130: 54-58.

Gienow U, Kunze G, Schauer F, Bode R, Hofemeister J (1990) The yeast genus *Trichosporon spec* LS3; Molecular characterization of genomic complexity. Zentralbl Mikrobiol 145: 3-12.

Glumoff V, Käppeli O, Fiechter A, Reiser J (1989) Genetic transformation of the filamentous yeast, *Trichosporon cutaneum*, using dominant selection markers. Gene 84: 311-318.

Glumoff V (1992) Heterologous gene expression in the soil yeast *Trichosporon cutaneum*. Doctoral thesis, Swiss Federal Institute of Technology Zürich.

Gritz L, Davies J (1983) Plasmid-encoded hygromycin B resistance: the sequence of hygromycin B phosphotransferase gene and its expression in *Escherichia coli* and *Saccharomyces cerevisiae*. Gene 25: 179-188.

Guého E, Tredick J, Phaff HJ (1984) DNA base composition and DNA relatedness among species of *Trichosporon* Behrend. A van Leeuwenhoek 50: 17-32.

Hara N, Tubota Y, Saito K, Suto T (1989) Biochemical characteristics and ubiquinone of 44 strains of *Trichosporon beigelii* and related organisms. J Gen Appl Microbiol 35: 1-10.

Hasegawa Y, Higuchi K, Obata H, Tokuyama T, Kaneda T (1982) A cyclohexanecarboxylic acid utilizing yeast: isolation, identification, and nutritional characteristics. Can J Microb 28: 942-944.

Hess PN (1988) Untersuchungen zur Leistungsfähigkeit aerober Suspensionskulturen in gerührten Bioreaktoren. Doctoral thesis, Swiss Federal Institute of Technology Zürich.

Hrmová M, Biely P, Vrsanska M, Petraková E (1984) Induction of cellulose- and xylan-degrading enzyme complex in the yeast *Trichosporon cutaneum*. Arch Microb 138: 371-376.

Hoy J, Hsu K-C, Rolston K, Hopfer RL, Luna M, Bodey GP (1986) *Trichosporon beigelii* infection: A review. Rev Infect Dis 8: 959-967.

Käppeli O, Fiechter A (1982) Growth of *Trichosporon cutaneum* under oxygen limitation: kinetics of oxygen uptake. Biotech Bioeng 24: 2519-2526.

Kemker BJ, Lehmann PF, Lee JW, Walsh TJ (1991) Distinction of deep versus superficial clinical and nonclinical isolates of *Trichosporon beigelii* by isoenzymes and restriction fragment length polymorphisms of rDNA generated by polymerase chain reaction. J Clin Microb 29: 1677-1683.

Kreger-van Rij NJW, Veenhuis M (1971) A comparative study of the cell wall structure of basidiomycetous and related yeasts. J Gen Microb 68: 87-95.

Laaser G, Möller E, Jahnke K-D, Bahnweg G, Prillinger H, Prell HH (1989) Ribosomal DNA restriction fragment analysis as a taxonomic tool in separating physiologically similar basidiomycetous yeasts. System Appl Microb 11: 170-175.

Mattern IE, Punt P (1988) A vector of *Aspergillus* transformation conferring phleomycin resistance. Fungal Genet Newslett 35: 25.

McBride ME, Kalter DC, Wolf JE Jr (1988) Antifungal susceptibility testing of *Trichosporon beigelii* to imidazole compounds. Can J Microb 34: 850-854.

Middelhoven WJ, van den Brink JA, Veenhuis M (1983) Growth of *Candida famata* and *Trichosporon cutaneum* on uric acid as the sole source of carbon and energy, a hitherto unknown property of yeasts. Antonie van Leeuwenhoek 49: 361-368.

Middelhoven WJ, Hoogkramer-Te Niet MC, Kreger-Van Rij NJW (1984) *Trichosporon adeninovorans* sp nov, a yeast species utilizing adenine, xanthine, uric acid, putrescine and primary n-alkylamines as the sole source of carbon, nitrogen and energy. Antonie van Leeuwenhoek 50: 369-378

Mörtberg M, Neujahr HY (1986) Transport and hydrolysis of disaccharides by *Trichosporon cutaneum*. J Bact 168: 734-738.

Neujahr HY (1990) Yeasts in biodegradation and biodeterioration processes. In: Verachtert H, De Mot R (eds) Yeasts: Biotechnology and Biocatalysis, M Dekker Inc, New York, pp 322-348

Neujahr HY, Gaal A (1973) Phenol hydroxylase from yeast. Purification and properties of the enzyme from *Trichosporon cutaneum*. Eur J Bioch 35: 386-400.

Ochsner UA, Glumoff V, Kälin M, Fiechter A, Reiser J (1991) Genetic transformation of the filamentous yeast *Trichosporon cutaneum* using homologous and heterologous marker genes. Yeast 7: 513-521.

Punt PJ, Oliver RP, Dingemanse MA, Pouwels PH, van den Hondel CAMJJ (1987). Transformation of *Aspergillus* based on the hygromycin B resistance marker from *Escherichia coli*. Gene 56: 117-124.

Ratledge C (1988) Yeasts for lipid production. Biochem Soc Trans 16: 1088-1091.

Samsonova IA, Böttcher F, Werner C, Bode R (1989) Auxotrophic mutants of the yeast *Trichosporon adeninovorans*. J Basic Microb 29: 675-683.

Schneider H, Biely P, Latta R, Lee H, Dorscheid D, Levy-Rick S (1990) Utilization by yeasts of D-glucarate, galactarate, and L-tartarate is uncommon and occurs in strains of *Cryptococcus* and *Trichosporon*. Can J Microb 36: 856-858.

Spånning Å, Neujahr HY (1987) Growth and enzyme synthesis during continuous culture of *Trichosporon cutaneum* on phenol. Biotech Bioeng 29: 464-468.

Stevens BJH, Payne J (1977) Cellulase and xylanase production by yeasts of the genus *Trichosporon*. J Gen Microb 100: 381-393.

Stüttgen E, Sahm H (1982) Purification and properties of endo-1,4-β-xylanase from *Trichosporon cutaneum*. Eur J Appl Microb Biotech 15: 93-99.

Varga JM, Neujahr HY (1970) Purification and properties of catechol 1,2-oxygenase from *Trichosporon cutaneum*. Eur J Bioch 12: 427-434.

Veenhuis M, Hoogkamer-Te Niet MC, Middelhoven WJ (1985) Biogenesis and metabolic significance of microbodies in urate-utilizing yeasts. Antonie van Leeuwenhoek 51: 33-43.

Veenhuis M, van der Klei IJ, Harder W (1986) Physiological role of microbodies in the yeast *Trichosporon cutaneum* during growth on ethylamine as the source of energy, carbon and nitrogen. Arch Microb 145: 39-50.

Weijman ACM (1979) Carbohydrate composition and taxonomy of *Geotrichum*, *Trichosporon* and allied genera. Antonie van Leeuwenhoek 45: 119-127.

West M, Emerson GW, Sullivan PA (1990) Purification and properties of two lactose hydrolases from *Trichosporon cutaneum*. J Gen Microb 136: 1483-1490.

Zimmermann M, Emeis CC (1989) Extracellular polysaccharidases from *Trichosporon beigelii*. In: Martini A, Vaughan Martini A (eds), Proceedings of the Seventh International Symposium on Yeasts, Wiley, Chichester, UK, S131.

Physiological functions of vacuoles in yeast: Mechanism of sequestration of metabolites and proteins into vacuoles

Y. Ohsumi
Department of Biology
College of Arts and Sciences
The University of Tokyo
3-8-1, Komaba, Meguro-ku
Tokyo 153
Japan

Vacuoles are the largest compartment in yeast cells, occupying about 25% of the total cell volumes. Vacuoles contain mainly low molecular weight solutes and ions, but quite small amount of proteins. Recent studies revealed that vacuoles are not inert organelles as previously thought, but play important roles on maintaining homeostasis of cytosol in many respects. Last 15 years we have been making efforts to understand the structure and function of vacuoles in yeast, *Saccharomyces cerevisiae*, by taking several different approaches, as described below.

Compartmentation of amino acids in the cell

Yeast cells have an intracellular organization similar to that of higher eukaryotic cells,and their intracellular homeostasis of ions and metabolites is regulated by the interaction of organelles. Yeast cells contain a large amount of amino acids and about half of them are metabolically less active and localized in the vacuoles. In order to understand metabolic regulation it is prerequisite to know about the dynamics and regulation of compartmentation of cellular constituents. We developed a widely applicable method to permeabilize plasma membrane specifically by treating cells with Cu^{2+} ion (Ohsumi *et al.*, 1988). By this method the vacuolar and cytosolic amino acid pools were differentially extracted and their dynamics was examined (Kitamoto *et al.*, 1988). The composition of cytosolic pool was found to be constant under various physiological conditions, while the vacuolar pool varied extremely with the growth media. When the cells were grown in minimal medium supplemented with various amino acids, the vacuolar pool of added the amino acid expanded, especially in the cases of histidine, lysine andarginine. Acidic amino acids were excluded from the vacuoles in any cases, and glycine distributed evenly both in the cytosol and vacuoles. When the cells which accumulated a large amount of arginine in the vacuoles were transferred to a nitrogen-free medium, arginine in the vacuoles disappeared rapidly. Mobilization of arginine accompaniedwithcountertransport ofK^+ from cytosol to vacuoles, which compensate osmotic pressure of the vacuoles.

NATO ASI Series, Vol. H 69
Molecular Biology and its Application to Medical Mycology
Edited by B. Maresca, G. S. Kobayashi and H. Yamaguchi

Since the total amount of solutes and ions within the vacuoles should have certain limitation, arginine came out of the vacuoles by addition of excess amount of lysine or histidine to the medium, or addition of high dose of mating pheromone, alpha factor, to ***MATa*** cells which induces influx of a large amount of Ca^{2+} (Ohsumi and Anraku, 1985). Compartmentation of amino acids is dynamic process and should be considered in the regulation of amino acid metabolism.

Transport systems in the vacuolar membrane

In order to understand the mechanism of transport of solutes and ions into vacuoles, first we established a procedure for large scale preparation of vacuolar membrane vesicles from highly purified vacuoles (Ohsumi and Anraku, 1981). Vesicles gave much more information about the mechanism of transport than intact whole vacuoles. The vacuoles obtained was proved to perfectlyretain their right-side-out orientation by freeze-fracture electron microscopy and their biochemical analysis. During the vesicle preparation we realized that ATPase activity was copurified with α mannosidase, a marker enzyme of vacuolar membrane, and found that the vacuolar membrane has its own energy transducing system, a third type of proton translocating ATPase (Kakinuma *et al.*, 1981, Uchida *et al.*, 1985, 1988). Vacuolar H^+-ATPase has been extensively investigated biochemically and molecular biologically (Anraku *et al.*, 1989). That is a complex enzyme consisted of more than ten different subunits and is not essential but quite important devise of yeast cell for keeping inside of vacuoles acidic. We also found various secondary active transport systems for amino acids (Ohsumi and Anraku, 1981, Sato *et al.* 1984) and Ca^{2+} (Ohsumi and Anraku, 1983) driven by a proton motive force donated by the H^+-ATPase. These vacuolar H^+/amino acids antiport systems were characterized with their low affinities and high capacities for their substrates. Kinetic analyses revealed that there are at least seven independent transport systems for total ten amino acids (Arg, Lys, His, Phe, Trp, Tyr, Gln, Asn, Ile, and Leu) in the vacuolar membranes. These H^+/amino acids antiporters may be a main transport system to sequester amino acids into vacuoles since they well explain the composition of vacuolar amino acid pool. As Ca^{2+}concentration of vacuolar sap is known to be high, the vacuoles are the main Ca^{2+} reservoir and keep Ca^{2+} concentration in the cytosol low by the Ca^{2+}/H^+ antiport system.
Further, we found a membrane-potential dependent cation channel in the vacuolar membrane by the artificial planar bilayer membrane fusion method using the vacuolar membrane vesicles (Wada *et al.* 1987). This ion channel conducts K^+ (single-channel conductance, 435 pS in 0.3 M KCl) and some other monovalent cations. This ion channel might play an essential role in maintaining the chemical and electrical potential differences across the vacuolar membrane (Ohsumi *et al.*, 1988). Patch clump method is now directly applicable to the isolated vacuoles and at least one another type of ion channel is detected in the vacuolar membranes (Yabe, personal communication).

Vacuole as detoxification compartment

Yeast cells accumulate several compounds in large amounts. *S*-adenonsyl-methionine (SAM) and polyphosphates are good examples, which are known to come to more than 10% of the dry mass under certain conditions. This kind of large capacity for accommodation is possible only when cells have large intracellular compartment, vacuoles, otherwise homeostasis of osmotic and ionic environment within cytosol may be spoiled. Since SAM was not taken up by the vacuolar membrane vesicles in the presence of ATP, the huge accumulation of SAM does not depend upon the proton motive force across the vacuolar membrane. It is not uncovered yet what is a driving force for this transport. We isolated a mutant lacking vacuoles entirely (Kitamoto *et al.* 1988), which could still grow in nutrient medium such as YEPD. This means that vacuoles may not be essential, at least for mitotic cell cycle progression in rich medium. However, the mutant (*slp1*) cells showed quite pleiotropic phenotypes. Its growth was sensitive to lysine, histidine, but not to arginine. Arginine is a good nitrogen source. While the former two amino acids are non-metabolizable in yeast cells, therefore, excess accumulation of these amino acids in the cytosol may have toxic effect. The mutant is also sensitive to high concentration of Ca^{2+} or heavy metal ions in the medium. These facts indicate that defect in compartment for storage causes unbalance of metabolites and ions in the cytosol. In addition to storage of excessive nutrients, vacuoles seem to segregate obstructive or harmful substances from cytosol. In this respect, vacuoles function as detoxification compartment. Many drugs are apparently accumulated in the vacuoles when introduced in the cell. Recently 5(6)-carboxyfluorescein diacetate is often used as a good fluorescence probe for vacuolar compartment because of its specific accumulation in the vacuoles by unknown mechanism. It is not unreasonable to suppose that antifugal drugs are also sequestered into vacuoles. In fact *slp1* mutant cells described above were supersensitive to many antibiotics such as geneticin, kanamycin, tetracycline, and methyglyoxal. While sensitivities to miconazole and aculeacin whose target sites are predicted to be on the plasma membrane, were not altered by the mutation. These facts support the idea that vacuolar function affect to the effective value of drug which has target in cytosol. Though it is important to uncover the molecular mechanism of transport phenomena of various foreign compounds across the vacuolar membrane, our knowledge is still in an early stage at this moment.

Vacuoles as lytic compartment

Vacuoles in yeast contain various kinds of hydrolases and their specific inhibitors are localized in the cytosol, so vacuoles have been postulated to be cellular digestive organelles that are homologous to lysosomes in animal cells (Matile, 1975; Wiemken *et al.*, 1979). Lots of information has been accumulated about the biochemical characteristics of these vacuolar enzymes and their biogenesis, especially molecular mechanism of targeting them to the vacuoles (Klionsky *et al.*, 1990). The genes for major vacuolar proteases proteinase A, proteinase B and carboxypeptidase Y have been cloned and their structure has been analyzed (Achstetter and Wolf, 1985); Klionsky *et al.*, 1988). Several specific roles of these proteases were proposed, but most of them were disproved. Recently Chiang and Schekman reported that vacuolar proteases may involved in the "catabolite inactivation", degradation of the key enzyme in gluconeogenesis, fructose 1,6-bisphosphatase, in the presence of good catabolite, glucose (Chiang and Schekman, 1991). The levels of these main vacuolar enzymes vary with the growth conditions. They also change with the growth stage, reaching maximum levels when the cells approach the stationary phase. There are several reports that activities of vacuolar hydrolases increase and drastic degradation of cellular proteins occurs during sporulation. Furthermore, cells lacking proteinase A and B shows sporulation minus phenotype. Many other vacuolar function defective mutants we obtained also cannot undergo the normal sporulation process. These findings strongly suggest that protein degradation in the vacuoles may have essential role in protein turnover in nutrient-deficient conditions. However, the mechanism of protein degradation in the vacuoles and its contribution to the total protein turnover are remained unsolved. Vacuolar protein degradation must be a precisely controlled process, involving many steps. To understand the whole process of protein degradation in the vacuoles, it is desirable to dissect it to elementary steps by molecular genetical approaches.

Sequestration of cytosol into vacuoles

We expected that if mutants cells deficient in vacuolar proteases show the autophagic response to nutrient starvation condition, cellular components sequestered in the vacuoles may not be degraded and may be detected under light microscope, because the vacuoles are the most prominent compartment and are easily distinguishable from other parts of the cell under light microscope. Then the morphological changes of vacuoles during nitrogen starvation were examined with multiple vacuolar proteases-deficient strains constructed by Jones (Jones, 1984). In a nutrient medium YEPD, these mutant cells grew well, and their vacuoles appeared normal. Their vacuoles were transparent and contained few detectable particles. When these cells were transferred to a nitrogen free medium, they stopped growing. After incubation for 1 h in nitrogen free medium, spherical bodies were detected in the vacuoles, moving ceaselessly by Brownian movement because of low viscosity of vacuolar sap. The number of these bodies gradually increased and in 3 h they filled the vacuoles almost completely. The accumulation of these bodies occurred synchronously in almost all cells, regardless of their phase in the cell cycle. Similar accumulation of spherical bodies was induced under carbon or sulfate free medium. Therefore, it is general response of yeast cells in adverse conditions. To characterize the bodies in the vacuoles, we examined their ultrastructure by electron microscopy using rapid freezing and freeze-substitution methods. Spherical bodies detected by light microscopy was found to be cytosol surrounded by a unit membrane. The contents of the bodies were morphologically indistinguishable from the cytosol, containing the same density of cytoplasmic ribosomes. To characterize the bodies biochemically, vacuoles were prepared from cells grown in carbon starvation medium. Isolated vacuoles still retained many bodies inside. Intact vacuoles showed a little activity of the cytoplasmic marker enzyme, G6PDH. However, the vacuoles exhibited G6PDH activity after freezing-thawing or treating them with 1% Triton X100. This latent G6PDH activity increased according to the accumulation of the bodies in the vacuoles. These data strongly suggested that the bodies are the intermediate form of cytosol sequestered into the vacuoles to be degraded, therefore, we named these structure "autophagic body".

Process of autophagy in yeast

Whole process of accumulation of autophagic bodies in the vacuoles were further analyzed by electron microscopy. Spherical double membrane structures enclosing cytosol were observed in the cytoplasm of carbon- or nitrogen starved cells. Fusion of these double membrane structure with their outer membrane to the vacuoles was also detected. As the result single unit membrane structure, autophagic body, was discharged into vacuole. These observations indicate that double membrane structure correspond to autophagosomes and that autophagic process in yeast is essentially the same as that in higher eukaryotic cells. Autophagosome in the process of formation, cup shape structure enclosing a part of cytosol, was also detected. The autophagosomal membrane was apparently thinner than those of rough endoplasmic reticulum, Golgi body, nucleus and vacuole. It was positively stained by polysaccharides staining. Autophagic bodies ranged from 200 to 900 nm in diameter, and contained ribosomes, rough endoplasmic reticulum, mitochondria, lipid granules, glycogen granules and membrane vesicles. But there was no indication that organelles were preferentially sequestered into the vacuoles. These findings suggest that non-selective and bulk protein degradation by autophagy is induced in various nutrient starvation conditions.

Conditions for induction of autophagy

Autophagy in yeast is obviously recognizable by the accumulation of autophagic bodies in the vacuoles using multiple proteases-deficient mutants. First starvation condition to induce autophagy was examined. Depletion of single auxotrophic amino acidwas found to be sufficient to induce autophagy in yeast. However, the extent of accumulation of autophagic bodies varied with the amino acid depleted. Depletion of methionine induced most extensive accumulation of autophagic bodies by unknown reason. In contrast uracil starvation did not cause any accumulation, although it stopped cell growth. Signal(s) for induction of autophagy in yeast is important problem remained to be solved. Dissipation of the pH gradient across the vacuolar membrane has been reported to cause mis-localization of vacuolar enzymes (Rothman *et al.*, 1989). Further recently it is reported that Bafilomycin A1, a potent inhibitor of vacuolar type H^+-ATPases, causes accumulation of autophagolysosomes in animal cells (Henomatsu *et al.*, 1990). When acidication of vacuoles was inhibited by Bafilomycin A1 or ammonium acetate, accumulation of autophagic bodies was still observed. We concluded that acidification is not necessary for autophagy in yeast. To determine which vacuolar protease is responsible for the accumulation of autophagic bodies, genetic analyses were performed, and revealed that cell lacking either proteinase A or B accumulated autophagic bodies under various starvation conditions. Proteinase A functions as a processing enzyme for many other vacuolar enzymes including proteinase B. Furthermore disruptant of *PRB1* gene was constructed by standard procedure, and was shown to accumulate the autophagic bodies similarly as multiple proteases-deficient cells. These results indicate that proteinase B may be essential to degrade the autophagic bodies incorporated in the vacuoles. Proteinase B is a kind of serine protease and is inhibited by PMSF irreversibly. Wild type cells grown in YEPD medium were incubated in various starvation medium containing 1 mM PMSF. In exactly the same manner as protease deficient mutants, they accumulated autophagic bodies in the vacuoles. When the wild type cells accumulated autophagic bodies were transferred to fresh YEPD medium, autophagic bodies in the vacuoles disappeared rapidly. These facts strongly suggest that the autophagic body is an intermediate form of the normal autophagy in yeast. Protein degradation during starvation was estimated by the release of radioactivities from ^{14}C-leucine labeled cells. From wild type cells radioactivities released into medium significantly increased during starvation. In contrast multiple proteases-deficient mutants or *prb1* disruptants showed low levels of the release. PMSF inhibited the release from wild type cells to the extent of *prb1*mutant.

Physiological roles of protein turnover by autophagy

The sporulation process in yeast is triggered bynitrogen starvation, therefore, this cell differentiation should completely depend upon the degradation of their own proteins. The data presented above strongly suggest that sporulationminus phenotype of *pra1* , *prb1*or *slp1* mutants is due to the defect of autophagy. Yeast cells are known to enter the G1 phase of cell cycle, when they are confronted with nitrogen or sulfate deficiency. However, multiple proteases-deficient cells or *slp1*mutant cells could not progress their cell cycle to arrest at G1 phase in such starvation media. This means that protein degradation in the vacuoles by autophagic process must be essential to supply materials under adverse conditions.

References

Achstetter T, Wolf DH (1985) Proteinases, proteolysis and biological control in the yeast *Saccharomyces cerevisiae*. Yeast 1:139-157.

Anraku Y, Umemoto N, Hirata R, Wada Y (1989) Structure and function of the yeast vacuolar membrane proton ATPase. J. Bioenerg. Biomembr. 21:589-603.

Henomatsu N, Yoshimori T, Moriyama Y, Futai M, Tashiro Y (1990) Effect of Bafilomycin A1 on intracellular distribution of prolactin in the GH3 rat pituitary tumor cells. Cell Struct. Funct. 15:447.

Jones EW (1984) The synthesis and function of proteinases in *Saccharomyces cerevisiae*: genetic approaches. Annu. Rev. Genet. 18:233-270.

Kakinuma Y, Ohsumi Y, Anraku Y (1981)Properties of H^+-translocating ATPase in vacuolar membrane of *Saccharomyces cerevisiae*. J. Biol. Chem. 256:10859-10863.

Kitamoto K, Yoshizawa K, Ohsumi Y, Anraku Y (1988) Dynamic aspects of vacuolar and cytosolic amino acid pools of *Saccharomyces cerevisiae* . J. Bacteriol. 170:2683-2686.

Kitamoto K, Yoshizawa K, Ohsumi Y, Anraku Y (1988) Mutants of *Saccharomyces cerevisiae* with defective vacuolar function. J. Bacteriol. 170:2687-2691.

Klionsky DJ, Herman PK, Emr SD (1990) The fungal vacuole: Composition, function, and biogenesis. Microbiol. Rev. 54:266-292.

Klionsky DJ, Banta LM, Emr SD (1988) Intracellular sorting and processing of a yeast vacuolar hydrolase: proteinase A propeptide contains vacuolar targeting information. Mol. Cell. Biol. 8:2105-2116.

Matile P (1975) The lytic compartment of plant cells. Cell Biology Monographs, Vol. 1, Springer-Verlag, Wien, New York.

Ohsumi Y, Anraku Y (1981) Active transport of basic amino acids driven by a proton motive force in vacuolar membrane vesicles of *Saccharomyces cerevisiae*. J. Biol. Chem. 256:2079-2082.

Ohsumi Y, Anraku Y (1983) Calcium transport driven by a proton motive force in vacuolar membrane vesicles of *Saccharomyces cerevisiae*. J. Biol. Chem. 258:5614-5617.

Ohsumi Y, Anraku Y (1985) Specific induction of Ca^{2+} transport activity in *MATa* cells of *Saccharomyces cerevisiae* by a mating pheromone a factor. J. Biol. Chem. 269:10482-10486.

Ohsumi Y, Kitamoto K, Anraku Y (1988) Changes induced in the permeability barrier of the yeast plasma membrane by cupric ion. J. Bacteriol. 170:2676-2682.

Rothman JH, Yamashiro CT, Raymond CK, Kane PM, Stevens TH(1989) Acidification of the lysosome-like vacuole and vacuolar H^+-ATPase are deficient in two yeast mutants that fail to sort vacuolar proteins. J. Cell. Biol. 109:93-100.

Sato T, Ohsumi Y, Anraku, Y (1984) Substrate specificities of active transport systems for amino acids in vacuolar membrane vesicles of *Saccharomyces cerevisiae*. J. Biol. Chem. 259:11505-11508.

Uchida E, Ohsumi Y, Anraku Y (1985) Purification and properties of H^+-translocating, Mg^{2+}-adenosine triphosphatase from vacuolar membranes of *Saccharomyces cerevisiae*. J. Biol. Chem. 260:1090-1095.

Uchida E, Ohsumi Y, Anraku Y (1988) Characterization and function of catalytic subunit a of H^+-translocating adenosine triphosphatase from vacuolar membranes of *Saccharomyces cerevisiae*. J. Biol. Chem. 263: 45-51.

Wada Y, Ohsumi Y, Tanifuji M, Kasai M, Anraku Y (1987) Vacuolar ion channel of the yeast, *Saccharomyces cerevisiae*. J. Biol. Chem. 262:17260-17263.

Wiemken A, Schellenberg M, Urech K (1979) Vacuoles: the sole compartments of digestive enzymes in yeast (*Saccharomyces cerevisiae*. Arch. Microbiol. 123:23-35.

Characterization of a *Candida albicans*-specific DNA fragment

Pati M. Glee, Jim E. Cutler
Department of Microbiology
Montana State University
Bozeman, Montana 59717
USA

Investigative efforts toward understanding the basic biology of *Candida* species, primarily *Candida albicans*, have increased in recent years. Not only do significant clinical questions exist concerning colonization versus infection of mucosal surfaces, pathogenesis, virulence characteristics, diagnosis and treatment of deep-seated disease, and epidemiology, but also, developmental components and control of *C. albicans* morphogenesis remain undefined (Odds, 1988; Graybill, 1990).

Analysis of *C. albicans* MspI fragment distribution

Our original interest in *C. albicans* molecular biology was to analyze methylation differences that may occur during yeast to mycelial morphogenesis. Evidence in other eukaryotic organisms indicated that some developmentally regulated genes undergo 5-methylcytosine changes that are associated with differential gene expression (Jones, 1984; Jones, Taylor, 1980; Doerfler, 1983; Felsenfeld, McGhee, 1982). Restriction enzyme isoschizomers *Msp*I and *Hpa*II, that recognize CCGG sites[1], were utilized to assay methylation differences between *C. albicans* yeast and mycelial form DNA.

Gel electrophoresis and ethidium bromide (EtBr) staining indicated that *Msp*I- or *Hpa*II-digested *Candida* DNA distributed over a broad range of sizes and yielded strikingly more high molecular weight fragments than similarly digested *Neurospora* DNA. The fluorescent intensity for some of the *Candida* fragments suggested repetitive DNA sequences even into higher molecular weights (Cutler *et al.*, 1988).

Though we have not determined any *Msp*I versus *Hpa*II fragment differences by visual inspection of EtBr bright bands in yeast and mycelial form DNA samples, two observations regarding the *Msp*I/*Hpa*II fragment distribution of *C. albicans* DNA were intriguing: radiolabeled ribosomal DNA (rDNA) hybridized to Southern blots of *Msp*I-digested *C. albicans* DNA exhibited signal sizes of $\leq$ 1.5 kbp; and *Msp*I-digested mitochondrial DNA (mtDNA) prepared by the method of (Wills *et al.*, 1984a) displayed fragment sizes smaller than

---

[1] *Msp*I will digest CCGG and C^{m}CGG, but not ^{m}CCGG; whereas *Hpa*II will digest CCGG and ^{m}CCGG, but not C^{m}CGG (McClelland, 1981).

NATO ASI Series, Vol. H 69
Molecular Biology and its Application to Medical Mycology
Edited by B. Maresca, G. S. Kobayashi and H. Yamaguchi

4.4 kbp. These observations suggested that *C. albicans MspI* fragments larger than 4.4 kbp, including the putative repetitive sequences mentioned above, would be non-rDNA and non-mtDNA.

The published estimates (Scherer, Magee, 1990) of 40 - 80 copies of the 12 - 14 kbp rDNA repeat and 30 copies of the 40 kbp mtDNA would account, by our calculations, for 1,680 - 2,320 kbp of repetitive DNA per cell. The reported genome size for *C. albicans* isolates ranges from 37 - 40 femptograms per cell [reviewed in (Riggsby, 1990)] or approximately 36 - 39 Mb. Reassociation kinetics of *C. albicans* DNA indicated 13% repetitive sequences in the genome (Riggsby *et al.*, 1982) or 4,680 - 5,070 kbp of repetitive DNA. These rough calculations suggest that at least half of the projected amount of repetitive sequences in *C. albicans* DNA would not be accounted for by mtDNA plus rDNA repeats.

Two methods were used to assess the relative proportion of high molecular weight *C. albicans MspI* fragments as separated by gel electrophoresis: radiolabeled genomic DNA was digested, electrophoresed, and gel slices counted; and Polaroid negatives of *MspI* digests were scanned with an image analyzer to determine the distribution of EtBr-DNA fluorescence intensity. The mean distribution of *MspI* radiolabeled pieces ≤4.4 kbp was 27% for both 0.8% and 1.4% agarose gels which was similar to the 23% value determined by image analysis of the negatives. These data indicated that numerous *MspI* fragments from *C. albicans*, including single copy and putative repetitive elements, are larger than 4.4 kbp. The large repetitive *MspI* fragments may contribute substantially to non-rDNA and non-mtDNA repetitive DNA sequences in *C. albicans*.

Analysis of a *C. albicans*-specific DNA fragment

Attempts to clone *MspI/HpaII* repetitive elements into the *ClaI* site of pBR322 led to our report of a genomic fragment which hybridized to DNA from *C. albicans* and *C. stellatoidea* isolates, but not to DNA from other species of *Candida, Saccharomyces cerevisiae, Neurospora crassa, Gaeumannomyces graminis*, or mammalian DNA (mouse and human) (Cutler *et al.*, 1988). The recombinant plasmid, designated pHpaCa7, was found to contain unexpected *MspI* sites and additional *C. albicans* fragments totaling about 5.5 kbp instead of 2.9 kbp as originally thought (Cutler *et al.*, 1988). *MspI* digestion of pHpaCa7 generated three fragments larger than similarly digested pBR322. Because ligation of the original *C. albicans HpaII* fragments into *ClaI*-digested pBR322 does not regenerate a *ClaI* or *MspI/HpaII* recognition site, digestion of pHpaCa7 with *MspI* would leave some pBR322 bases attached to the 5' and 3' sides of any insert piece(s).

Additional restriction enzymes and Southern blots were done to determine the relative position of the *Candida* DNA insert pieces in pHpaCa7 and identify which fragments contained the pBR322 5' or 3' flanking DNA. The large (L) fragment contained only *C. albicans* DNA and displayed the species-specific hybridization; the 5'-flank was the small (S) *MspI* fragment; and the 3'-flank was the medium (M) *MspI* fragment. Size estimates for the L, M, and S fragments were determined from at least three separate 1 - 2% agarose gels and are listed in Table 1. The approximate amount of *C. albicans* DNA contained in the M and S fragments are 1.45 and 0.33 kbp, respectively.

Table 1
Fragment Length Estimates

Fragment	Kbp mean size (SD)	
Large (L)	3.80	(0.18)
Medium (M)	1.59[a]	(0.04)
Small (S)	0.77[a]	(0.06)

SD - standard deviation of the mean
[a] - Size includes pBR322 bases and *C. albicans* DNA

Because the large (L) fragment size of 3.8 kbp was similar to putative *Msp*I genomic repeats and *Msp*I-digested mtDNA fragments in the 3.6 to 4.2 kbp region, hybridization analysis to establish copy number and compartment origin was done. The L fragment indicated 1 - 2 copies per cell in DNA dot blots, and hybridized to whole cell DNA extracts but not mtDNA.
Labeled L fragment, used to probe blots of *Msp*I-digested genomic DNA from 47 *C. albicans* and 3 *C. stellatoidea* isolates, indicated four different signal patterns (Table 2). The 50 DNA samples included isolates from serogroups A (18 isolates) and B (20 isolates) with no apparent correlation of serotype and hybridization pattern. L fragment patterns displayed by *C. albicans* and *C. stellatoidea* isolates were identical. The original L fragment was cloned from *C. albicans* 9938 which gave a single band (pattern III) at 3.8 kbp. The other band sizes in patterns I, II, and IV were within 400 bp of the 9938 band size. The basis for L fragment restriction length differences among various *C. albicans* isolates remains undefined.
Results from *Msp*I-digested genomic blots probed with radiolabeled S fragment indicated numerous restriction length polymorphisms (RFLPs) among *C.*

Table 2
L Fragment Hybridization Patterns

Band Pattern	Occurrence [Serotype]
I (doublet)	25/50 [6 A, 15 B, 3 ND]
II (single, large)	6/50 [2 A, 2 B, 2 ND]
III (single, med.)	12/50 [6 A, 4 B, 2 ND]
IV (single, sm.)	7/50 [4 A, 3 ND]

ND - not determined

Table 3
Donated Fungal and Protozoan DNA

Gaeumannomyces graminis varieties - Joan Henson, Montana State University
N. crassa 74-OR-23-1A - Peter J. Russell, Reed College
Cryptococcus neoformans H99 - John R. Perfect, Duke University
Mucor racemosus 1216B - John E. Linz, Michigan State University
Aspergillus flavus H125 - David M. Geiser, University of Georgia
Trichomonas vaginalis S-981 - Don Riley, University of Washington
Histoplasma capsulatum G22B ATCC 26034, *Blastomyces dermatitidis* CR, *Sepedonium chrysospermum* ATCC 16084, and *Chrysosporium keratinophilum* ATCC 14803 - from Elizabeth J. Keath, Washington University

albicans isolates that exhibited 5 - 10 *Msp*I bands, depending on the strain. The S fragment indicated higher copy number than the L fragment in chromosomal hybridization analysis and may be useful as a DNA typing tool, but S RFLP pattern stability over multiple generations was not determined. Southern blot hybridization analysis for the putative species-specific L fragment was extended to include other fungal and protozoan DNA samples which were graciously donated by individuals noted in Table 3. The L fragment did not hybridize to DNA from *Aspergillus, Histoplasma, Sepedonium, Chrysosporium, Mucor, Cryptococcus, Coccidioides, Blastomyces, Torulopsis,* or *Trichomonas vaginalis*.

In contrast, a mixture probe of radiolabeled M and S fragments hybridized to all of the DNAs tested, including human and mouse; and additional blots suggested that the S fragment may account for most of the observed cross-hybridization. The M and S fragment hybridization was not due to the attached pBR322, because labeled pBR322 only cross-hybridized to a single band of restriction digested *Chrysosporium* DNA and not to other *Candida* species, other fungi, or human and mouse DNAs.

Northern blots of *C. albicans* RNA extracts probed with the L fragment were positive and suggested that at least a portion of L fragment sequences are transcribed. The M and S fragments also indicated positive Northern hybridization signals. Partial nucleotide information from double-stranded sequencing of all three *C. albicans* DNA fragments indicated many open reading frames and other features which are summarized in Table 4.

Table 4
Sequence Information

Fragment	Data
L	1.42/3.8 kbp sequenced; 35% G + C; no large regions of significant homology (GenBank)
M	0.54/1.6 kbp sequenced; 37% G + C; region 31 - 171 bp gave 74 - 80% homology to *S. cerevisiae*, *S. carlsbergensis*, and *M. racemosus* rDNA
S	0.33/0.79 kbp sequenced; 61% G + C; 13 RGYCAG tandem repeats (where R=A or G, Y=C or T); 9 other CAG repeats present; CAG repeat region was homologous to many human and mouse proteins

Summary

C. albicans has a significant number of high molecular weight *MspI* fragments which are likely A-T rich, lack CCGG sequences, and some sequences may be multi-copy elements that contribute to the non-rDNA and non-mtDNA repetitive portion of the *Candida* genome. A 3.8 kbp CCGG-poor sequence, denoted the L fragment, appears to be highly specific for *C. albicans*, has low copy number, and indicated transcription capacity.

References

Cutler JE, Glee PM, Horn HL (1988) *Candida albicans*-specific DNA fragment. J Clin Microbiol 26: 1720-1724.

Doerfler W (1983) DNA methylation and gene activity. Ann Rev Biochem 52: 93-124

Felsenfeld G, McGhee J (1982) Methylation and gene control. Nature 296: 602-603.

Graybill JR (1990) Therapeutic strategies for management of Candidiasis. In: Second Conference on *Candida* and Candidiasis. 1 - 4 April 1990. Abst.S4.

Jones PA (1984) Gene activation by 5-azacytidine. In: Razin A, Cedar H, Riggs A (eds) DNA Methylation: Biochemistry and Biological Significance. Springer-Verlag New York, Inc., New York, NY, pp 165-187.

Jones PA, Taylor SM (1980) Cellular differentiation, cytidine analogs and DNA methylation. Cell 20: 85-93.

McClelland M (1981) Purification and characterization of two new modification methylases: MClaI from *Caryophanon latum* MTaqI from *Thermus aquaticus* YTI. Nuc Acids Res 9: 6795-6804.

Odds, FC (1988) *Candida* and candidosis. 2nd edition. Bailliere Tindall, London

Riggsby WS (1990) Physical characterization of the *Candida albicans* genome. In: Kirsch DR, Kelly R, Kurtz MB (eds) The Genetics of *Candida*. CRC Press, Inc., Boca Raton, FL, pp 125-146.

Riggsby WS, Torres-Bauza LJ, Wills JW, Townes TM (1982) DNA content, kinetic complexity, and the ploidy question in *Candida albicans*. Mol Cell Biol 2: 853-862.
Scherer S, Magee PT (1990) Genetics of *Candida albicans*. Microbiol Rev 54: 226-241.
Wills JW, Lasker BA, Sirotkin K, Riggsby WS (1984a) Repetitive DNA of *Candida albicans*: nuclear and mitochondrial components. J Bacteriol 157: 918-924.

Recent advances in the molecular biology of *Cryptococcus neoformans*

Jeffrey C. Edman
Hormone Research Institute &
Department of Laboratory Medicine
University of California
San Francisco, CA 94143-0534
USA

Introduction

Cryptococcus neoformans is the causative agent of cryptococcosis. Once a rare disease seen only in the terminally ill, it became much more common with the advent of immunosuppressive regimens for the treatment of neoplastic disease and for transplant maintenance. The disease has taken on added importance as one of the most common opportunistic infections seen in patients with AIDS. Anti-cryptococcal therapy is difficult in AIDS patients due to decreased cure rates and increased relapse rates. This increased incidence of cryptococcosis has lead to increased interest in the mechanism of pathogenesis and virulence.

A number of molecular biological studies have been performed on *C. neoformans*. These have recently been reviewed and are indicative of the substantial progress on this organism. Over a dozen genes have been isolated by various investigators with the aims of developing transformation systems, understanding the molecular basis of virulence, and addressing the phylogenetic placement of *C. neoformans*. As these studies have been reviewed (Kwon-Chung and Edman, 1992), this review will focus on published and unpublished observations on the nature of DNA-mediated transformation in *C. neoformans*. A brief introductory section on biology and genetics will serve to demonstrate the utility of *C. neoformans* as an experimental organism.

Biology and genetics

C. neoformans was originally felt to be a member of the Fungi Imperfecti. The demonstration of its perfect or sexual state (*Filobasidiella neoformans*) by Kwon-Chung in the 1970s led to its placement in the Basidiomycetes (Kwon-Chung, 1975). The organism is probably heterothallic (asexual forms are of stable mating types), though self-fertile (and presumably diploid) strains have been described (Kwon-Chung, 1977; White and Jacobson, 1985). Its life cycle has been elucidated and is shown in Figure 1.

Under appropriate conditions, opposite mating types fuse to form a filamentous dikaryon. Meiosis then occurs in the terminal basidia and chains of basidiospores are formed. Each of the basidiospores can then give rise to the asexual state (yeast form) which will undergo continuous mitosis and

NATO ASI Series, Vol. H 69
Molecular Biology and its Application to Medical Mycology
Edited by B. Maresca, G. S. Kobayashi and H. Yamaguchi

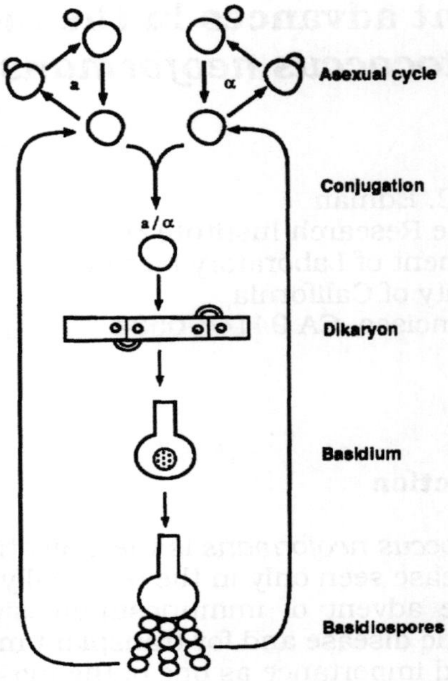

Figure 1. Schematic of the life cycle of *Cryptococcus neoformans*. Adapted from Whelan, 1987.

division via budding until encountering a cell of the opposite mating type. Mating type is controlled by a single locus, two allele system (**a** and a; ref. Kwon-Chung, 1980; Whelan, 1987). Similar to the well-studied *S. cerevisiae*, the life cycle of *C. neoformans* allows for extremely powerful genetic studies. Only limited numbers of such studies have been described, but they provide the bases for the genetic dissection of this organism (Whelan and Kwon-Chung, 1986; Still and Jacobson, 1983). Auxotrophic mutants are easily obtained from appropriate selective media (Whelan and Kwon-Chung, 1986). Conjugation and sporulation have been shown to result in the Mendelian segregation of phenotypes in basidiospores (Kwon-Chung, 1980; Jacobson *et al.*, 1982). Microdissection of spore masses from individual basidia allows random spore analysis to be carried out (Still and Jacobson, 1983).

C. neoformans can be propagated on standard yeast media. It grows rapidly and forms distinctive glistening round colonies on solid media. It can be replicated plated and mutagenized by standard techniques. Mating can be perform on semi-starvation media (either V8 juice agar or synthetic media) and basidiospores develop within three days. Techniques for the isolation of proteins, carbohydrates, and nucleic acids have been reported (Edman and Kwon-Chung, 1990; Bhattacharjee *et al.*, 1979, Cherniak, 1988; Polacheck *et al.*, 1982). A mouse model of cryptococcosis allows for the study of the role of various genes in virulence.

Genes

The study of gene structure and function in *C. neoformans* can be divided into two purposes. One is the isolation and characterization of genes that may be useful for transformation and gene disruption/replacement. The second is the characterization of genes which are thought to encode functions that are necessary for the virulence of *C. neoformans*. The set of genes in the second class include those responsible for melanin production, capsule production, mating type determination, and thermotolerance (Fromtling *et al.*, 1982; Kwon-Chung *et al.*, 1982; Kwon-Chung and Rhodes, 1986; Littman and Tsubura, 1959; Kwon-Chung *et al.*, 1992).

Transformation

Transformation of *C. neoformans* has been reported by Edman and Kwon-Chung using the *C. neoformans URA5* gene and *ura5* mutants (Edman and Kwon-Chung, 1990; Varma *et al.*, 1992). Transformation has also be achieved with the *ADE1* and *ADE2* genes of *C. neoformans* and has similar characteristics to that reported with *URA5* (unpublished observations).
Transformation of fungi with exogenous DNA can be broadly separated into two outcomes: the presence of autonomously replicating extrachromosomal transforming DNA or integration of transforming DNA into chromosomal DNA. Integration events can either be ectopic (inserting into DNA regions with no apparent sequence similarity to the transforming DNA) or homologous (inserting into regions with sequence similarity). Homologous insertions can be of two types: addition/substitution events or replacement/conversion events (Fincham, 1989).
The fate of transforming DNA in most fungal systems is dependent on the presence of autonomous replicating sequences (ARS) in the vector. If an ARS is present, then vector tends to be maintained as extrachromosomal circle. ARS sequences also allow for high copy number and increased transformation frequencies. The addition of telomere sequences will allow for the maintenance of linear plasmids. Centromeric sequences improve mother-daughter segregation of vectors while slightly decreasing copy number.
When *C. neoformans ura5* mutants are transformed with supercoiled *URA5* vectors, transformation frequencies of approximately 10-50 per μg are obtained. Two types of transformants are obtained: stable (retain uracil prototrophy after growth on rich media) and unstable. Between 60-90% of the total are unstable. Southern analysis of stable transformants reveal that all have integrated *URA5* vector sequences and exclusively in ectopic locations (i.e. the restriction pattern of the resident *ura5* locus is unaltered). Unstable transformants, on the other hand, harbor extrachromosomal DNA molecules. Since the transforming vector in this case is supercoiled DNA, it was first assumed that the vector contained an ARS sequence and that the DNA was present as a circular molecule in the unstable transformants. Restriction mapping of the extrachromosomal molecule, however, was inconsistent with a circular structure. It appeared that the molecules were being randomly linearized and maintained in linear state. The use of linearized vectors and analysis of their transformants confirmed this (see below). In summary, supercoiled *URA5* vectors transform at low frequency and result in two types of transformants: a majority of unstable transformants harboring apparently linear extrachromosomal DNA in high copy number and a minority of stable transformants with ectopically integrated vector sequences.

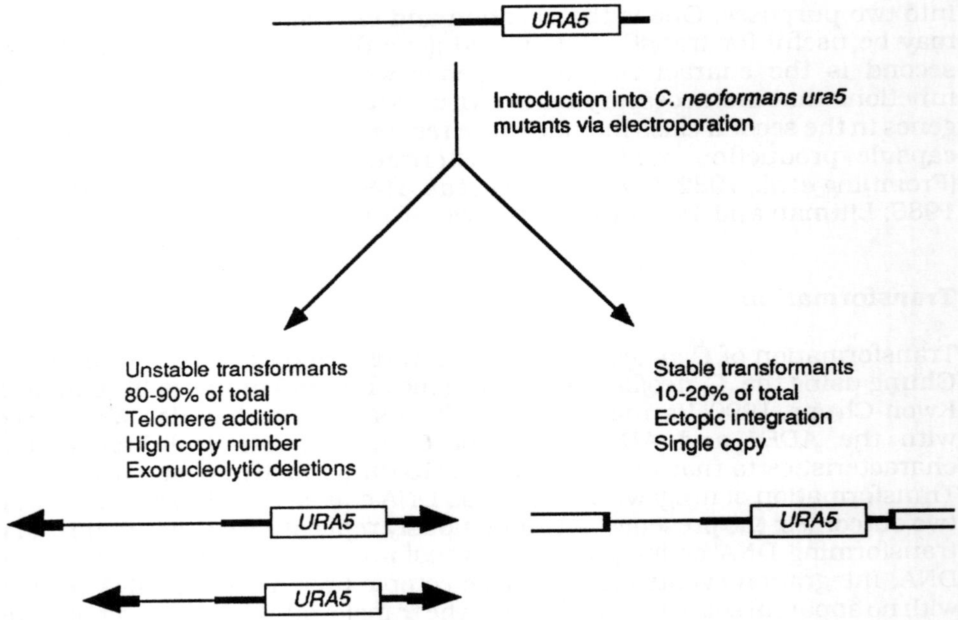

Figure 2. Summary of transformation of *C. neoformans* with linear *URA5* vectors.

When linearized *URA5* vectors or restriction fragments containing the *URA5* gene are used to transform *C. neoformans*, the frequency of transformation increases five to ten-fold. In *S. cerevisiae*, linearization of non-ARS containing vectors leads to an increased transformation frequency and an increase in homologous integration events. However, when the *C. neoformans* transformants were characterized by Southern analysis, it was shown that linearization did not result in an increase in homologous events. Rather, a generalized increase in transformation was evident. A similar ratio of stable (integrants) to unstable (extrachromosomal) transformants to that of supercoiled DNA was seen. Southern analysis of the unstable transformants revealed that most contained extrachromosomal vector sequences of a size that was similar to that of the transforming DNA. A minority of the unstable transformants harbored extrachromosomal DNA that was smaller than the transforming DNA. Rare transformants were also seen that contained extrachromosomal *URA5* sequences with sizes greater than the input DNA. Restriction mapping indicated a linear structure very similar to that of the linearized transforming DNA. In those cases where the extrachromosomal DNA is smaller than the input, it appears that deletions from the ends of the molecule were taking place rather than gross structural alterations. A summary of transformation of *C. neoformans* with linearized vectors is presented in Figure 2.

The linear nature of extrachromosomal *URA5* sequences in unstable *C. neoformans* transformants was confirmed by exonuclease treatment (Edman, 1992). DNA from transformants was treated with Bal31 nuclease or T7 gene 6 exonuclease and could be shown to be sensitive. In order for linear DNA to be maintained at a constant size, a mechanism for the maintenance of chromosomal ends must be available. Specialized sequences (telomeres) and enzymatic machinery (telomerases) are present in all eukaryotic cells to achieve this maintenance (Blackburn, 1990; Zakian, 1989). As the extrachromosomal DNA is linear in *C. neoformans*, it could be predicted that the ends have undergone telomere addition. In order to prove this, it was necessary to recover the extrachromosomal DNAs in *E. coli* and subject the putative telomeres to sequence analysis. As linear molecules are not propagated in *E. coli*, it was necessary to recircularize the *URA5* vector sequences prior to transformation. Direct recircularization of the molecules present in *C. neoformans*, however, would lead to an inverted repeat of telomere sequences. As inverted repeats are known to be unstable in *E. coli* (Collins, 1981), an alternate procedure involving the prior removal of single telomere and then recircularization was employed (Figure 3).

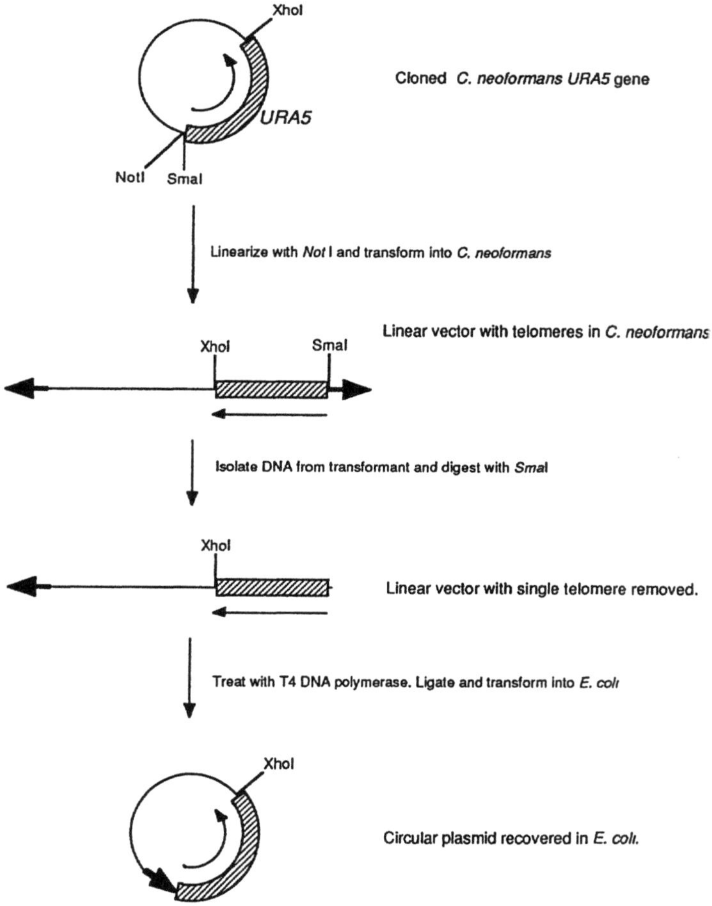

Figure 3. Strategy for the recovery of telomeres from extrachromosomal vectors of *C. neoformans*.

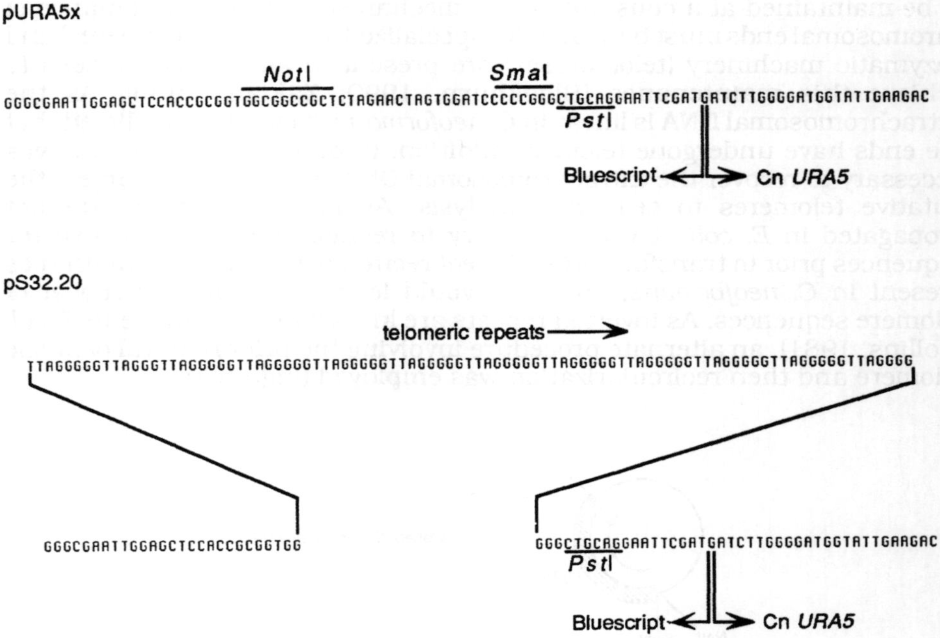

Figure 4. Sequence of transforming vector pURA5x and recovered vector with telomeric sequences.

DNA from cells transformed with the *URA5* vector was digested with *Sma*I (which would remove one telomere), treated with T4 DNA polymerase to create "blunt" ends on the telomere, and ligated into circular molecules with T4 DNA ligase. The ligation was used to transform *E. coli* and plasmids were sequenced across the putative telomere addition site. Figure 4 shows a typical result.

In all cases, a repetitive sequence characteristic of telomere sequences seen in other organisms has been added at the *Not*I site. The repeats are the octanucleotide AGGGGGTT. The number of repeats varies among the individual recovered plasmids.

The effect of telomere sequences on transformation frequencies was tested by introduction of telomere containing vectors into *C. neoformans*. As the recovered plasmids contained only one telomere sequence and we wanted to test the effect of have telomeres on both ends of the transforming DNA, a vector was constructed *in vitro* that possessed two telomeres separated by the kanamycin resistance gene from Tn903. In this construct, the kanamycin fragment could be removed from the vector by *Pst*I digestion leaving a vector with two telomeres in the appropriate orientation (Figure 5).

When telomere-containing *URA5* vectors are introduced into *C. neoformans*, a dramatic increase in transformation frequency is observed. As noted above, vectors without telomeres transform at frequencies of about 100-500 per μg. Telomere-containing vectors transform at frequencies of 50,000 - 100,000 per μg. When transformants from telomere-containing vectors are

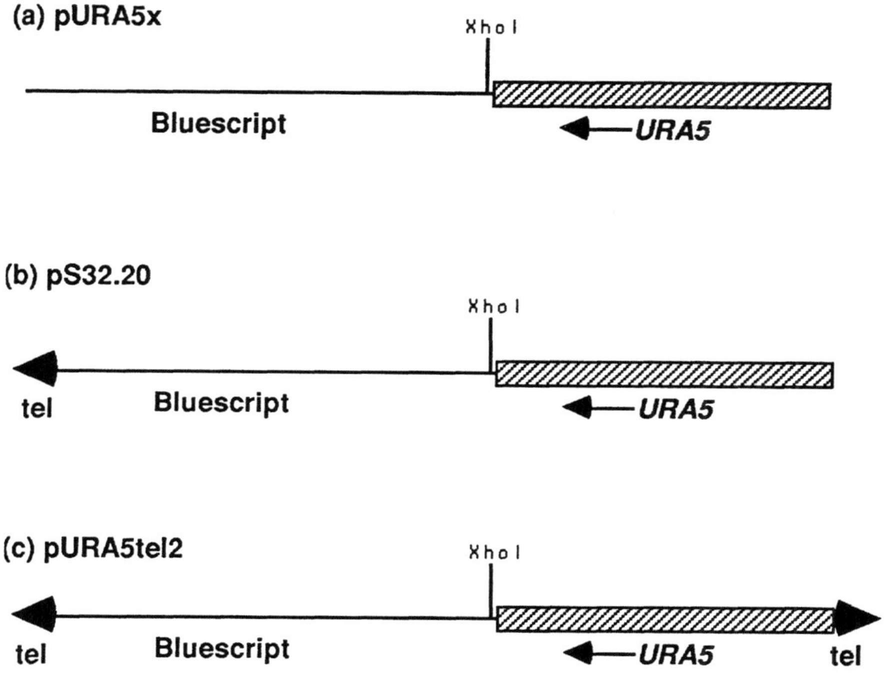

Figure 5. Telomere containing vectors of *C. neoformans*.

analyzed by Southern blot for the presence of *URA5* sequences, it can be seen that most contain extrachromosomal DNA that is virtually identical to the introduced vector. When only one telomere is present (as in S32.20), some deletions still occur on the *URA5* end of the vector. Transformants from a double telomere vector (pURA5tel2) appear to have remained intact. No deletions, rearrangements, or integrations were seen among twenty independent transformants. These data suggest that not only does the presence of telomeres dramatically enhance transformation efficiencies, but in addition protect the transforming DNA from exonucleolytic degradation, rearrangements, and integration.

The effect of telomeres on transformation frequency is reminiscent of the effect that ARS sequences have on *S. cerevisiae* and other fungal vectors. From Southern blotting estimation, it can be shown that the copy number of extrachromosomal *URA5* is on the order of hundreds per cell. Therefore, irrespective of whether or not the transforming vector had telomeres prior to transformation. This suggests that the URA5 sequences are being replicated efficiently in the extrachromosomal state. The enhancement of transformation frequency seen with telomere containing vectors could either be due to the telomere providing some sort of ARS function or protection from degradation. It has already been noted that the most common rearrangement seen with linear vectors are simple exonucleolytic deletions. Telomeres could presumably provide a means of escaping such degradative processes. Telomere addition and high copy number extrachromosomal transformation are inseparable in transformation with linear *C. neoformans* vectors. It may be difficult to prove experimentally that the telomeres are providing some sort of replication origin, but it is clear that the presence of telomeres on *C. neoformans* linear vectors markedly increase transformation efficiencies and protect the transforming DNA from deletions, rearrangements, and integration events.

Summary and prospects for the future

The development of transformation system for *C. neoformans* has allowed the beginnings of the molecular dissection of this pathogen. The ability to complement existing mutations of *C. neoformans* with genomic DNA sequences should allow the molecular isolation of genes that have been associated with virulence by classical genetic studies (i.e. capsule, melanin, and mating type genes). Currently lacking from the transformation systems of *C. neoformans* is a method for gene replacement or conversion. This would require a higher frequency of homologous integrations that is seen with the *URA5* gene and *ura5* mutants. Preliminary data suggest that homologous integration can occur at much higher frequencies using the *ADE2* gene and serotype A *C. neoformans ade2* mutants (Toffaletti, D.L., Perfect, J.R., Rude, T.H., Johnston, S.A., and Durack, D.T., personal communication). As all the studies described to date have involved the use of serotype D *ura5* mutants, it will be interesting to see if the effect is due to the *ADE2* gene or is serotype specific. As noted above, there are significant karyotypic differences between these serotypes and perhaps serotype A strains possess a much more effective mechanism for homologous recombination. The ability to perform homologous will eventually allow the manipulation of the *C. neoformans* genome in a similar fashion to that of *S. cerevisiae*. Such studies will expand our knowledge of the basic biology of *C. neoformans* and to begin to dissect the role of various genes in virulence.

References

Bhattacharjee AK, Kwon-Chung KJ and Glaudemans CPJ (1979) The structure of the capsular polysaccharide of *Cryptococcus neoformans* serotype D. Carbohydr Res 73: 183-192.

Blackburn EH (1990) Telomeres: structure and synthesis. J Biol Chem 265: 5919-5921.

Cherniak R (1988) Soluble polysaccharides of *Cryptococcus neoformans*. Curr Top Med Mycol 2: 40-54.

Collins J (1981) Instability of palindromic DNA in *E coli*. Cold Spring Harbor Symp Quant Biol 45: 409.

Edman JC (1992) Isolation of telomere-like sequences from *Cryptococcus neoformans* and their use in high-efficiency transformation. Mol Cell Biol 12:2777-2783.

Edman JC and Kwon-Chung KJ (1990) Isolation of the *URA5* gene from *Cryptococcus neoformans* var *neoformans* and its use as a selective marker for transformation. Mol Cell Biol 10: 4538-4544.

Fincham JRS (1989) Transformation in Fungi. Microbiol Rev 53: 148-170.

Fromtling RA, Shadomy HJ and Jacobson ES (1982) Decreased virulence in stable, acapsular mutants of *Cryptococcus neoformans*. Mycopathologica 79: 23-29.

Jacobson, ES, Ayers DJ, Harrell AC and Nicholas CC (1982) Genetic and phenotypic characterization of capsule mutants of *Cryptococcus neoformans*. J Bacteriol 150: 1292-1296.

Kwon-Chung KJ (1975) A new genus, *Filobasidiella*, the perfect state of *Cryptococcus neoformans*. Mycologia 67: 1197-1200.

Kwon-Chung KJ (1977) Heterothallism vs self-fertile isolates of *Filobasidiella neoformans* (*Cryptococcus neoformans*) in The black and white yeasts Pan American Health Org Sci Pub 356 Proceedings of the IV Int Conf on Mycoses Washington, DC pp 204-213.

Kwon-Chung KJ (1980) Nuclear genotypes of spore chains in *Filobasidiella neoformans* (*Cryptococcus neoformans*). Mycologia 72: 418-422.

Kwon-Chung KJ and Edman JC (1992) Recent advances in the molecular genetics of *Cryptococcus neoformans* in New Strategies in Fungal Disease. Churchill Livingstone London pp 195-207.

Kwon-Chung KJ, Edman JC and Wickes BL (1992) Genetic association of mating types and virulence in *Cryptococcus neoformans*. Infect Immun 60: 602-5.

Kwon-Chung KJ, Polachek I and Popkin TJ (1982) Melanin-lacking mutants of *Cryptococcus neoformans* and their virulence for mice. J Bacteriol 150: 1414-1421.

Kwon-Chung KJ and Rhodes JC (1986) Encapsulation and melanin formation as indicators of virulence in *Cryptococcus neoformans*. Infect Immun 51: 218-233.

Littman ML and Tsubura E (1959) Effect of degree of encapsulation upon virulence of *Cryptococcus neoformans*. Proc Soc Exp Biol Med 101: 773-777.

Polacheck I, Hearing VJ and Kwon-Chung KJ (1982) Biochemical studies of phenoloxidase and utilization of catecholamines in *Cryptococcus neoformans*. J Bacteriol 150: 1212-1220.

Still CN and Jacobson ES (1983) Recombinational mapping of capsule mutations in *Cryptococcus neoformans*. J Bacteriol 156: 460-462.

Varma A, Edman JC and Kwon-Chung KJ (1992) Molecular and genetic analysis of *URA5* transformants of *Cryptococcus neoformans*. Infect Immun 60: 1101-1108.

Whelan WL (1987) The genetics of medically important fungi.Crit Rev Microbiol 14: 99-170.

Whelan WL and Kwon-Chung KJ (1986) Genetic complementation in *Cryptococcus neoformans*. J Bacteriol 166: 924-929.

White CW and Jacobson ES (1985) Occurrence of diploid strains of *Cryptococcus neoformans*. J Bacteriol 161: 1231-1232.

Zakian VA (1989) Structure and function of telomeres. Ann Rev Genet 23: 579-604.

Molecular biology of *Kluyveromyces lactis*

L. Frontali, M.M. Bianchi, L. Fabiani, C. Falcone, C. Mazzoni,
A. Ragnini, M. Saliola, C.Wilson
Dept. of Cell and Developmental Biology
University of Rome "La Sapienza"
P.le A. Moro, 5
00185 Rome
Italy

The yeast *Saccharomyces cerevisiae* is probably the most thoroughly understood amongst eukaryotic organisms and an excellent model for the study of eukaryotic cells in general; indeed, the term "yeast" is often used as a synonim for this species. Recently, however, other yeasts have attracted the attention of researchers as a result of their distinct biological and metabolic properties which open up new possibilities of biological utilization. Among these so called "non-conventional" yeasts, *Kluyveromyces lactis* has attracted a special interest because of several peculiar characteristics, including the easiness of mass cultivation, its status of safe organism and its very good secretion properties. *K.lactis* is a budding yeast and, like *S.cerevisiae*, lends itself easily to genetic analysis. It differs, however, from *S.cerevisiae* for several important metabolic properties and is a "petite negative" yeast, i.e. a species in which no mitochondrial respiratory deficient mutants have been found. It has a considerably smaller number of chromosomes than *S.cerevisiae*, which are however larger in size. Pulsed field gel electrophoresis reveals six DNA bands ranging in size roughly between 1.2 and 2.8 megabasepairs (Sor and Fukuhara, 1989). Several genes are being mapped by hybridization on the chromosomes and results are being compared with linkage maps of known mutations obtained by tetrad analysis.
An other important property that *K.lactis* shares with *S.cerevisiae* is the capability for homologous recombination, which makes it possible to use gene disruption techniques to study gene function. In recent years the taxonomy of the genus *Kluyveromyces* has been the object of extensive discussions which have lead to repeated taxonomical modifications. The possibility of crosses between strains previously thought to be different species had lead to the unification of as many as 13 entities, which had been defined as species in 1970 (Van der Walt 1970), into the single species *Kluyveromyces marxianus* (Van der Walt and Johannsen 1984). Subsequent work based on reassociation kinetics (Fuson *et al.*, 1987; Vaughan and Martini, 1987) and on other molecular characteristics, summarized in table I, consistently supports the fact that *K.lactis* and *K.marxianus* are distinct taxonomical entities. An interesting aspect of the results shown in table I are the widely different molecular characteristics, such as the existence of GC clusters in mtDNA and the stable replication of some vectors (see below), which provide the same indications as those derived from reassociation kinetics.

NATO ASI Series, Vol. H 69
Molecular Biology and its Application to Medical Mycology
Edited by B. Maresca, G. S. Kobayashi and H. Yamaguchi

Plasmids

K.lactis strains can harbour two kinds of plasmids: a) a circular plasmid called pKD1 which bears a strong structure resemblance to the 2μ circles from *S.cerevisiae*, but has no sequence homology (Falcone *et al.*, 1986; Chen *et al.*, 1986; Bianchi *et al.*, 1987); b) linear DNA plasmids responsible for a killer phenotype due to the secretion of a plasmid encoded toxin which inhibits growth of sensitive yeasts (Stark *et al.*, 1990). The biology of the above mentioned plasmids exhibits some very interesting characteristics which we will review here very briefly. Circular plasmids have been found in several yeast species: *S.cerevisiae*, *K.lactis* (*var. drosophilarum*) and a certain number of osmophilic yeasts belonging to the genus *Zygosaccharomyces* (Volkert *et al.*, 1989). All of them share the same structural organization: two inverted repeats separate two unique regions in which three or four open reading frames can be found. In all studied circular plasmids functional analysis (Broach *et al.*, 1982, Jayaram *et al.*, 1983, Kikuchi 1983, Araki *et al.*, 1985; Toh-e and Utatsu 1984) has shown that two genes are involved in correct plasmid partitioning at cell division, while the third open reading frame encodes a site-specific recombinase which catalyzes intramolecular recombination at the inverted repeats and causes isomeric forms to be

Table I

Species in the genus *Kluyveromyces* (Van der Walt 1970)	%G+C (l)	Estimate.N° of chromosomes (2)	Transformability by pKD1 based vectors (3)	SacII Sites in mtDNA (4)
K.aestuari	40	7	+/-	-
K.africanus	38	14	-	-
K.blattae	34	4-5	-	n.t.
K.delphensis	40	9	-	-
K.loddarae	36	13	-	-
K.marxianus	41	8-12	-	-
K.bulgaricus	41	7-10	-	-
K.wikenii	41			
K.dobjanskii	43	6	+	+
K.lactis	40	6	++	+
K.drosophilarum	40	6	++	+
K.phaseolosporus	40	6	++	+
K.vanudenii	40	6	++	+
K.phaffii	35	11	-	-
K.polysporus	35	9	-	-
K.wickerhamii	42	8	-	+
K.thermotolerans	46.2	7	+	-
K.waltii	45.6	5	+	-

(l) From Fuson *et al.* 1987 and Vaughan and Martini 1987 (2) From Sor and Fukuhara 1989 (3) From Chen *et al.* 1989 (4) From Ragnini and Fukuhara 1988
n.t. = not tested

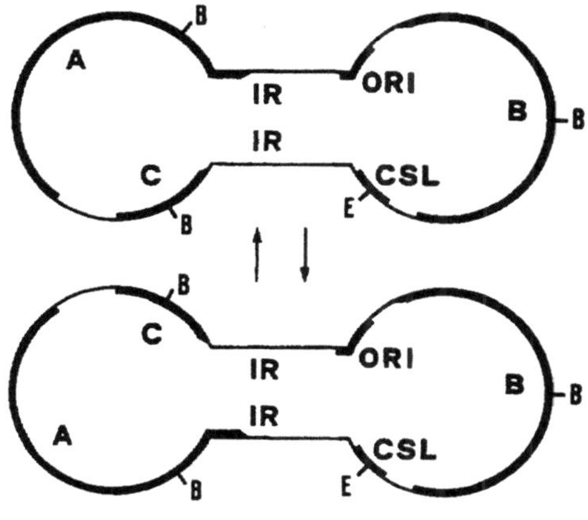

Fig. 1 The two isomeric forms of pKD1 are shown. The three genes (A, B and C), the two inverted repeats (IR), the putative replication origin (ORI) and the cis-acting stability locus (CSL) are indicated. BamHI (B) and EcoRI (E) restriction sites were used for the construction of vectors described in Table II.

present in the cells in equimolar amounts (see Fig. 1). This isomerization seems to be involved in plasmid amplification and hence in copy number control (Futcher 1986, Volkert and Broach 1986). Extensive regulatory circuits among plasmid genes have been shown to exist in *S.cerevisiae* (Murray *et al.*, 1987; Som *et al.*, 1988; Veit and Fangman 1988), while nothing is known in regard to the regulation of the other circular plasmids.

In all studied circular plasmids, correct partitioning at cell division requires, the products of two plasmid genes and the integrity of a short sequence (called stability locus) bearing several direct or inverted repeats (Jayaram *et al* 1985; Jearnpipatkul *et al.*, 1987, Bianchi *et al.*, 1991) which probably acts as an attachment site for the mitotic apparatus.

What is very peculiar about these plasmids is the fact that they do not confer any apparent advantageous phenotype to their hosts, and that all their genetic information seems to be involved only in proper partitioning and copy number maintenance. The absence both of advantageous phenotypes and of any sequence homology among these plasmids makes their evolutionary origin a very puzzling issue.

If, while taking into account the lack of sequence homology, we assume that the circular plasmids found in different yeasts had independent origins, then the common mechanism for plasmid partitioning and copy number control must be the result of an evolutionary convergence. Alternatively one would think that a very strong divergence from a common ancestor may have been the result of species-specific co-evolution of plasmid and nuclearly-encoded factors. As regards the killer phenotype, several *K.lactis* strains secrete a toxin that kills a variety of sensitive strains. The killer character is determined by two linear DNA plasmids, one of which encodes the toxin. The plasmids have their own transcription and replication systems (Stark *et al.*, 1990).

Gene manipulation in *K. lactis*

In general, the possibility of establishing host vector transformation systems for non conventional yeasts is based on three distinct alternatives: i) the use of integrative vectors, which in some yeasts give rise to multiple integrations;

ii) the use of replicative plasmid based vectors; iii) the use of replicative ARS-based vectors. The establishing of a very efficient transformation system (Bianchi *et al.*, 1987) has been a key element in the development of both the basic and applied aspects of *K. lactis* molecular biology.

Table II

Vector		Stability a)	
		pKD1°	pKD1⁺
	Inactivated gene		
- pE1	none	87.3	66.1
- pAA	A	69.6	55.3
- pBB	B	1.1	66.2
- pCC	C	2.4	65.8
	Elements present		
- pRO	ori	4.2	7.7
- pROCS2	ori,CSL	6.1	68.1
- pKRF4	KARS	3.3	3.3
- pACS2	KARS,CSL	4.2	66.0

a) % of ura⁺ cells after 7-8 generations in complete medium
Vector pEl was constructed by inserting the integrative vector YIp5 bearing the *URA3* selection marker into the unique EcoRI site of pKD1. Vectors pAA, pBB and pCC were constructed by inserting YIp5 into one of the three BamHI sites located inside each of the three open reading frames of pKD1 (see Fig. 1). vectors pRO, pROCS2, pKRF4, pACS2 were constructed by inserting into YIp5 DNA fragments containing the pKD1 replication origin (RO), the same plus stability locus (CS2), a chromosomal ARS sequence (KRF4), the same plus CS2. Details of the constructions are reported in Bianchi *et al.*, (1991).

pKDl-based vectors can transform *K. lactis* with a very high efficiency (5-10 x 10^4 transformants per μg DNA). Transformants are stable and, in general, contain 50-100 copies per cell of the transforming plasmid.
Stable vectors may contain the entire pKD1 sequence or the region containing the replication origin and the stability sequence (Bianchi *et al.*, 1991). In the latter case, stability is only obtained in pKD1⁺ cells, i.e. in cells bearing a resident plasmid, which supplies *in trans* the gene products necessary for correct partitioning at cell division. The *URA3* gene from *S.cerevisiae* or the G418 resistance gene from Tn903 have been used as selection markers and cloning, expression and secretion vectors have been constructed. Examples of the stabilities of different vectors are reported in Table II.
The species specificity of pKD1 based vectors is very narrow (Chen *et al.*, 1989), as indicated in Table I. This is probably the result of a defective interaction between plasmid and nuclearly - encoded factors involved in plasmid partitioning. pKD1 is actually capable of autonomous replication in *S. cerevisiae* but transformants are very unstable.

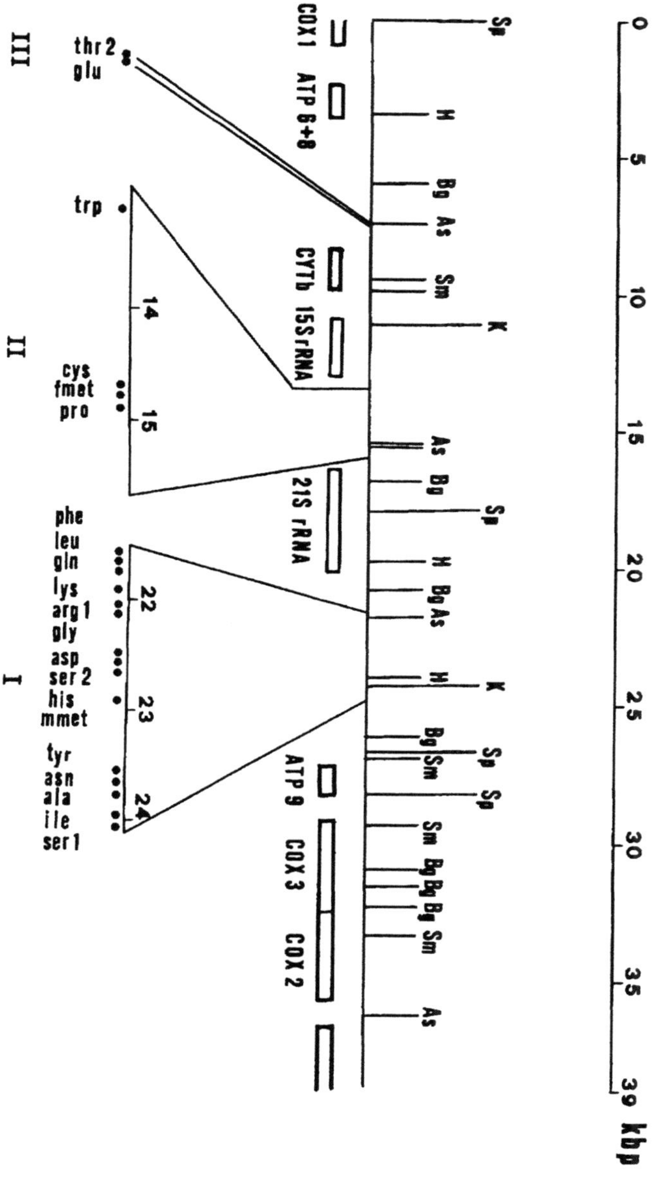

Fig. 2 Linearized map of the mitochondrial genome of *K.lactis*. Approximate positions of mitochondrial genes are indicated with respect to the major restriction sites. Genes are indicated by open bars or by dots (tRNA genes) Sp, SpeI; H, HindIII; Bg, BglII; As, AsuII; Sm, SmaI; K, KpnI.

Chromosomal fragments bearing ARS sequences have also been used for the construction of vectors (Das and Hollenberg 1982; Fabiani *et al.*, 1990). Some of the fragments manifest ARS activity also in *S.cerevisiae*, and the corresponding vectors can transform both yeasts. A comparison of sequences having ARS function in both *K.lactis* and *S.cerevisiae* or in only one of the two species was performed to obtain a detailed characterization of *K.lactis* chromosomal DNA replication origins.

ARS-based vectors are rather unstable in *K.lactis*, unless a fragment containing

the pKD1 stability sequence is inserted, in which case high stability is obtained in transformed cells bearing a resident plasmid (Bianchi *et al.*, 1991).
Transforming vectors have also been obtained from a linear killer plasmid in which the gene encoding the killer toxin had been substituted with a foreign gene. Since the killer plasmids have their own transcription system, one of the promoters of the killer plasmid itself must be used to obtain gene expression (Tanguy-Rougeau *et al.*, 1990).

Mitochondria

As we have mentioned before, mitochondrial "petite" mutants have never been found in *K.lactis*, so an analysis of the mitochondrial genome of this yeast might have provided some clues on the phenomenon of "petite negativity". However the informational content of *K.lactis* mtDNA was found to be quite the same as that of other studied yeasts. The gene order is reported in Fig. 2. In some strains an important translocation was found (Skelly *et al* 1991).
Other interesting characteristics of the *K.lactis* mitochondrial genome are listed below. Its size is around 39.000bp; as in *S.cerevisiae*, the genes are separated by AT-rich sequences in which several GC clusters are scattered. These clusters generally contain SacII sites. Hence, SacII digestion of *K.lactis* mtDNA yields many more fragments than are found in other *Kluyveromyces* species (See Table I).
Some genes contain non-coding sequences: the CoxI gene may contain one or four introns, depending on the origin of the strain, all of them bearing ORFs in frame with the preceding exons (Hardy and Clark-Walker 1991). One of the intervening sequences is a groupII intron and crossing experiments have revealed that it is very frequently (89%) transferred to mitochondrial genomes lacking the intron (Skelly *et al.*, 1991).
The large ribosomal subunit (LSU) rRNA gene contains an intron which is highly homologous to the one present in the corresponding gene of *S.cerevisiae.* In the latter yeast the intron contains an ORF encoding an endonuclease which permits intron transposition to the intron-less LSU rRNA genes in crosses with strains not containing the intron. In the case of *K.lactis*, however, the same intron does not contain the ORF, thus suggesting that the ORF itself might be a transposable element (Wilson and Fukuhara, 1991).
All of the above mentioned results point to a very high variability of yeast mitochondrial genomes, based both on translocations and on intron mobility.

Respiration-fermentation relationships

At present only a dozen genes, mainly involved in metabolism have been cloned from *K.lactis* genome. Their sequence revealed a high identity with the corresponding genes of *S.cerevisiae*, whereas upstream regions diverge and gene expression might be differently regulated. In fact *K.lactis* and *S.cerevisiae* exhibit important metabolic differences especially in regard to the regulation of fermentative and respiratory metabolisms. The ability to shift from respiration to fermentation seems to be rare among yeast species and, in this

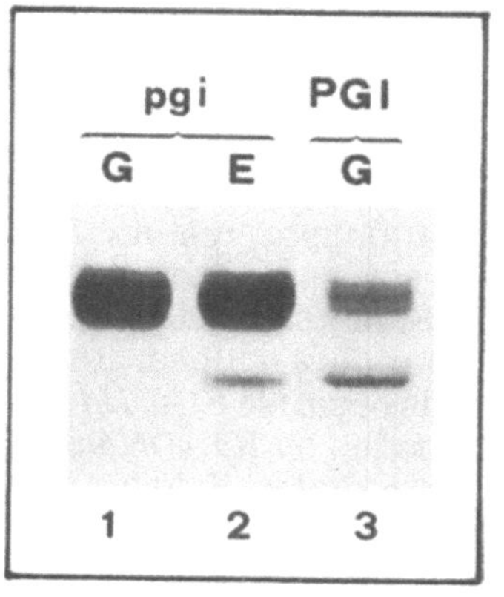

Fig. 3. Electrophoretic analysis of ADH activities in ethanol producing and non-producing strains of *K.lactis*. Samples of cell extracts, corresponding to 10 μg of protein, were electrophoresed on non-denaturing polyacrylamide gels and the ADH isozymes were revealed as bands by staining the gels for the enzyme activity (Lutstorf and Megnet, 1968). Extracts from ethanol nonproducing cells (*pgi* mutant strain) grown on 7% glucose (lane 1) and 2% ethanol (lane 2). Extracts from ethanol producing cells (*PGI* strain) grown on 7% glucose (lane 3).

respect, *S.cerevisiae* is exceptional. In this yeast, glucose represses respiratory enzymes and cells grow mainly by fermentation even in the presence of oxygen. By contrast, when *K.lactis* is grown on glucose under aerobic conditions, less than 30% of the sugar is metabolized by fermentation (for a review on yeast metabolism, see Gancedo and Serrano, 1989). Both the fermentative and respiratory metabolisms co-exist in cells of *K.lactis*, a fact which indicates that this yeast can escape the repression effect of glucose. It has in fact been reported that glucose does not affect mitochondrial respiration in this yeast (De Deken, 1966). As a result of their predominantly aerobic metabolism, *Kluyveromyces* species are not good producers of ethanol compared to *Saccharomyces* strains. Despite the low fermentative activity, four genes encoding alcohol dehydrogenase (ADH) activities are present in *K. lactis* (Saliola *et al*., 1990). The four ADH genes of *K.lactis*, named and numbered *KlADH1* to *KlADH4*, have been cloned and their nucleotide sequences have been determined (Saliola *et al*., 1990 and 1991; Denis C, personal communication). The four genes show strong sequence identities (near to 80%) with each other and with the corresponding genes of *S.cerevisiae* both at the nucleotide and the amino acid level. It has been reported that the expression of the ADH genes in *K. lactis* cells is under the regulatory control of the carbon source (Saliola *et al*., 1990). The electrophoretic analysis of the ADH isozyme pattern, in fact, revealed that one activity is present only in extracts from cells grown on ethanol. The analysis of mutants obtained by gene disruption techniques allowed to identify *KlADH4* as the gene encoding such ethanol-induced activity (Mazzoni *et al*., submitted).

Interestingly, *KlADH4* can be used as an indicator of the fermentative and aerobic metabolism of *K.lactis* cells. This gene is up regulated at the transcriptional level by ethanol and not by other respiratory carbon sources, and its expression is not sensitive to glucose repression. The induction of *K1ADH4* can be observed only in strains which do not produce ethanol, such as those harbouring a mutation in the *PGI* gene (Goffrini *et al*., 1991) which encodes the glycolytic enzyme phosphoglucose isomerase (Fig. 3, lanes 1 and 2). In such mutants, glucose is probably metabolized through the pentose phosphate pathway; the reduced amount of pyruvate is essentially utilized aerobically and ethanol is not produced by cells.

This is confirmed by the observation that the expression of *KlADH4* is constitutive in wild-type strains which can produce ethanol endocellularly

(Fig. 3, lane 3). From these data one can conclude that relevant differences in the regulation of the ADH system exist between *S.cerevisiae* and *K.lactis*. In *S.cerevisiae*, most of glucose is converted to ethanol by the product of the *ADH1* gene while the expression of *ADH2*, the gene involved in ethanol utilization, is repressed by the sugar and a predominantly fermentative metabolism results. Only after glucose has been consumed, *ADH2* is derepressed and the cells can utilize ethanol aerobically.
In *K.lactis* on the contrary respiration and fermentation can coexist and the ethanol produced during growth on glucose can be utilized aerobically by cells because *KlADH4* is not sensitive to glucose repression. Since the activity encoded by *K1ADH4* is located in mitochondria, it seems that the utilization of ethanol in *K. lactis* cells has been compartmentalized in these organelles.

Expression of heterologous genes in *K.lactis*.

S. cerevisiae has been the eukaryotic organism of choice for the production of heterologous proteins in the past few years. However, this yeast is not a good secretor and presents some disadvantages, such as the hyperglycosylation of proteins, which when used for therapeutic purposes could lead to immunological reactions and a diminished activity of the product. Moreover, as a result of its metabolic properties, *S.cerevisiae* accumulates ethanol during growth on glucose and this results in reduced biomass yields which could be of some importance in scaling-up of fermentors. For these reasons, more and more interest has been directed towards non-*Saccharomyces* yeasts (for a recent review, see Reiser *et al.*,, 1990 and Buckholz *et al.*,, 1991).
K. lactis is one of the very few yeasts that can use lactose as a carbon and energy source. Because of this property this yeast has been used in the dairy industry for the production of cell biomass and β-galactosidase. More recently, increasing interest has been shown in *K.lactis* as an alternative host to *S.cerevisiae* for the expression of heterologous genes because of the possibility of introducing foreign DNA into this organism and to the availability of stable self-replicating vectors. Not much is known at present on the glycosylation properties of *K.lactis*. In any case, available data strongly indicate that *K.lactis* can be preferred to *S.cerevisiae* because of its better secretion properties. van den Berg and colleagues (1991) integrated into the genome of both yeasts a vector carrying the prochymosin gene fused to the *S.cerevisiae* α-factor leader and found that 95% of the protein was secreted from *K.lactis* compared with 10% from *S.cerevisiae*. Moreover, in comparable culture conditions the total yield (endocellular plus secreted) of prochymosin was nearly twenty-fold higher for *K.lactis*. Efficient secretion of prochymosin is not related to any peculiarity of this protein and seems to depend, instead, on a natural property of *K.lactis*. In fact good secretion has been obtained also for human serum albumin (HSA) and human interleukin-1β (IL-1β) (Fleer *et al.*,, 1991 a and b). The difference in the yield of secreted proteins obtained for *K.lactis* and *S.cerevisiae* could reflect a different saturation level of the secretion system in the two yeasts.
The availability of inducible promoters of the *K.lactis LAC4* (β-galactosidase) and *KlADH4* genes could be very useful for the production of heterologous proteins in a regulated fashion.

Conclusions

Knowledge on yeast molecular biology has been for a long time limited to *S.cerevisiae* and, recently, to *Schizosaccharomyces pombe*. However more and more interest has recently been focused on other yeasts either because of their pathogenic importance or for the possibility of biotechnological application. Comparison of results and methodologies concerning these yeasts will be very important for further knowledge at the molecular level. The general conclusions we draw from the study of *K.lactis* are mainly concerned with a) the importance of a molecular taxonomy b) the importance of establishing an efficient transformation system for the study of the molecular characteristics of a yeast species c) carbon metabolism and the related regulatory circuits have been essentially studied in the "fermentative" yeast *S cerevisiae*. *K.lactis* could represent a simple eukaryotic model for such studies in an essentially aerobic organism.

Acknowledgments

This work was supported by C.N.R. Progetto Finalizzato Biotecnologia e Biostrumentazione and Progetto Finalizzato Ingegneria Genetica.

References

Araki H, Jearnpipatkul A, Tatsumi H, Sakurai T, Ushio K, Muta T and Oshima Y (1985) Molecular and functional organization of yeast plasmid pSRl J. Mol. Biol. 182: 191-203.

Bianchi MM, Falcone C, Chen XJ, Wésolowki-Louvel M, Frontali L, and Fukuhara H (1987) Transformation of the yeast *Kluyveromyces lactis* by new vectors derived from the 1.6 μm circular plasmid pKDl. Curr Genet 12: 185-192.

Bianchi MM, Santarelli R, Frontali L (1991) Plasmid functions involved in the stable propagation of the pKDl circular plasmid in *Kluyveromyces lactis*. Curr Genet 19: 155-161.

Broach JR, Guarascio VR and Jayaram M (1982) Recombination within the yeast plasmid 2μ circle is site - specific. Cell 29: 227-234.

Buckholz RG, Gleason MAG (1991) Yeast systems for the commercial production of heterologous proteins. Bio/Technology 9: 1067-1072.

Chen XJ, Bianchi MM, Suda K, Fukuhara H (1989) The host range of the pKDl-derived plasmids in yeast. Curr Genet 16: 95-98.

Das S, Hollenberg CP (1982) A high frequency transformation system for the yeast *Kluyveromyces lactis*. Curr Genet 6: 123-128.

De Deken RH (1966) The crabtree effect and its relation to the petite mutation. J Gen Microbiol 44: 157-165.

Fabiani L, Aragona M, Frontali L (1990) Isolation and sequence análysis of a *K.lactis* chromosomal DNA element able to autonomously replicate in *S.cerevisiae* and in *K.lactis*. Yeast 6: 69-76.

Falcone C, Saliola M, Chen XJ, Frontali L and Fukuhara H (1986) Analysis of a 1.6-μm circular plasmid from the yeast *Kluyveromyces drosophilarum*: structure and molecular dimorphism. Plasmid 15: 248-252.

Fleer R, Yeh P, Amellal N, Fournier A, Bacchetta F, Baudel P, Jung G, L'Hôte H, Becquart J, Fukuhara H and Mayaux JF (1991) Stabel multicopy vectors for high-level secretion of recombinant human serum albumin by *Kluyveromyces* yeasts. Biotechnology 9: 968-975.

Fleer R, Chen XJ, Amellal N, Yeh P, Fournier A, Guinet F, Gault N, Faucher D, Folliard F, Fukuhara H, Majaux JF (1991) High level secretion of correctly processed recombinant human interleukin-lβ in *Kluyveromyces lactis*. Gene 107: 285-295.

Fuson GB, Presley HL, Phaff HJ (1987) Deoxyribonucleic acid base sequence relatedness among members of the yeast genus *Kluyveromyces*. Int J Syst Bact 37: 371-379

Futcher AB (1986) Copy number amplification of the 2 μm circle plasmid of *Saccharomyces cerevisiae*. J Theor Biol 119: 197-204.

Cancedo C, Serrano R (1989) Energy-yelding methabolism. In: Rose AH, Harrison JS (eds) The Yeasts. Academic Press Inc., San Diego, pp 205-259.

Goffrini P, Wêsolowski-Louvel M, Ferrero I (1991) Phosphoglucose isomerase gene is involved in the Rag phenotype of the yeast *Kluyveromyces lactis*. Mol Gen Genet 228: 401-409.

Hardy CM, Clark Walker (1991) Nucleotide sequence of the COXl gene in *Kluyveromyces lactis* mitochondrial DNA: evidence for recent horizontal transfer of a group II intron. Curr Genet 20: 99-114.

Jayaram M, Li YY, Broach JR (1983) The yeast plasmid 2 μm circle encodes componenents required for its high copy propagation. Cell 34: 95-lO4.

Jayaram M, Sutton A, Broach JR (1985) Properties of REP3 : a cis-acting locus required for stable propagation of the Saccharomyces cerevisiae plasmid 2 μm circle. Mol Cell Biol 5: 2466-2475.

Jearnpipatkul A, Hutacharoen R, Araki H, Oshima Y (1987) A cis-acting locus for the stable propagation of yeast plasmid pSRl. Mol Gen Genet 207: 355-360.

Kikuchi Y (1983) Yeast plasmid requires a cis-acting locus and two plasmid proteins for its stable maintenance. Curr Genet 119: 197-204.

Lutstorf U, Megnet R (1968) Multiple forms of alcohol dehydrogenase in *S.cerevisiae*.I. Physiological control of ADH-2 and propeties of ADH-2 and ADH-4. Archives Biochem Biophys 126: 933-944.

Murray JAH, Scarpa M, Rossi N, Cesareni G (1987) Antagonistic controls regulate copy number of the 2μ plasmid. EMBO J 6: 4205-4212.

Ragnini A, Fukuhara H (1988) Mitochondrial DNA of the yeast *Kluyveromyces*: guanine-cytosine rich sequence clusters. Nucl Acid Res 16: 8433-8442.

Reiser J, Glumoff V, Kalin M, Ochsner U (1990) Transfer and Expression of heterologous genes in yeasts other than *Saccharomyces cerevisiae*. Adv Bioch Eng Biotech 43: 75-102.

Saliola M, Gonnella R, Mazzoni C, Falcone C (1991) Two genes encoding putative mitochondrial alcohol dehydrogenase are present in the yeast *Kluyveromyces lactis*. Yeast 7: 401-412.

Saliola M, Shuster JR, Falcone C (1990) The alcohol dehydrogenase system in the yeast *Kluyveromyces lactis*. Yeast 6: 193-204.

Skelly PJ, Hardy CM, Clark Walker (1991) A mobile group II intron of a naturally occurring rearranged mitochondrial genome in *K.lactis*. Curr Genet 20: 115-120.

Som T, Armstrong KA, Volkert FC, Broach JR (1988) Autoregulation of 2 μm circle gene expression provides a model for maintanance of of stable plasmid copy levels. Cell 52: 27-37.

Sor F, Fukuhara H (1989) Analysis of chromosomal DNA patterns of the genus *Kluyveromyces*. Yeast 5: 1-10.

Stark MJR, Boyd A, Mileham AJ, Romanos MA (1990) The plasmid - encoded killer system of *Kluyveromyces lactis*: A Review. Yeast 6: 1-29.

Tanguy-Rougeau C, Chen XJ, Wêsolowski-Louvel M, Fukuhara H (1990) Expression of a foreign KmR gene in linear killer DNA plasmids in yeast. Gene 91: 43-50.

Toh-e A and Utatsu I (1985) Physical and functional structure o a yeast plasmid, pSB3, isolated from *Zygosaccharomyces bisporus*. Nucl Acids Res 13: 4267-4283.

Van den Berg JA, van der Laken KJ, van Ooyen AJJ, Renniers TCHM, Rietveld K, Schaap A, Brake AJ, Bishop RJ, Schultz K, Moyer D, Richman M, Shuster JR (1990) *K.lactis* as a host for heterologous gene expression: expression and secretion of prochymosin. Bio/Technology 8: 135-139.

Van der Walt JP (1970) *Kluyveromyces* van der Walt emend. van der Walt In: Lodder J (ed) The Yeasts: A taxonomic study, 2 Ed. North Holland Publishing Co., Amsterdam, pp 316-378.

Van der Walt JP, Johanssen E (1984) *Kluyveromyces* van der Walt emend. van der Walt in Kreger van Rij NJW (ed) The yeasts, a taxonomic study. 3rd Ed. Elsevier Science Publisher, Amsterdam, pp 224-251.

Vaughan-Martini A, Martini A (1987) Taxonomic revision of the genus *Kluyveromyces* by nuclear deoxyribonucleic acid reassociation. Int J Syst Bact 37: 380-385.

Veit BE, Fangman WL (1988) Copy number and partition of the *Saccharomyces cerevisiae* 2 µm plasmid controlled by transcription regulators. Mol Cell Biol 8: 4949-4957.

Volkert FC, Broach JR (1986) Site specific recombination promotes plasmid amplification in yeast. Cell 46: 541-550.

Volkert FC, Wilson DW, Broach JR (1989) Deoxyribonucleic acid plasmids in yeast. Microbiol Rev 53: 299-317.

Wilson C, Fukuhara H (1991) Distribution of mitochondrial rl type introns and the associated open reading frame in the yeast genus *Kluyveromyces*. Curr Genet 19: 163-167.

Gene regulatory circuits in *Saccharomyces cerevisiae* as a tool for the identification of heterologous eukaryotic regulatory elements

John F. Pulitzer*° and Alessandra Pollice*
*Dept. of Genetics, General and Mol. Biology
University of Naples
Via Mezzocannone, 8
80134 Naples
Italy

Introduction

The yeasts *S. cerevisiae* and *Schizosaccharomyces pombe* are non-pathogenic, prevalently haploid unicellular fungi that grow and form colonies on defined media. Both organisms have been for decades favoured objects of investigation, first by classical genetics and more recently by molecular techniques. Knowledge accumulated, specially over the last few years, has revealed that both yeasts, not only clarify mechanisms of genetic transmission (gene expression, mitosis and meiosis) and signal transduction common to higher eukaryotes but, unexpectedly, also illuminate many processes crucial to the development of multicellular organisms such as cell differentiation and cell-cell interactions. Many elements (proteins, cis acting elements) of the control circuits of the cell are conserved from yeast to mammals; remarkably, in many cases, yeast control elements can be replaced with elements of heterologous control circuits. It is thus possible to isolate genes from higher eukaryotes by complementation and to scrutinize their function by the powerful genetics now available in both yeasts.
We will briefly discuss some experiments on the introduction of conserved heterologous elements into yeast control circuits. Although the described experiments concern higher eukaryotes, the same experimental rationale might be profitably applied to other fungi.

Cell cycle networks

Parallel work on cell cycle circuits has been carried out on two different yeasts: *S.pombe* and *S.cerevisiae*. In both cases the convergence of classical genetic and molecular techniques has brought about striking progress in the last few years. Morphological criteria were used to identify *landmark* events in the progression through the yeast cell cycle and in the subsequent

---

°International Institute of Genetics and Biophysics, C.N.R. Via Marconi, 12 - 80125 Naples, Italy

NATO ASI Series, Vol. H 69
Molecular Biology and its Application to Medical Mycology
Edited by B. Maresca, G. S. Kobayashi and H. Yamaguchi

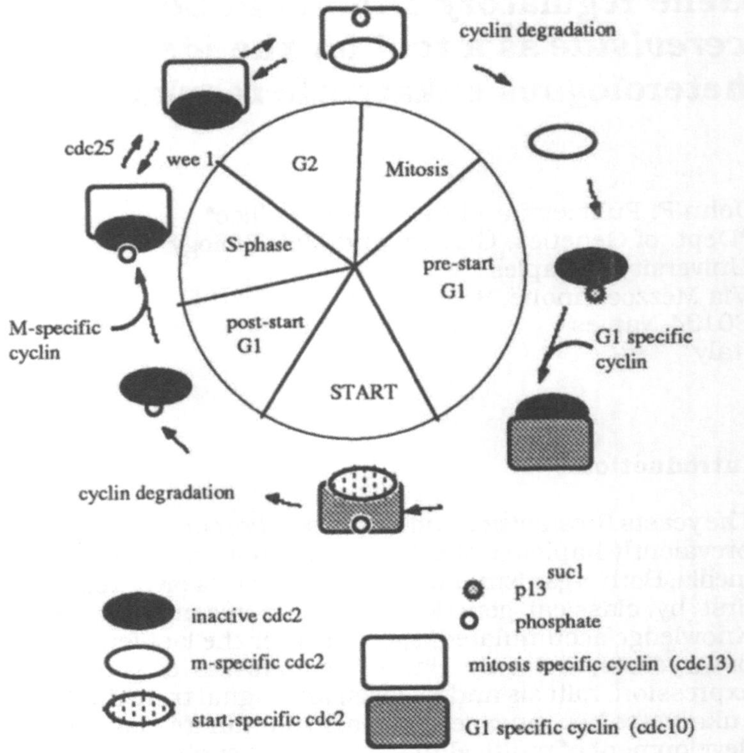

Figure 1.

screening of mutants affecting this progression; *epistatic relationships* between mutants were then used to order the execution points of the mutant alleles in a temporal map correlated to the succession of landmark events (Pringle and Hartwell, 1981) (Lee and Nurse, 1988). We outline below the cell cycle as seen in *S.pombe* (note that in *S.cerevisiae* the equivalent genes have different names!).

Central to this circuit is the mauration-promoting factor (MPF). MPF is adimeric complex of a *protein kinase* p34^{cdc2} and a larger regulatory subunit known as cyclin. The kinase is activated twice in the cycle: at entry into S-*phase* when chromosomes are replicated and at *M-phase* when chromosomes are segregated· It is thought to function by phosphorylating key proteins involved in M-phase and S-phase events. In turn, p34^{cdc2} kinase activity is controlled by a *tyrosine phosphatase* (p80^{cdc25}) (W. G. Dunphy and A. Kumagai, 1991).

A human homologue of the *S.pombe cdc2* gene was cloned by functional complementation: a human cDNA library was expressed in fission yeast and clones that could complement a mutant of *cdc2* were selected. The predicted protein sequence of the human homologue turned out to be very similar to that of the yeast *cdc2* gene.

Similarly, a p80^{cdc25} mutant is complemented by human c-DNA coding for a protein tyrosine phosphatase (Gould et al., 1990).

The realization that the molecular mechanisms that control certain steps in the cycle may be universal, suggested the possibility of cloning other higher eukaryote cell cycle genes by cross-species complementation of yeast mutants. *Drosophila melanogaster* homologues of fission yeast *S. pombe* cell division control (*cdc*) genes were cloned by the complementation of temperature sensitive mutations. The *Drosophila* genes were expressed in *S.pombe* as cDNAs from a pooled pre-existing *Drosophila* embryonic cDNA library adapted for propagation in fission yeast by introduction of an ars1-LEU2 DNA fragment. This library was introduced into *S.pombe cdc2* and *cdc25* mutants, and plasmids isolated carrying cDNAs that complement these mutations. The gene that encodes the *Drosophila cdc2* homologue mapped to a single locus in the *Drosophila* genome. Two different cDNAs that complement *cdc25-22* were isolated. One corresponds to a transcript of *string*, previously described as the *Drosophila* homologue of *cdc25*, and the other to a gene that has not been previously characterized. The product of latter gene is likely to be an allele specific suppressor of the ts *cdc25* mutant and may thus represent the *Drosophila* equivalent of another element of the circuit. Three, redundant G1 specific cyclins have been identified in *S.cerevisiae* and are coded for by three genes: CLN1, CLN2 and CLN3. Mutation of all three leads to cell-cycle arrest in G1. *Human* and *Drosophila* c-DNA clones rescue the yeast cyclin deficiency; two of these share homology to each other and are related to known cyclins (Leopold, P. and O'Farrell, P. H., 1991).

Caution

Of course, the possibility of evolutionary divergence imposes some limitations on the use of functional complementation of yeast mutants; in fact, a homology screen yielded two genes from *Drosophila* that are structurally related to *cdc2* but only one complements the pombe mutant. Also, even though a given heterologous gene will complement a yeast mutant may not necessarily imply that this gene performs the same function in yeast and in the native host. Thus, the human calmodulin gene (Calcium binding protein) complements the lethal disruption of the yeast homologue; however, point mutants that disrupt the Ca binding domain do not affect the capacity of the human gene to complement the yeast function (Geiser *et al.*, 1991).

Oncogenes

The *ras* proteins are a family of GTP binding, membrane-associated proteins involved in the regulation of a variety of cellular processes. A substantial number of novel *ras*-related proteins have been identified and, from yeast to man, appear to be involved in controlling a diverse set of essential cellular functions including growth, differentiation, cytoskeletal organization, intracellular vesicle transport and secretion (Hall, 1990). *Ras* proteins are thought to exist in two states: active *ras* bound to GTP and inactive *ras* bound to GDP. GTP is bound in an exchange reaction while GDP is produced by the intrinsic GTPase activity of *ras*. GTPase activity is greatly stimulated by interaction with a protein called GAP, present in all mammalian cells. Active *ras* mediates cAMP synthesis.

Mutant forms of cellular *ras* genes are often found in human cancers. Mutant *ras* genes transform NIH3T3, an established murine cell line. The oncogenic

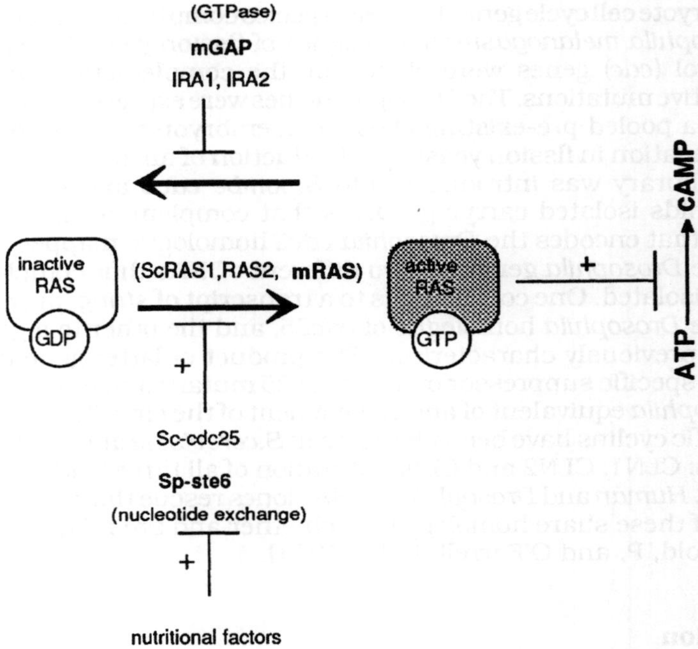

Figure 2.

mutations prevent GAP from stimulating the GTPase activity of *ras* which remains in the active GTP bound conformation.

An exciting development in yeast molecular biology was the discovery that *S.cerevisiae* encodes homologs of mammalian *ras* genes and that the product of these yeast genes function in regulating adenylate cyclase activity. This discovery raised the possibility of developing in yeast a model system which could lead to a better knowing of the physiological role of mammalian *ras* genes and their oncogenics alleles.

S.cerevisiae has two *ras* genes, RAS1 and RAS2, identified by hybridization techniques using a human DNA probe (DeFeo-Jones *et al.*, 1983); the products of these genes , somewhat larger than the human homologues, are about 50% homologous to the mammalian *ras* proteins.The *ras* genes play a key role in the biology of yeast cells: disruption of RAS1 and RAS2 is lethal to the cell. Yeast RAS1 and RAS2 function can be replaced by complementation with chimeric mammalian/yeast *ras* genes or even the mammalian H-*ras* gene. (Kataoka *et al.*, 1985).

The isolation of mutants in yeast has provided a more detailed knowledge of the function of these genes; for example the RAS2val19 mutation, which is equivalent to the oncogenic *ras*val12 mutation in mammals, greatly increases adenylate cyclase activity and cyclic AMP levels. (Toda *et al.*, 1985). These results clearly show that a major role of *ras* proteins in *S.cerevisiae* is to activate adenylate cyclase activity. Several components of the RAS-adenylate cyclase pathway have been identified in the yeast *S.cerevisiae*.

In *S.cerevisiae* the CDC25 products has been identified as the upstream regulator of RAS protein function (Broek *et al.*, 1987).

Genetic analysis indicates that activation is probably effected by direct protein-protein interaction between RAS and CDC25 (Powers *et al.*,1989); mutant forms of CDC25 have been isolated that constitutively activate the RAS-cAMP pathway, showing that CDC25 is a regulatory molecule.
CDC25 is also defined as a START gene because of its involvement in the control of progression through the START point in the G1 phase of the yeast cell cycle.
The signal(s) that activates the *ras* pathway(s) is not known. Other components of the *ras*-cyclic AMP pathway in *S.cerevisiae* are IRA1 and IRA2, negative regulators of *ras* genes (Tanaka *et al.*, 1990). As mentioned earlier, the human H-*ras* gene can substitute for yeast *ras*; the relationship is sufficiently close that dominant H-*ras* mutations that inhibit CDC25 in yeast have been found. Furthermore, these dominant H-*ras* mutation have the appropriate phenotype in mammalian cells, suggesting the presence of a CDC25-like protein. This underscores the possibility of using yeast genetics (functional complementation) as a tool to understand mechanisms common to other eukaryotes.

Caution

S. pombe has only one *ras* gene, *ras*1+, that does not appear to function in activating adenylate cyclase; neither disruption of the gene nor mutations corresponding to the mammalian oncogenic counterparts, affect cAMP levels in the cell. Surprisingly, even though the function of *Ras* protein in fission yeast differs from that in budding yeast of fission yeast, at least part of the regulatory pathway is conserved: the ste6 gene an upstream regulator of RAS protein function is a homologue of CDC25: the ste6 gene product and the CDC25 gene product have significant amino-acid similarity in their C-terminal regions. Probably, ste6 and CDC25 activate Ras protein through a common mechanism, perhaps by promoting GDP-GTP exchange (Hughes *et al.*, 1991). Again, the possibility of evolutionary divergence raises the possibility of picking up elements unpredicted by the rationale of heterologous complementation of yeast mutants.

Transcriptional control

In the last few years it has become apparent that, although details of promoter structure may vary from organism to organism as well as the rules governing the exact sites at which transcription initiate, the components of the transcriptional machinery (RNA polymerase subunits, transcription factors) are remarkably conserved from lower to higher eukaryotes.

Promoter structure

In the yeast *Saccharomyces cerevisiae*, efficient transcription initiation by RNA polymerase II (B) is dependent on three cis-acting elements: the initiation site/s itself (*IR*), the *TATA* box, and the upstream activation site (*UAS*) (Struhl, 1987a).

The *IR* sites (unlike their higher eukaryote counterparts, are preferred sequences (Chen & Struhl, 1985; Nagawa and Fink, 1985). The *TATA* box elements restrict initiation to potential *IR* sequences situated within an initiation window 60-100 bp downstream. The *UAS* elements, situated 100 bp or more upstream, determine the level of transcription and often, via interaction with cognate *TATA* boxes, which start sites are used. *UAS* elements may overlap regulatory sites (*URS*) or be constitutive (eg. *polydAT* sequences). Some *TATA* boxes respond to *URS* elements and some only to constitutive *UAS* elements. Yeast *UAS* elements have been inverted and their distance upstream (but not downstream) varied extensively without appreciable effect on transcription. This behaviour is reminiscent of that of mammalian promoter elements, such as the viral enhancers (Struhl, 1987a).

RNA polymerase II components

There is also conservation in the proteins that bind to promoter elements. The formation of a preinitiation complex by RNA polymerase II involves docking of the large subunits (B220, B150) of RNA polymerase on to the basic transcription factors TFIID,TFIIA, TFIIE and TFIIB. Yeast TFIID is a DNA binding protein that initiates formation of the basal complex at the TATA box; it is also, with TFIIA and TFIIB the target contacted by transcriptional activators (see below) (Pugh and Tjian 1992). TFIID and TFIIA isolated from yeast, can (at least in part) functionally replace their mammalian counterparts in the formation of an *in vitro* preinitiation complex (Hahn *et al.*, 1988).

Transcriptional activators

For effective transcription, the pre-initiation complex needs to be activated by transcriptional activators that bind at UAS elements. Also in this case there are numerous instances of conservation between lower and higher eukaryotes. We will consider two prototypical cases: GCN4 and HAP2/HAP3. GCN4, a key element in the regulation of aminoacid biosynthesis in yeast is a transcriptional activator that binds to a specific sequence found in several copies upstream of aminoacid biosynthetic genes (Hope and Struhl 1985, Arndt and Fink, 1986).
The JUN oncoprotein, which causes sarcomas in chickens, shows significant homology to the DNA-binding domain of GCN4. The GCN4 and JUN proteins bind the same DNA sequences, consensus ATGA(C/G)TCAT, even though the DNA-binding domains are only 45% identical in amino acid sequence. The JUN protein appeares to represent the oncogenic version of the normal AP-1 transcription factor, suggesting an evolutionary relationship between yeast and vertebrate activator proteins. JUN efficiently activates transcription in yeast either through its own or a heterologous DNA-binding domain (Struhl, 1987). As is the case for yeast activator proteins, transcriptional stimulation by JUN requires an acidic activation region distinct from the DNA-binding domain.
The *S.cerevisiae* HAP2 and HAP3 gene products regulate the transcription of the cytochrome-C gene CYC-1 by binding to CCAAT-containing transcription elements as a HAP2/HAP3 multisubunit heteromer (Hahn and Guarente,

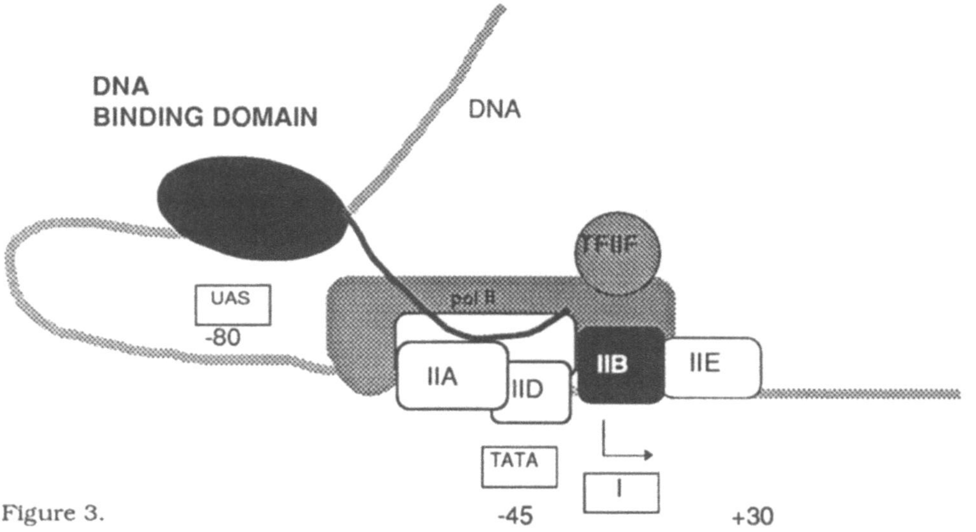

Figure 3.

1988). The human sequence-specific DNA-binding protein, CP1, also recognizes CCAAT-containing transcription elements as a heteromer. CP1 has DNA-binding properties that are virtually identical to the yeast HAP2/HAP3 complex. These two proteins bind to the same target sequences, make the same DNA contacts, and are affected in a similar manner by mutations in the CCAAT element. Furthermore the subunits of CP1 and HAP2/HAP3 are functionally interchangeable and yeast/human hybrid complexes retain the ability to specifically recognize CCAAT elements (Becker DM, Fikes JD, Guarente L 1991). The functional interchangeability between yeast and vertebrate transcription factors is a strong indication of basic similarity in the molecular mechanism of eukaryotic transcriptional activation.

It is possible by using a human c-DNA library cloned in a yeast expression vector to clone the human equivalent of HAP2, by functional complementation of a hap2 mutation. The cDNA encoding the human HAP2 homolog encodes a protein of 257 amino acids that has a 62-amino acid carboxyl-terminal region 73% identical to the essential core region of HAP2. This experiment shows that the general approach of cloning by complementation succesfully used in identifying mammalian cell-cycle genes can be applied in isolation of heterologous transcription factors for which corresponding yeast mutations exist.

Viral enhancers in yeast

A mammalian promoter element, the polyoma enhancer, is capable to activate transcription in yeast by productive interaction with native promoter elements (Ciaramella *et al.*, 1988). This finding is consistent with the conservation of elements of the transcriptional machinery (RNA polymerase subunits, transcription factors) from lower to higher eukaryotes.

Yeast proteins that bind to the different domains of the polyoma enhancer have been identified. In particular we found that a multifunctional yeast regulatory protein (RAP1) that binds to a conserved DNA element of the polyoma-virus enhancer is responsible for enhancer activity in yeast. RAP1 is unusual in that it is a DNA binding global control element, playing, according to context, at least four quite different roles: transcriptional activation, transcriptional silencing, telomere elongation and (at least in one context) enhancement of recombination; perhaps it carries out these different roles by an ability to bind other proteins that play a more specific role in all these processes. Furthermore, it is of particular interest that RAP1 is mostly localized in the nuclear scaffold.
A RAP1 binding sequence is conserved in SV40 and BPV enhancers as well, suggesting that this site could have a functional role in mammals. We are currently studying proteins from mammalian cells that bind to this site in the hope of finding a mammalian equivalent of RAP1.

References

Arndt K and Fink G (1986) GCN4 protein, a positive transcription factor in yeast, binds general control promoters at all 5' TGACTC 3' sequences. Proc. Natl. Acad. Sci. USA 83: 8516-8520.

Becker DM, Fikes JD and Guarente L (1991) A cDNA encoding a human CAAT-binding protein cloned by functional complementation in yeast. Proc. Natl. Acad. Sci. USA 88: 1968.

Broek D, Toda T, Michaeli T, Levin L, Birchmeier C, Zoller M, Powers S and Wigler M (1987) The *S. cerevisiae* CDC25 gene product regulates the RAS/adenylate cyclase pathway. Cell 48: 789-799.

Buratowski S, Hahn S, Sharp PA and Guarente L (1988) Function of a yeast TATA element-binding protein in a mammalian transcription system. Nature 334: 37-42.

Chen W and Struhl K (1985) Yeast mRNA initiation sites are determined primarily by specific sequences, not by the distance from the TATA element. EMBO J. 4: 3273-3280.

Ciaramella M, Sacco M and Pulitzer JF (1988) Foreign transcriptional enhancers in yeast. I. Interactions of papovavirus transcriptional enhancers and a quiescent pseudopromoter on supercoiled plasmids. Nucl. Acids Res. 16: 8847-8868.

DeFeo Jones D, Scolnick EM, Koller R and Dhar R (1983) *ras*-related gene sequences identified and isolated from *Saccharomyces cerevisiae*. Nature 306: 707-709.

Dunphy WG and Kumagai A (1991) The *cdc25* protein contains an intrinsic phosphatase activity. Cell 67: 189-196.

Geiser JR, van Tuinen D, Brockerhoff SE, Neff MM and Davis TN (1991) Can calmodulin function without binding calcium? Cell 65: 949-959.

Gould KL, Moreno S, Tonks NK and Nurse P (1990) Complementation of the mitotic activator, p80*cdc25*, by a human protein-tyrosine phosphatase. Science 250: 1573-1576.

Hahn S and Guarente L (1988) Yeast HAP2 and HAP3: transcriptional activators in a heteromeric complex. Science 240: 317-321

Hall A (1990) The cellular functions of small GTP-binding proteins. Science 249: 635-40.

Hope I and Struhl K (1985) GCN4 protein, synthesized *in vitro*, binds HIS3 regulatory sequences: implications for general control of amino acid biosynthetic genes in yeast. Cell 43: 177-188.

Hughes DA, Fukui Y, Yamamoto M (1990) Homologous activators of *ras* in fission and budding yeast. Nature 344 6264: 355-357.

Kataoka T, Powers S, Cameron S, Fasano O, Goldfarb M, Broach J and Wigler M (1985) Functional homology of mammalian and yeast RAS genes. Cell 40: 19-26.

Lee MG and Nurse P (1988) Complementation used to clone a human homologue of the fission yeast cell cycle control gene *cdc2*. Nature 327: 31-35.

Leopold P and O'Farrell PH (1991) An evolutionarily conserved cyclin homolog from *Drosophila* rescues yeast deficient in G1 cyclins. Cell 66: 1207-1216.

Nagawa F and Fink GR (1985) The relationship between the "TATA" sequence and transcription initiation sites at the HIS4 gene of *Saccharomyces cerevisiae*. Proc. Natl. Acad. Sci. USA, 82: 8557-8561.

Powers S, O' Neill K and Wigler M (1989) Dominant yeast and mammalian RAS mutants that interfere with the *CDC25*-dependent activation of wild-type RAS in *Saccharomyces cerevisiae*. Mol. Cell. Biol. 9: 390-395.

Pringle JR and Hartwell LH (1982) The *Saccharomyces cerevisiae* life cycle. In The Molecular Biology of the Yeast *S. Cerevisiae* Cold Spring Harbor Lab. (J.N. Strathern *et al.* Eds) p. 97-142.

Pugh BF and Tjian R (1992) Diverse transcriptional functions of the multisubunit eukaryotic TFIID complex. J. Biol. Chem. 267: 679.

Struhl K (1987a) The DNA-binding domains of the jun oncoprotein and the yeast GCN4 transcriptional activator protein are functionally homologous. Cell 50: 841-846.

Struhl K (1987b) Promoters, activator proteins, and the mechanism of transcriptional initiation in yeast. Cell 49:295-297.

Toda T, Uno I, Ishikawa T, Powers S, Kataoka T, Broek D, Cameron S, Broach J, Matsumoto K and Wigler M (1985) In yeast, RAS proteins are controlling elements of adenylate cyclase. Cell 40: 27-36.

Acknowledgement

The experiments reported in this review were supported in part by the following grants from CNR, P.J. Biotecnolgia e Biostrumentazione and P.F. Ingegneria Genetica.

Molecular Biology of Filamentous Fungi

Molecular breeding in filamentous fungi with emphasis on *Aspergilli*

Y. Iimura and K. Gomi
National Research Institute of Brewing
2-6-30 Takinogawa, Kita-ku
Tokyo 114
Japan

Introduction

Filamentous fungi such as *Aspergillus* and *Penicillium* are extensively used in industries which produce alcohol, organic acid, and antibiotics. So far breedings of these useful fungi have been mainly achieved by methods of mutagenesis. Recently, the application of recombinant DNA technology to filamentous fungi has allowed strain improvement. Since *Aspergilli* produce a large amount of enzymes to decompose starch and proteins to sugars and amino acids, respectively, it is possible that these fungi serve useful secretion systems with strong expression promoter for efficient productions of heterologous proteins. Among the filamentous fungi, *Aspergillus nidulans* and *Neurospora crassa* have been most extensively investigated in molecular genetics. Based on the knowledge of these fungi, the molecular genetic systems of several industrial species, including *Aspergillus niger* and *Aspergillus oryzae*, have been developed and allowed strain improvement by increasing the number of genes encoding enzymes of importance.
Here, we discuss the transformation system of *Aspergilli*, especially of *A. oryzae*, and the expression control of the α-amylase gene in *A. oryzae* using the molecular genetic system.

Transformation procedure of filamentous fungi

In order to introduce the exogenous DNA into the filamentous fungal cells, the use of protoplast is required. Protoplasts are prepared by treating the mycelia with cell wall lytic enzymes, which contain chitinase and β-1,3-glucanase. Commercial enzyme preparations are produced mainly from *Trichoderma* sp. For protoplast formation, 0.8 M NaCl or KCl is used as an osmotic stabilizer. After mixed with transforming DNA, protoplasts are fused in the presence of 40% polyethylene glycol (PEG) and 10-50 mM $CaCl_2$. Fused protoplasts are embedded in regeneration agar medium. Recently, *A. niger* was transformed by electroporation (Ward, Kodama and Wilson, 1989), but *A. oryzae* has not been transformed by this method.

NATO ASI Series, Vol. H 69
Molecular Biology and its Application to Medical Mycology
Edited by B. Maresca, G. S. Kobayashi and H. Yamaguchi

Host-vector systems in *Aspergillus oryzae*

Little work has been done to develop transformation system of *A. oryzae*, because it has no sexual life cycle and the multinucleate conidia and hyphae make mutant isolation difficult. Based on the molecular genetic systems of *A. nidulans* and *N. crassa*, host-vector systems for several industrial species, including *A. oryzae* and *A. niger*, have been developed. Table 1 and Figure 1 show the selectable markers and the plasmid vectors used for transformation of *A.oryzae*, respectively.

As in *A. nidulans*, trasformation system in *A. oryzae* have been developed by complementation of nutritional requirements. The successful transformation of *A. oryzae* by complementation of a methionine-auxotrophic mutation was first reported by Iimura *et al.* (1987). A genomic library of *A. oryzae* was screened for a DNA fragment complementing the auxotrophic mutation. Consequently, a plasmid carrying a 3.5 Kbp *Bam*HI fragment of *A. oryzae* was isolated. The resulting plasmid was recovered in *E. coli*, suggesting that it might exist, in part, as a free plasmid in the transformants. The enzyme encoded by the isolated DNA fragment has not been determined. However, these results have opened the way to investigate *A. oryzae* at the molecular level.

Furthermore, Gomi, Iimura and Hara (1987) and Mattern *et al.* (1987) developed the transformation systems using an *argB* gene encoding ornithine carbamoyltransferase (OCTase) of *A. nidulans* and a *pyrG* gene encoding orotidine-5'-phosphate decarboxylase of *A. niger*, respectively, as selectable markers. The recipient *argB* mutants can be easily distinguished among many arginine-auxotrophic mutants by their growth phenotype on a medium containing citrulline, though there is no available positive selection procedure for isolating the *argB* mutants. In contrast, the *pyrG* mutants having uracil or uridine auxotrophy can be isolated by positive selection for 5-fluoro-orotic acid resistance. In these cases, the two gene used as selectable markers were derived from *A. nidulans* and *A. niger*, species related to *A. oryzae*, and could function in *A. oryzae*.

Features of transformation with *argB* gene

Transformation using *argB* or *pyrG* occurred by integration of the transforming

Table 1.
Selectable markers used in transformation systems of *Aspergillus oryzae*

Maker	Encoded enzyme	Source	Reference
argB	Ornithine carbamoyltrasferase	*A.nidulans*	Gomi, Iimura and Hara (1987) Hahm and Batt (1988)
pyrG	Orotidine-5'-phosphate decarboxylase	*A. niger* *A. oryzae*	Mattern *et al.* (1987) de Ruiter-Jacobs *et al.* (1989)
niaD	Nitrate reductase	*A. oryzae*	Unkles *et al.* (1989)
amdS	Acetamidase	*A. nidulans* *A. oryzae*	Christensen *et al.* (1988) Gomi, Kitamoto and Kumagai (1991)
met	Undetermined	*A. oryzae*	Iimura *et al.* (1987)

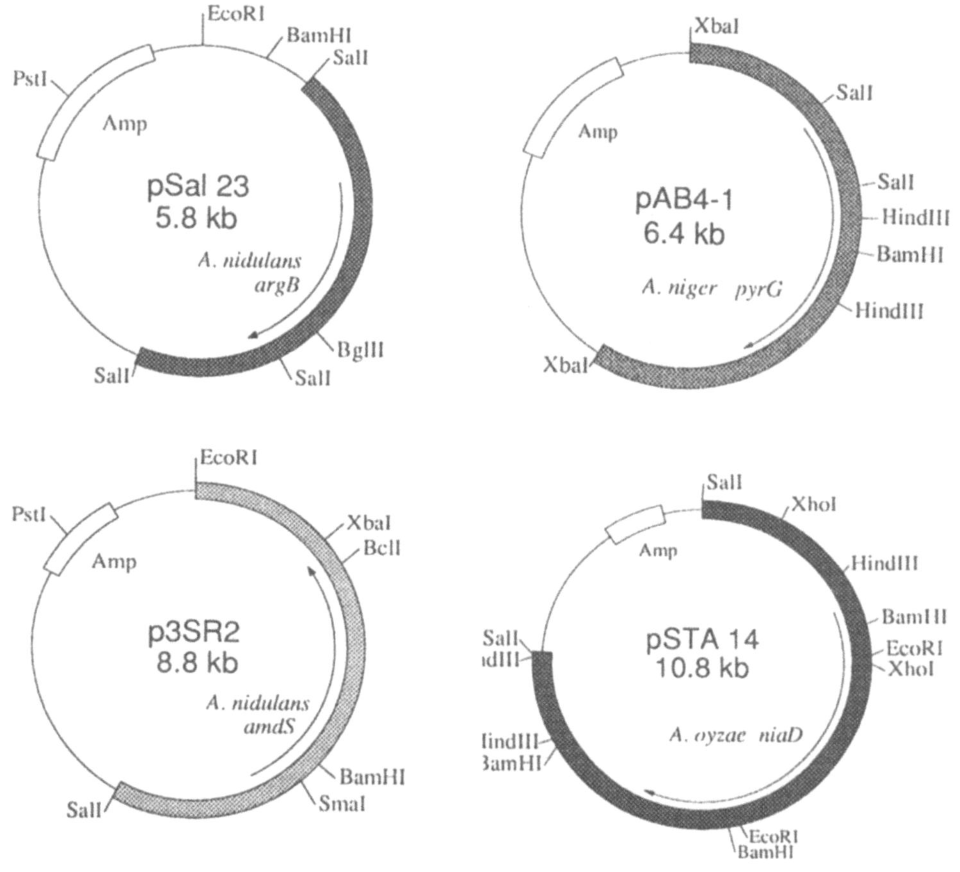

Figure 1. The structures of the plasmid vectors used for transformation of *A. oryzae*

plasmid into the chromosome. Southern blot analysis revealed that the genes did not have a high degree of homology to the corresponding genes of *A. oryzae* (Gomi, Iimura and Hara, 1987). Therefore it is likely that the transformation occurred by non-homologous integration. This might account for the low frequency of transformation (in the transformation with *argB*, less than 10 transformants per mg of transforming DNA). Gomi, Iimura and Hara(1987) estimated the copy number in the integrated *argB* gene in the transformants, one of which contained at least 60 copies and the others also had more than 15 copies, showing that the transformants generally contained many copies of the foreign gene in tandem arrays on the chromosome. These phenotypes were extremely stable for many generations under non-selective conditions. The activity of OCTase encoded by the *argB* gene in the transformants increased in proportion to the copy number of the integrated gene (Table 2), suggesting that the transformation system would allow improvement in the productivity of enzymes, in addition to the possibility of the production of heterologous proteins. Recently, de Ruiter-Jacobs *et al.* (1989) established the transformation system using a *pyrG* gene isolated from *A oryzae*, showing that this gene was integrated to the chromosome by homologous recombination, which allowed the transformation frequency to increase.

Transformation using dominant selectable marker

Transformation systems using complementation of the auxotrophic mutation seem to be unsuitable for the breeding of the useful strain for industries, because it is possible that the treatment of mutagenesis causes damage to the genes important for the use. To avoid this disadvantage, positive transformation system is necessary. *A. oryzae* is resistant to most antibiotics used for positive selection in yeast (*S. cerevisiae*) and other filamentous fungi, but is sensitive to benomyl (1-2 μg/ml) and phleomycin (50 μg/ml). However, transformation system using the genes responsible for resistance against these compound is not yet available.

Another dominant selective marker, acetamidase, which is encoded by *amdS* gene, has been widely used for the transformations of filamentous fungi, such as *A. niger* (Kelly and Hynes, 1985), *Penicillium chrysogenum* (Beri and Turner, 1987) and *Trichoderma reesei* (Penttila *et al*, 1987), which are unable to assimilate acetamide as nitrogen source. The *amdS* gene is not effective for selective marker in transformation system of *A. oryzae*, which is able to grow on a medium containing acetamide as a sole nitrogen source. However, in *A. nidulans*, the introduction of multiple copies of *amdS* caused the transformant more vigorous growth on the medium containing acetamide as a sole nitrogen source (Kelly and Hynes, 1987). Similar transformation systems using *amdS* gene from *A. nidulans* or *A. oryzae* have been developed (Christensen *et al*, 1988, Gomi, Kitamoto and Kumagai, 1991). Additionally, *amdS* gene of *A. oryzae* has proved to be applicable for the transformation by the homologous integration to the chromosome of the wild type of *A. oryzae* (Gomi *et al.*, unpublished).

Table 2.
Specific activity of ornithine carbamoyltransferase in the transformants
(Gomi, Iimura and Hara, 1987)

Strain	Czapek-Dox medium (Selective condition)		Dextrin-pepton medium (Non-selective condition)	
	Spec.act.[a]	Ratio [b]	Spec. act. [a]	Ratio [b]
A.oryzae				
FN-16 [c]	17.1	1.0	2.8	1.0
Transformant-1	469.9	27.5	71.3	25.6
Transformant-2	904.6	52.9	133.7	47.9
Transformant-3	514.3	30.1	42.3	15.2
A. nidulans				
IAM2006	22.2	1.3	3.4	1.2

(a) Specific activity; μg citrulline /min/mg protein; (b) Specific activity of FN-16 is expressed as 1.0; (c) Parent strain of the recipient, M-2-3.

Improvement of transformation frequency

An efficient transformation system of *A. oryzae*, which mediated through the gene involving the nitrate assimilation, was developed by Unkles *et al.*(1989). In this system, *niaD* gene encoding nitrate reductase was used as a selectable marker, while the *niaD* mutant is unable to assimilate nitrate as a sole nitrogen source. This transformation system has the following advantages. Since the transformation using a plasmid carrying the *niaD* gene of *A. oryzae* occurred only by homologous integration at a resident locus on the host chromosome, the transformation frequency was much higher than that by heterologous integration. Using this homologous integration system, the control of the expression in the genes of interest can be analyzed. Furthermore, since *niaD* mutants can be easily isolated as strains resistant to chlorate without any induction procedures of mutation, there is little damage on the genes necessary for industrial use.

Transformation using autonomously replication plasmids

Autonomously replication plasmids will allow the increase of the transformation frequency and re-isolation of plasmids carrying the gene complementing the mutation. Many attempts have been made to develop autonomously replication plasmids in filamentous fungi. In the zygomycete fungus *Mucor circinelloides*, a chromosome sequence which confers autonomous replication of plasmids was found (van Heeswijeck,1986). Recently, a sequence designated *AMA1* conferring autonomous replication in *Aspergillus* was also isolated from *A. nidulans* (Gems, Johnstone and Clutterbuck,1991). The transformation frequency was enhanced over 200-fold by the plasmid containing *AMAI* sequence, which was maintained extrachromosomally in the transformants of *A. nidulans*. Moreover, the sequence was also able to function in similar fashion in *A. oryzae*. Thus, *AMA1* will be a prominent tool for the cloning of gene from *A. oryzae* by complementation.

Cloning of α-amylase gene from *A. oryzae*

A. oryzae produces many kinds of hydrolases, the most important of which for industrial use are α-amylase, glucoamylase and protease. The α-amylase of *A. oryzae*, which is known as Taka-amylase A (TAA) has been investigated extensively. This enzyme is a glycoprotein consisting of a single polypeptide chain of 478 amino acid residues (Toda, Kondo and Narita,1982). In contrast to biochemical and biophysical analyses, there was little information regarding the organization and regulation of the genes encoding α-amylase. *A. oryzae* has the ability secreting amounts of TAA, the synthesis of which is inducible by oligosaccharides, such as maltose and starch. To elucidate the function of TAA at the molecular level, the TAA gene was cloned and the nucleotide sequence was determined (Tada *et al*,1989). The TAA genes were isolated from a λ EMBL3 genomic library of *A. oryzae* RIB40, by plaque hybridization using two kinds of 26 mer oligonucleotide probes, which were synthesized according to the amino acid sequence reported by Toda, Kondo and Narita (1982).

Structure of the TAA gene

Genomic Southern blot analysis using the cloned TAA gene as a probe showed the existence of at least three TAA genes, which were designated as *amyA*, *amyB* and *amyC*, in *A. oryzae*. The nucleotide sequences of *amyB* and *amyC* were completely identical in the coding region and the 5' and 3' non-coding regions. Another gene (*amyA*) had almost the same sequence as the others except two nucleotides in the coding region, but differing significantly in the 3' non-coding region. The nucleotide sequence showed that the TAA-coding region consisted of eight introns and nine exons, coding 499 amino acids (Tada *et al.*, 1989). All eight introns start with GT and terminate with AG sequence, showing general feature of introns. In addition, a sequence homologous to the internal consensus sequence found in fungi, PuCTPuAC, was observed in the introns.(Gurr, Unkles and Kinghorn, 1987).
Compared with the amino acid sequence of the mature TAA, the deduced amino acid sequence consisting of additional 21 amino acids having a hydrophobic character was observed at the N-terminus of the cloned TAA gene. This sequence is thought to be a signal sequence for secretion. Furthermore, one insertion, one deletion and 10 substitutions were found throughout the whole sequence in two identical genes, *amyB* and *amyC*. It is likely that these changes are due to a strain variation. Additionally, when introduced into *A. oryzae*, each of three genes is functional in the resulting transformants. Independently, the TAA genes from *A. oryzae* were cloned and sequenced by other groups (Gines Dove and Seligy, 1989, Wirsel *et al.*, 1989). Wirsel *et al.* reported that there are three TAA genes, nucleotide sequences of which are almost same. According to detection of the corresponding mRNAs, at least two genes are thought to be expressed.
To elucidate the location of the TAA genes on the chromosome of *A. oryzae*, physical mapping has been done by Southern blot analysis using orthogonal-field-alteration gel electrophoresis (OFAGE). As a result, the three TAA genes were located on the three different chromosomes. *Aspergillus sojae*, which have lower productivity of TAA than *A. oryzae*, had only one chromosome hybridized with the TAA gene (Gomi *et al.*, unpublished).

Regulational control of the TAA gene expression

By introduction of the cloned TAA gene into *A. oryzae* using the integrative transformation system with a plasmid (pSal23) containing the *A. nidulans argB* gene, high-α-amylase-producing strain was constructed (Tada *et al*, 1989). As shown in Table 3, the transformant appeared to have two- to sixfold higher TAA activity.
Northern blot analysis of the TAA gene was performed in order to analyze the regulational control of the gene expression. Total RNA was extracted from the mycelia of *A. oryzae* grown in the medium containing glucose or starch as a sole carbon source, and subjected to Northern blot analysis using the TAA gene as a probe. Consequently, though ethidium bromide staining showed that almost the same amounts of RNA were present in each sample, mRNA of the TAA was prevalent in mycelia grown in the presence of starch as a sole carbon source. This result confirmed that the production of TAA is controlled at a transcriptional level, consistent with the result of an *in vitro* translation

Table 3.
TAA activity produced by transformants (Tada *et al*,1989)

Strain	Plasmid	W. D. M[(a)]	TAA activity[(b)]	Ratio[(c)]
FN-16[(d)]		3.9	8,800	107
M-2-3[(e)]	pSal23	2.3	8,200	100
M-2-3	pSal23 +*amyB*	1.9	32,100	391
M-2-3	pSal23 +*amyB*	3.6	21,400	261
M-2-3	pSal23 +*amyB*	1.9	49,000	598

(a) Weight of dry mycelia (gram).
(b) Specific activity (units/ gram dry mycelia)
(c) Specific activity of M-2-3 transformed with pSal23 is taken as 100.
(d) Parent strain of *argB* mutant, M-2-3.
(e) Control.

experiments (Erratt *et al*, 1984).

As described above, molecular analyses of the TAA genes have shown that the genes form a multi-gene family in *A. oryzae*. Nucleotide sequences of these structural genes as well as their 5' flanking regions are almost identical. However, little is known about the sequences necessary for high-level expression and regulation of the TAA gene.

In order to elucidate the mechanism of transcriptional regulation in the TAA gene expression, fusion experiments, in which *E. coli uidA* gene encoding β-glucuronidase (GUS) was linked to the promoter of the TAA gene, have been done. A 613 bp promoter sequence was sufficient to confer TAA gene regulation on the heterologous reporter gene. To locate functional regions responsible for regulation and high-level expression within the upstream region of the TAA gene, Tada *et al.*(1991) prepared a series of deletions within 613 bp of the promoter region. Sequential deletions were introduced into the *A. oryzae* recipient using the integrative transformation system with a plasmid (pSal23) containing the *A. nidulans argB* gene. The resulting prototrophic transformants were grown in the medium containing maltose or glucose as a sole carbon source. Their GUS activity was determined(Figure 2).

From the 5 deletion experiment , Δ377 did not alter the level of GUS activity substantially, while Δ290 significantly reduced the level of GUS activity. Interestingly, even Δ61 showed this level of GUS activity, though it lost the transcription start points. These results indicate that distinct DNA sequence elements from -377 to -290 are required for the high-level expression and the induction of the TAA gene. There are several elements which are reported to be involved in promoter function, such as TATA and CAAT elements (Gurr, Unkles and Kinghorn, 1987). Two CAAT element exist at -377 bp and - 192bp. A TATA element is at -100 bp. Only about 90 bp (from -377 to -290) sequence of the promoter region might be involved in high-level expression and regulation of the TAA gene. One of putative CAAT elements was located within this region of 90 bp. Thus, it is interest whether this element is responsible for the transcriptional regulation. Further experiments should permit identification of *cis*-acting elements such as UAS, enhancer or receptor sequences within this 90 bp region and detection of DNA binding protein interacting with those elements.

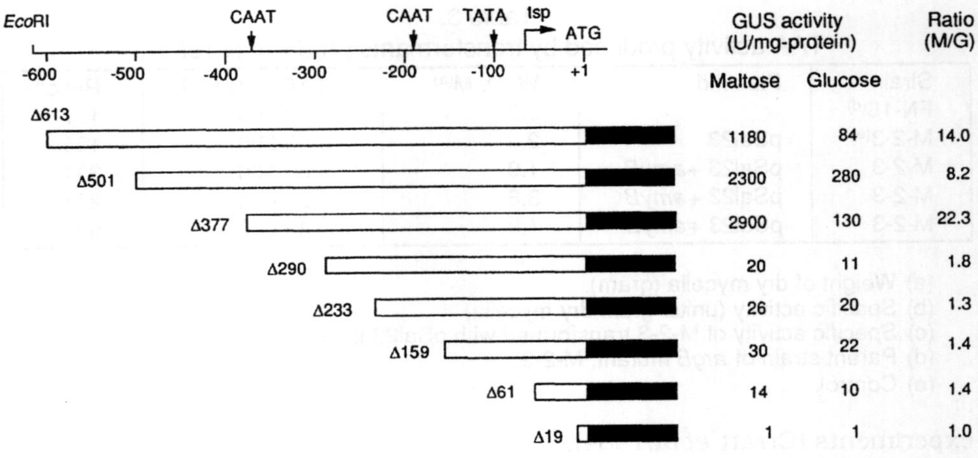

Figure 2. GUS activities in TAA promoter deletion mutants grown in medium containing maltose or glucose as a sole carbon source.
Numbers for deletion mutants denote the distance in base pairs from the initiation ATG codon at +1 to the deletion end point. Putative TATA box exists at -100 bp, and CAAT boxes exist at -192 bp and -377bp. Major transcription start points (tsp) are located at -66, -64 and -61 bp.
Promoter region of the TAA gene is indicated by an open box and *E. coli uidA* coding region is indicated by a solid box.

Similarly, high glucoamylase producing strain were constructed by the transformation of *A. oryzae* with the cloned glucoamylase gene (Hata *et al*, 1991). In these transformants, though the glucoamylase activity was about eightfold higher than the parent strain, the α-amylase activity was about one-tenth lower. The mechanism of this effect is unclear. Furthermore, when the gene encoding glucoamylase of *Aspergillus usamii* mut. *shirousamii*, which has the ability to digest raw starch, was introduced into *A. oryzae*, the resulting transformants showed the increase of raw starch digesting activity (Shibuya *et al*, 1990). These results show that the increase in the copy number of gene results in the increase in expression level of corresponding enzyme activity in *A. oryzae*, suggesting that this method will be a powerful tool for the breeding of *Aspergilli*.

References

Beri RK and Turner G (1987) Transformation of *Penicillium chrysogenum* using *A. nidulans amdS* gene as a dominant selective marker. Curr. Genet. 11: 639-641.
Christensen T, Woeldike H, Boel E, Mortensen SB, Hjortshoej K, Thim L and Hansen MT (1988) High level expression of recombinant genes in *Aspergillus oryzae*. Biol.Tech. 6: 1419-1422.
de Ruiter-Jacobs YMJT, Broekhuijsen M, Unkles SE, Campbell EI, Kinghorn JR, Contreras R, Pouwels PH and van den Hondel CAMJJ (1989). A gene transfer system based on the homologous *pyrG* gene and efficient expression of bacterial genes in *Aspergillus oryzae*. Curr. Genet. 16: 159-163
Erratt JA, Douglas P, Morrannelli F and Seligy VL (1984) The induction of α-amylase

by starch in *Aspergillus oryzae*. evidence for controlled mRNA expression. Can. J. Biochem. Cell Biol. 62 : 678-690.

Gems D, Johnstone IL and Clutterbuck AJ (1991) An autonomously replicating plasmid transforms *Aspergillus nidulans* at high frequency. Gene 98: 61-67.

Gines MJ, Dove M J and Seligy VL (1989) *Aspergillus oryzae* has two nearly identical Taka-amylase genes, each containing eight introns. Gene 79: 107-117.

Gomi K, Iimura Y and Hara S (1987) Integrative transformation of *Aspergillus oryzae* with a plasmid containing the *Aspergillus nidulans argB* gene. Agric. Biol. Chem. 51: 2549-2555.

Gomi K, Kitamoto K and Kumagai C (1991) Cloning and molecular characterization of the acetamidase-encoding gene *amdS* from *Aspergillus oryzae*. Gene 108: 91-98.

Gurr S, Unkles SE and Kinghorn JR (1987) The structure and organization of nuclear genes of filamentous fungi. In Gene Structure in Eukaryotic Microbes (JR Kinghorn, Ed.) pp.93-139. IRL Press, Oxford.

Hahm YT and Batt CA. (1988) Genetic transformation of *argB* mutant of *Aspergillus oryzae*. App. Env. Microbiol. 54: 1610-1611.

Hata Y, Tsuchiya K, Kitamoto K, Gomi K, Kumagai C, Tamura G and Hara S (1991) Nucleotide sequence and expression of the glucoamylase-encoding gene (*glaA*) from *Aspergillus oryzae*. Gene 108: 145-150.

Iimura Y, Gomi K, Uzu H and Hara S (1987) Transformation of *Aspergillus oryzae* through the complementation of methionine auxotrophic mutation. Agric. Biol. Chem. 51: 323-328.

Kelly JM and Hynes MJ (1985) Transformation of *Aspergillus niger* by the *amdS* gene *Aspergillus nidulans*. EMBO J, 4: 475-479.

Kelly JM and Hynes MJ (1987) Multiple copies of the *amdS* gene of *Aspergillus nidulans* causes titration of *trans*-acting regulation proteins. Curr. Genet. 12: 21-31.

Mattern IE, Unkles SE, Kinghorn JR, Pouwels PH and Van den Hondel CAMJJ (1987) Transformation of *Aspergillus oryzae* using the *A. niger pyrG* gene . Mol. Gen. Genet. 210: 460-461.

Penttila M, Nevalainen H, Ratto M, Salminen E and Knowles J (1987) A versatile transformation system for the cellulolytic filamentous fungus *Trichoderma reesei*. Gene 61: 155-163.

Shibuya I, Gomi K, Iimura Y, Takahashi K, Tamura G and Hara S (1990) Molecular cloning of the glucoamylase gene of *Aspergillus shirousamii* and its expression in *Aspergillus oryzae*. Agric. Biol. Chem. 54: 1905-1914.

Tada S, Iimura Y, Gomi K, Takahashi K, Hara S and Yoshizawa K (1989) Cloning and nucleotide sequence of the genomic Taka-amylase A gene of *Aspergillus oryzae*. Agric. Biol. Chem. 53: 593-599.

Tada, S., Gomi, K., Kitamoto, K., Kumagai, C., Tamura, G. and Hara, S. (1991) Identification of the promoter region of the Taka-amylase A gene required for starch induction. Agric. Biol. Chem. 55: 1939-1941.

Toda H, Kondo K and Narita K (1982) The complete amino acid sequence of Taka-amylase A. Proc. Jap. Acad. 58B: 208-212.

Unkles SE, Campbell EI, de Ruiter-Jacobs YMJT, Broekhuijsen M, Macro JA, Carrez D, Contreras R, Van den Hondel CAMJJ and Kinghorn JR (1989) The development of a homologous transformation system for *Aspergillus oryzae* based on the nitrate assimilation pathway: A convenient and general selection system for filamentous fungal transformation. Mol. Gen. Genet. 218: 99-104.

Van Heeswijck R (1986) Autonomous replication of plasmids in *Mucor* transformants. Carlsberg Res. Comm. 51: 433-443.

Ward M, Kodama KH and Wilson LJ (1989) Transformation of *Aspergillus awamori* and *A.niger* by electroporation. Exp. Mycol. 13: 289-293.

Wirsel S, Lachmund A, Wildhardt G and Ruttkowski E (1989) Three α-amylase genes of *Aspergillus oryzae* exhibit identical intron-exon organization. Mol. Microbiol. 3: 3-14.

TubB α-tubulin is essential for sexual development in *Aspergillus nidulans*

N. Ronald Morris and Karen E. Kirk
Department of Pharmacology
University of Medicine and Dentistry of New Jersey
Robert Wood Johnson Medical School
Piscataway, N.J. 08854
U.S.A.

Abstract

A. nidulans has two genes encoding α-tubulin, *tubA* and *tubB*, which differ by 28% in their predicted amino acid sequences. Mutational analysis has shown that *tubA is* essential for mitosis and nuclear migration. We have also deleted the *tubB* gene and studied the effect of this deletion on growth, development and fertility. *tubB* deletion strains grow normally, produce normal asexual spores and undergo the early stages of sexual reproduction. However, they are infertile because production of ascospores is completely abolished. We have studied the cytology of sexual development in wild-type strains and *tubB* deletion strains. The *tubB* deletion strains develop normally to the stage of ascus formation, but are defective for later stages. Either the two zygotic haploid nuclei are blocked in karyogamy or the resulting diploid nucleus is blocked in meiosis I. Introduction of extra copies of the *tubA* coding region into *tubB* deletion strains partially reverses the fertility defect and leads to the production of viable spores.

Introduction

The major structural protein of microtubules (MTs), tubulin, has two subunits, α- and β-tubulin, which form a dimer that polymerizes to form spindle and cytoplasmic MTs (Cleveland and Sullivan, 1985). Each type of subunit comprises a multigene family with different numbers of members in different organisms. Why these proteins are encoded by multigene families is still unclear, but two general explanations have been proposed. One is that the different tubulin subunit isotypes are functionally different and are incorporated into different classes of MTs (Fulton and Simpson, 1976). The other is that the multiplicity of tubulins is primarily for regulatory purposes, such that different tubulin isotypes can be expressed at different times and in different cell types during development (Raff *et al.*, 1987). Experiments designed to show that different tubulin isotypes are incorporated into unique types of MTs have generally supported the conclusion that any exogenously introduced subunit can be incorporated into any MT. Some recent papers have presented evidence for differential partitioning of MT isotypes into different classes of MTs (Oka *et al.*, 1990; Arai and Matsumoto, 1988). However,

NATO ASI Series, Vol. H 69
Molecular Biology and its Application to Medical Mycology
Edited by B Maresca, G. S. Kobayashi and H. Yamaguchi

the only convincing evidence for functional differences among different tubulin isotypes is that one of the β-tubulin isotypes of *Drosophila melanogaster* has been shown to be essential for spermatogenesis (Kimble *et al.*, 1990). In other organisms deletion of one tubulin subunit isotype may have an effect, but this can be compensated by overproduction of a sister isotype. For example, the effect of disrupting one of the two α-tubulin subunits of *Saccharomyces cerevisiae* can be repaired by overexpression of the other subunit (Schatz *et al.*, 1986) and similar results have been obtained with respect to the β-tubulins of *A. nidulans* (May, 1989).

Our laboratory has used a molecular genetic approach to look for evidence of tubulin functional specificity in the fungus *Aspergillus nidulans*. This organism has several advantages for this type of study. In comparison to most higher eukaryotes *Aspergillus's* collection of tubulin subunit isotypes is relatively uncomplicated: it has two genes for α-tubulins and two for β-tubulins (Doshi *et al.*, 1991; May *et al.*, 1987). It has a sophisticated classical genetic system and convenient mutations for studying tubulin isotype genetics. In fact, tubulin genetics was initially developed in *A. nidulans* (Sheir-Neiss *et al.*, 1978; Morris *et al.*, 1979). Genes can easily be cloned and manipulated. They can be deleted, replaced with other genes, and downregulated and upregulated by site specific, DNA mediated transformation. This has allowed us to examine each of the α-andβ-tubulin isotypes to determine whether they have a particular function in growth and/or development and whether these functions are isotype specific. *A. nidulans* also provides more opportunities than the budding and fission yeasts for identification of isotype specific functions because it has an extensive and interesting developmental program (Timberlake and Marshall, 1988). Our agenda has been to inactivate each of the tubulin subunit isotypes to determine the effect on growth and/or development and, if there is an effect, to see if the effect can be reversed by substitution of the other isotype in the pair. All four tubulin isotypes have now been inactivated and studied in this way.

The tubulin isotype first examined was the β-tubulin encoded by the *benA* gene (Sheir-Neiss *et al.*, 1978). *benA* β-tubulin is several fold more abundant during vegetative growth than *tubC*, the other β-tubulin isotype. Certain mutations in *benA* that cause resistance to the antiMT drug benomyl also cause conditional lethality at high temperatures (Morris *et al.*, 1979; Oakley and Morris, 1981). Thus, the *benA* β-tubulin appears to be required for vegetative growth and the *tubC* β-tubulin in these cells can not remedy the defect. This could be either because the normal, relatively low, *tubC* β-tubulin level in vegetative cells is insufficient for MT polymerization or function or because *benA* β-tubulin has a specific role for which *tubC* β- tubulin can not substitute. This was tested by transforming A. *nidulans* with a plasmid that simultaneously disrupted the *benA* gene and introduced a *tubC* structural gene under the control of the *benA* promoter (May, 1989). Strains generated in this way contained only *tubC* β-tubulin but were viable and exhibited no apparent MT disfunction. These data indicate that *tubC* β-tubulin has functional capabilities similar to those of *benA* β-tubulin. Interestingly, although *tubC* β-tubulin can complement a defect in *benA*, *its* normal function is obscure, since inactivation of *tubC* β-tubulin *per se* has no apparent effect on vegetative growth or on the development of asexual spores (conidia) or sexual spores (ascospores). Notwithstanding this lack of a requirement for

tubC during asexual spore formation, the *tubC* β-tubulin is upregulated during conidiogenesis (May and Morris, 1988).
The purpose of this report is to describe similar experiments designed to elucidate the role of the two α-tubulin isotypes in *A. nidulans* (see also Kirk and Morris, 1991). As noted above, *A. nidulans* has two α-tubulin genes, *tubA* and *tubB*. *tubA* was initially identified as a second site suppressor of a temperature sensitive *benA* mutation that allowed cells to grow at restrictive temperature. *tubA* mutations often exhibit vegetative supersensitivity to benomyl and are cold sensitive, indicating that *tubA* α-tubulin is essential for growth (Morris *et al.*, 1979; Oakley *et al.*, 1987). *tubB* was identified first as a cloned DNA sequence, and subsequently a deletion mutation was generated *(tubB)*.

Results

The *tubA* and *tubB* genes were cloned by conventional hybridization from a lambda genomic library using a chicken α-tubulin probe (Doshi *et al.*, 1991). The genes were then sequenced and found to be among the most divergent α-tubulin isotypes known, being 28% different from each other in predicted amino acid sequence. This extreme sequence diversity suggested that the *tubA* and *tubB* tubulin isotypes might be functionally differentiated. To determine whether the two genes have different functions, each has been inactivated by site specific, DNA-mediated transformation and the effect on growth and differentiation determined.
The *tubA* gene was disrupted by site specific integration of an internal sequence of the gene (Doshi *et al.*, 1991). The effect of this was to generate two incomplete, nonfunctional copies of this gene. Disruption of *tubA* caused a block in nuclear division and was vegetatively lethal, confirming previous evidence from cold sensitive mutations which indicated that *tubA* was an essential gene. Disruption of *tubB* produced slow growing, morphologically abnormal germlings with misshapen nuclei; however, in a different set of experiments a nearly complete deletion of *tubB* failed to have any effect on vegetative growth (Kirk and Morris, 1991). The N-terminal 80% of the *tubB* coding region was replaced by the *N. crassa pyr4* gene by site-specific integration of the plasmid at the *tubB* locus. This deletion mutation is henceforth referred to as *tubBΔ*. The site of plasmid integration in five transformants was determined by Southern blot analysis, which confirmed that *pyr4* had replaced the tubB gene in these transformants.
All five transformants were indistinguishable from wild type with respect to vegetative growth and conidia formation, but when allowed to continue to the stage of sexual spore (ascospore) formation were found to be sterile (Table 1). The deletion experiment is obviously epistatic to the disruption experiment with respect to vegetative growth and asexual spore production; therefore, we may conclude that the *tubB* gene is not needed for these functions. Strains carrying *tubBΔ* differentiated to form Hulle cells (large thick walled spore-like cells) and morphologically normal cleistothecia (the structure in which ascospores are made), but these cleistothecia contained no ascospores. *tubBΔ* strains are therefore sterile.
To determine why the *tubBΔ* mutants were sterile, it was first necessary to characterize the cytology of sexual development in wild type cells, since this

Table 1.
Effect of *tubBΔ* on sexual fertility

	Self-fertilization		*Cross-fertilization*
	tubB$^+$	*tubBΔ*	*tubB* x *tubBΔ*
Ascospore yield	3×10^4	0	7×10^4
Viability	93%	-	81%

developmental process had not previously been very well documented. The first sign of sexual differentiation in A. *nidulans* is the emergence, after about three days of vegetative growth, of minute clusters of Hülle cells on a mycelial lawn. This is soon followed by the appearance of cleistothecia, the protective spherical structures in which the ascospores develop. One of the earliest stages in sexual development is the establishment of ascogenous dikaryotic hyphae within the cleistothecia. Each segment of the dikaryotic ascogenous hyphae contains two haploid nuclei, which subsequently fuse to form a diploid meiotic nucleus in an ascus mother cell. This is recognizable

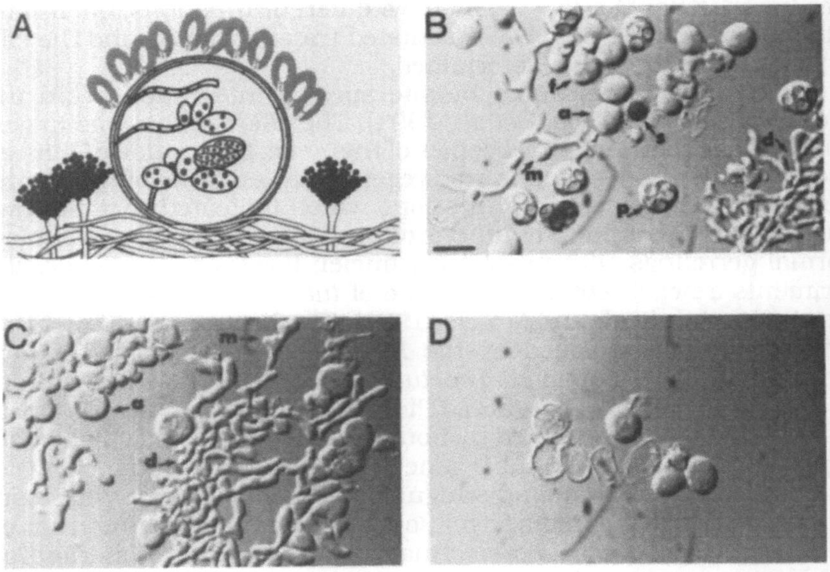

Figure 1. Cytology of sexual development. (A) Schematic diagram illustrating the events that take place inside a growing wild-type cleistothecium and result in ascospore formation. The structures inside the cleistothecium are drawn to approximate scale relative to one another but are about two to three times larger than the external scale. Nomarski images of contents of a developing wild-type cleistothecium (B), an early developing *tubB* cleistothecium (C), and a later developing *tubB* cleistothecium (D) are shown. The arrows indicate dikaryotic hyphae (d), ascus mother cell (m), ascus (a), freecell formation (f), early ascospore pigmentation (p), and mature released ascospore (s). Bar, 10μm.

as a small bud or more often as a swelling at the tip of the ascogenous hyphae. The ascus mother cell then expands to form an ascus. The diploid nucleus of the ascus mother cell undergoes a first and second meiotic division followed quickly by a mitotic division to generate eight haploid, spore progenitor nuclei. Each nucleus is then enclosed within a spore wall that becomes heavily pigmented, and the spore nuclei undergo one more mitotic division. Finally, the ascus degenerates to release binucleate spores which are still contained within the cleistothecium. When a mature, wild type cleistothecium is crushed, it releases tens of thousands of bright red, free, binucleate ascospores.

Since A. *nidulans is* homothallic, i.e. self-fertile, virtually any strain can be mated with itself. This makes it possible to study recessive loss of function mutations that affect mating directly, because the mutant strain will mate with itself. Thus we were able to study sexual development in self mated strains containing the recessive *tubBΔ* mutation (Kirk and Morris, 1991). The *tubBΔ* strain undergoes the initial stages of sexual development normally. Hülle cells appear, and cleistothecial structures are made that appear normal; but when these cleistothecia are opened, no spores are released. To determine at which step ascospore formation goes awry we dissected and examined *tubBΔ* mutant cleistothecia. Cytological investigation showed that morphologically normal ascogenous hyphae and ascus mother cells are made. The ascogenous hyphae have binucleate segments and are otherwise indistinguishable from the wild type. The first sign of abnormality is seen in the developing ascus. In wild type strains the asci contain one, two, four or eight nuclei depending on their stage of development. The *tubBΔ* asci exhibit only a single nuclear mass, which persists as the ascus grows to full size. Unfortunately, the resolution of the light microscope does not allow us to distinguish between two very closely apposed haploid nuclei and a single diploid nucleus and we have encountered technical problems with electron microscopy of cleistothecia; therefore the best we can presently say is that either the *tubBΔ* asci fail to undergo karyogamy or they fail to undergo normal meiotic and post meiotic divisions. In either case it is clear that *tubB* α-tubulin is required in the ascus mother cell at this specific developmental stage.

Why is there a requirement for *tubB* α-tubulin at this stage? There are two contrasting, general explanations for the *tubB* atubulin requirement during ascosporogenesis. The first is that in the absence of *tubB* α-tubulin the concentration of *tubA* atubulin in the ascus is too low to sustain MT polymerization or function and therefore meiosis can not occur. The second is that the ascus MTs have a specific requirement for *tubB* α-tubulin that can't be fulfilled by the *tubA* isotype. We have tried to distinguish between these possibilities by using DNA mediated transformation to introduce additional copies of *tubA* α-tubulin into the *tubB* deletion strain to determine whether raising the level of *tubA* α-tubulin sustains ascospore formation. Two *tubA* gene-containing plasmids, differing with respect to the promoters upstream of the *tubA* structural gene, were constructed and transformed into the *tubBΔ* strain. One used the normal *tubA* promoter. In the second the *tubB* promoter was fused to the *tubA* structural gene in the hope of generating levels of *tubA* α-tubulin in the *tubBΔ* ascus mother cell similar to the levels of *tubB* α-tubulin normally found in wild type strains. The transformants were then allowed to self-mate and the resulting cleistothecia analyzed for ascospore number and viability. The *tubA(p)tubA* transformants produced an ascospore with about

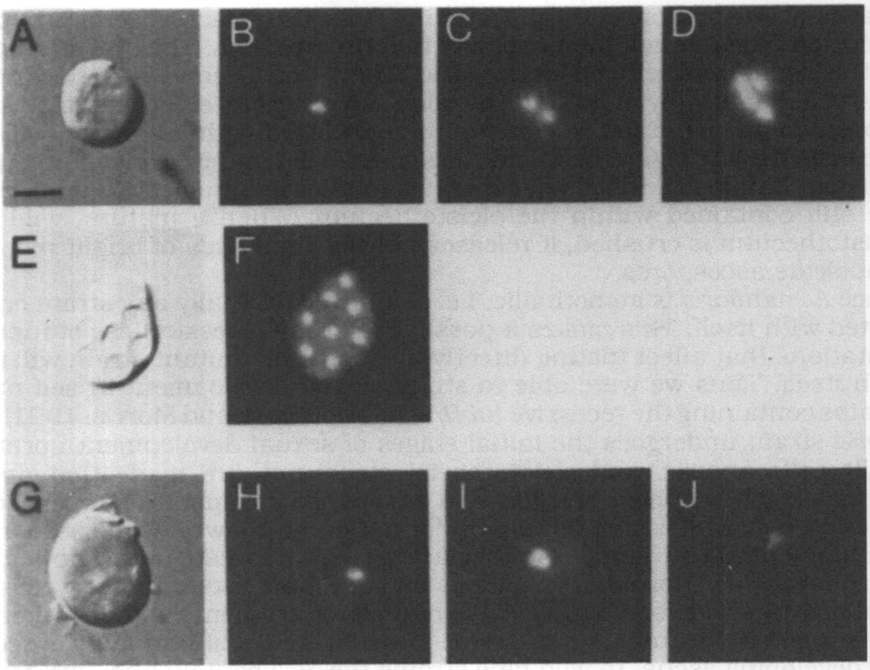

Figure 2. Cytology of ascus maturation in wild-type (A-F) and the *tubB* mutant. (G-J). (A,E and G) were photographed with Nomarski optics. (B-D, F and H-J) were stained with the DNA specific stain DAPI and photographed by fluorescence microscopy to reveal nuclei.

ten times fewer ascospores than wild type strains, but with three logs more than *tubBΔ* strains. The *tubB(ptubA* transformants produced an ascospore number comparable to wild type. However, in both cases the ascospores were morphologically abnormal and viability was greatly decreased. Cytological studies showed that the ascus mother cell was not tightly blocked in meiosis in these transformants. Later stages of ascospore formation were grossly abnormal. Some asci contained fewer than or more than eight ascospores, and nuclei sometimes remained unincorporated into ascospores in ascus mother cells. Ascospore viability was tested and found to be very low in both *tubA(ptubA* and *tubB(ptubA* transformants. However, when we constructed strains carrying two copies of *tubB(ptubA*, spore number and viability approached that of the wild type.
Since the *tubA* isotype was capable of functionally replacing the *tubB* isotype during formation of ascospores, we were interested to know whether the *tubB* isotype could replace the *tubA* isotype requirement during mycelial growth. A strain carrying the *tubA4* mutation, which causes mycelial growth to be cold sensitive, was transformed with a plasmid carrying the *tubB* structural gene under the control of the *tubA* promoter. These transformants were able to grow at restrictive temperature (K. Kirk and N.R. Morris, unpubl. data.)

Discussion

To determine why deletion of *tubB* affects fertility it was first necessary to describe ascospore development in wild type strains of *A. nidulans*. Although this has produced no surprises, it has given us a close understanding of the process of ascosporogenesis in *A. nidulans* and established a foundation for understanding the morphological effects of *tubB* deletion.
The fundamental question asked by this work is whether *tubB* α-tubulin has a function that is qualitatively different from that of *tubA* α-tubulin. To answer it we have improved upon a previous gene disruption experiment by deleting almost all of the *tubB* gene. Deletion of the *A. nidulans tubB* gene for α-tubulin has now shown unequivocally that *tubB* α-tubulin is not required for vegetative growth, conidiation, Hulle cell production, formation of dikaryotic ascogenous hyphae or appearance of cleistothecia. Whatever requirements these cell types and developmental processes may have for α-tubulin, they are presumably satisfied by *tubA* α-tubulin.
Only in the developing ascus does deletion of the *tubB* gene have a significant consequence with implications for the biology of the organism. Deletion of *tubB* completely prevents ascospore formation. This occurs at a very early stage in ascospore development. The dikaryotic nuclei are able to come together, but no further nuclear division occurs, and no ascospores are formed. Interestingly, the developmental program of the ascus mother cell is unaffected. The ascus mother cell continues to swell as it would in the wild type fungus. It forms an ascus, which matures to full size, produces its characteristic red pigment and then degenerates on the same time schedule that it would follow in the wild type organism, except that in the *tubB* deletion mutant no ascospores are there to be released. Thus the developmental program of the ascus is independent of both *tubB* α-tubulin and the ascospore developmental program.
The effect of *tubB* deletion on ascospore formation could be interpreted either to mean that *tubB is* normally the only α-tubulin present at a concentration consistent with MT formation and function in the developing ascus or to mean that it has a specific function which can not be replaced by *tubA* α-tubulin. To distinguish between these possibilities we introduced additional copies of the *tubA* gene into *a tubBΔ* mutant. The gene was introduced either with its own promoter or under the control of a *tubB* promoter. The result of adding one copy of *tubA* was that the mutant phenotype was partially complemented. The diploid nucleus could now undergo meiosis and form ascospores; but the number of ascospores in an ascus and their morphology was abnormal, and their viability was greatly diminished. Introduction of a second copy of *tubA* further ameliorated the defect. Since *tubA* α-tubulin can partially substitute for the deleted *tubB* α-tubulin, the most likely explanation for the effect of *tubB* deletion on ascospore formation is that either *tubA* α-tubulin is not made in the ascus mother cell or its concentration is too low to substitute for the missing *tubB* α-tubulin. It is also obvious that *tubA* α-tubulin can not upregulate to compensate for a *tubB* α-tubulin deficiency.
The converse experiment to determine whether the *tubB* isotype was functionally competent to rescue a defective *tubA* isotype gave a similar result. Putting the *tubB* structural gene under the control of the *tubA* promoter, remedied the growth defect caused by a cold sensitive mutation in the *tubA* gene. Thus, even though the structural genes for the two *A. nidulans* α-tubulins

are among the most divergent of any organism with respect to amino acid sequence, they appear to be functionally equivalent.

Acknowledgments

This research was supported by a National Institutes of Health grant to N.R.M. (GM29228) and by a Merck predoctoral fellowship to K.E.K.

References

Arai, T and Matsumoto, G (1988) Subcellular localization of functionally differentiated microtubules in squid neurons: regional distribution of microtubule-associated proteins and beta-tubulin isotypes. J. Neurochem. 51:1825-1838.

Cleveland, DW and Sullivan, KF (1985) Molecular biology and genetics of tubulin. Annu. Rev. Biochem. 54:331-365.

Doshi, P, Bossie, CA, Doonan, JH, May, GS, and Morris, NR (1991) The two α-tubulin genes of *Aspergillus nidulans* encode divergent proteins. Mol. Gen. Genet. 225:129-141.

Fulton, C and Simpson, PA (1976) Selective synthesis and utilization of flagellar tubulin. The multi-tubulin hypothesis. In *Cell Motility*. R. Goldman, T. Pollard, and J. Rosenbaum, editors. Cold Spring Harbor Laboratory, Cold Spring Harbor, N.Y. 987-1005.

Kimble, M, Dettman, RW, and Raff, EC (1990) The β3-tubulin gene of *Drosophila melanogaster is* essential for viability and fertility. Genetics. 126:991-1005.

Kirk, KE and Morris, NR (1991) The tubB gene α-tubulin is essential for sexual sporulation in *Aspergillus nidulans*. Genes Dev. 5:2014-2023.

May, GS, Tsang, ML-S, Smith, H, Fidel, 5, and Morris, NR (1987) *Aspergillus nidulans* β-tubulin genes are unusually divergent. Gene 55:231-242.

May, GS and Morris, NR (1988) Developmental regulation of a conidiation specific β-tubulin in *Aspergillus nidulans*. Dev. Biol. 128:406-414.

May, GS (1989) The highly divergent β-tubulins of *Aspergillus nidulans* are functionally interchangeable. J. Cell Biol. 109:2267-2274.

Morris, NR, Lai, MH, and Oakley, CE (1979) Identification of a gene for alpha tubulin in *Aspergillus nidulans*. Cell 16:437-442.

Oakley, BR and Morris, NR (1981) A beta tubulin mutation in *Aspergillus nidulans* that blocks microtubule function without blocking assembly. Cell 24:837-845.

Oakley, BR, Oakley, CE, and Rinehart, JE (1987) Conditionally lethal tubA α-tubulin mutations in *Aspergillus nidulans*. Mol. Gen. Genet. 208:135-144.

Oka, MT, Arai, T, and Hamaguchi, Y (1990) Heterogeneity of microtubules in dividing sea urchin eggs revealed by immunofluorescence microscopy: Spindle microtubules are composed of tubulin isotypes different from those of astral microtubules. Cell Motil. Cytoskeleton 16:239-250.

Raff, EC, Diaz, B, Hoyle, HD, Hutchens, J, Kimble, M, Raff, RA, Rudolph, JE, and Subler, M (1987) The origins of multiple gene families: are there both functional and regulatory restraints. In Development as an evolutionary process. R.A. Raff and E.C. Raff, editors. Alan R.Liss, New York. 203-238.

Schatz, PJ, Solomon, F, and Botstein, D (1986) Genetically essential and nonessential α-tubulin genes specify functionally interchangeable proteins. Mol. Cell Biol. 6:3722-3733.

Sheir-Neiss, G, Lai, MH, and Morris, NR (1978) Identification of a gene for beta tubulin in *Aspergillus nidulans*. Cell 15:639-647.

Timberlake, WE and Marshall, MA (1988) Genetic regulation of development in *Aspergillus nidulans*. Trends. Genet. 4:162-169.

The heat shock response of *Neurospora crassa*

Nora Plesofsky-Vig and Robert Brambl
Department of Plant Biology
The University of Minnesota
Saint Paul, Minnesota 55108
USA

Introduction

Nearly every type of cell in all organisms responds to the stress of sublethal heat shock with a major redirection of gene expression that helps to protect the cell against still higher, lethal temperatures (Lindquist, 1986; Lindquist & Craig, 1988). The molecular basis of this heat shock response, as well as responses to certain other physical stresses, centers about rapid synthesis of a unique set of mRNAs and their translation into the characteristic heat shock proteins. At the same time, normal gene transcription becomes repressed, and translation of normal mRNAs is reduced or discontinued. For most plants and microorganisms, this response to high temperature is transient, and they resume normal transcription and translation after a certain period of time at the heat shock temperature, presumably after accumulating sufficient amounts of the heat shock proteins (Lindquist, 1986). Numerous experiments have shown that cells that are exposed to a sublethal heat shock become more resistant to lethally high temperatures, as well as certain other stresses, and the proteins that cells synthesize during heat shock appear, in most cases, to be required for their increased thermotolerance (Plesofsky-Vig & Brambl, 1985).

The most prominently synthesized heat shock proteins fall into four separate classes: a group of proteins of about 70,000 daltons (hsp70), proteins between 82,000 and 90,000 daltons (hsp83), a protein of about 100,000 daltons (hsp100), and a group of low molecular weight proteins of 15,000 to 30,000 daltons that have a region of homology to vertebrate α-crystallin (Ingolia *et al.*, 1982). *Escherichia coli* produces heat shock proteins that are homologous to hsp70 (Bardwell *et al.*, 1984) and hsp83 (Bardwell *et al.*, 1987), with about 40-50% identity, but the only counterparts to the α-crystallin-related heat shock proteins observed in prokaryotes is the hsp18 of *Mycobacterium* (Nerland *et al.*, 1988). This class of heat shock proteins also displays much more variability among different eukaryotic organisms than do hsp70 and hsp83. In addition to these very highly induced heat shock proteins, several normal cellular proteins increase moderately (several-fold) in response to heat shock. These include ubiquitin (Bond & Schlesinger, 1985), which is active in cellular proteolysis, and hsp60, an organellar protein with mitochondrial and chloroplast representatives that are homologous to the prokaryotic GroEL protein of *E. coli* (Cheng *et al.*, 1989; Hemmingsen *et al.*, 1988).

A goal of many research groups has been to understand the cellular functions

NATO ASI Series, Vol. H 69
Molecular Biology and its Application to Medical Mycology
Edited by B. Maresca, G. S. Kobayashi and H. Yamaguchi

of proteins synthesized during heat shock. One productive approach has been the functional analysis of these or closely related proteins during cellular growth under non-stress conditions. For example, hsp70 has been found to be a member of a family of proteins (Lindquist & Craig, 1988), some of which are constitutively expressed, some induced by heat shock, and others induced by conditions such as glucose-deprivation (Munro & Pelham, 1986). Furthermore, hsp70 itself is both cytoplasmic and nuclear in location (Pelham, 1984), but other representatives of the hsp70 family are located within the endoplasmic reticulum (Munro & Pelham, 1986), in mitochondria (Craig *et al.*, 1989), or in chloroplasts (Marshall *et al.*, 1990).

The hsp70 group of proteins appears to aid in dissociating peptides or peptide domains from one another and in the disassembly of protein complexes (Pelham, 1986); this is apparently accomplished by hsp70 binding to peptides and subsequently releasing them as it hydrolyzes bound ATP (Flynn *et al.*, 1989). Constitutive hsp70 in mammalian cells is active in the disassembly of clathrin cages from coated vesicles (Chappell *et al.*, 1986), and in yeast cells it facilitates the translocation of proteins into both the endoplasmic reticulum and mitochondria (Deshaies *et al.*, 1988). The GroEL-related proteins are required in mitochondria and chloroplasts for the folding of proteins (Ostermann *et al.*, 1989) and for the assembly of subunit peptides into oligomeric enzyme complexes (Cheng *et al.*, 1989; Hemmingsen, *et al.*, 1988). The term chaperonin has been applied to both hsp60 and hsp70, because both activities are apparently required to guide translocated peptides into their proper state of assembly. Hsp83 also binds to other proteins, apparently when they are in an inactive conformation, such as unoccupied steroid hormone receptors (Sanchez *et al.*, 1985). In the water mold *Achlya*, where the mating interaction depends upon steroid hormone signals, the hsp83 heat shock protein reversibly interacts with the steroid receptor, perhaps to activate it, just as the same hsp83 does in animals (Riehl *et al.*, 1985). There are other regulatory proteins, like the receptor for dioxin (Wilhelmsson *et al.*, 1990) and the unassembled tyrosine kinases, key oncogenic proteins in Rous sarcoma virus-infected cells (Oppermann *et al.*, 1981), that also may be activated by hsp83 binding.

It is not known whether these heat shock proteins perform identical functions under heat shock conditions that they or their relatives perform under conditions of normal growth, but their functions would presumably be similar. Some of the heat shock proteins also are synthesized at specific points in cell development, such as fungal sporulation (Kurtz *et al.*, 1986), and we do not yet understand their involvement in these developmental processes.

No clear function has been identified for the α-crystallin class of heat shock proteins, and the low degree of similarity among proteins within this class suggests that they may have diverse functions in the cellular response to heat shock. These proteins as a class exhibit a low degree of sequence similarity, but they do have a similarity in their overall structural, or hydropathy profiles, with a highly conserved 33-amino acid segment shared with α-crystallin (Lindquist & Craig, 1988) which, it has been suggested, may mediate their assembly into oligomeric particulate complexes (Arrigo *et al.*, 1988) or it could contribute to their structural stability at high temperature (Maiti *et al.*,1988). Although members of this class of small heat shock proteins are clearly translocated into chloroplasts in plants (Kloppstech *et al.*, 1985; Vierling *et al.*, 1988), other related proteins have more variable locations. They have been reported to be nuclear (Arrigo *et al.*, 1988) or perinuclear (Leicht *et al.*, 1986) in animal cells, and in plants they appear in cytoplasmic granules (Nover *et al.*, 1983); other small heat shock proteins are found associated with mitochondria (Plesofsky-Vig & Brambl, 1985; 1990; Chou *et al.*, 1989; Pinelli & Shapira, 1990). Furthermore, the intracellular location of these proteins changes when the temperature of cell exposure is lowered or, for yeast, when the carbon source is varied (Rossi & Lindquist, 1989). Deletion of the gene for hsp26, the dominant small heat shock protein of yeast, did not affect yeast growth or survival at high temperature (Petko & Lindquist, 1986); constitutive expression of hsp26 during normal growth produced a small increase in cell survival at lethal temperature (Susek and Lindquist, 1989).

In this chapter we describe studies undertaken in our laboratory to characterize the responses of *Neurospora crassa* to heat shock. We will review briefly a molecular analysis of one of the individual heat shock proteins of this organism, a protein whose synthesis is specific to the heat shock response and that associates with the mitochondrial membranes. We also will describe what we have learned recently about another, high molecular weight heat shock protein that is enriched in the microsomes, and we summarize some recent investigations into the unusual heat shock response of the ascospores of *Neurospora*.

Heat shock protein synthesis in *Neurospora*

Neurospora exhibits a typical heat shock response, with the heat shock proteins most prominently synthesized by germinating conidia falling into the canonical classes: a protein of 67,000 daltons, a protein of 83,000 daltons, a protein of 98,000 daltons, and a small protein of about 30,000 daltons (whose sequence predicts a size of 25,000 daltons). Mycelial cells synthesize two additional heat shock proteins of 34,000 and 38,000 daltons. More extensive fractionation of cellular constituents reveals that other proteins, much less prominent in total cell extracts, also are synthesized during heat shock (see below).

We assayed the pattern of protein synthesis in *Neurospora* over a range of temperatures from 30°C, the normal growth temperature, up to 48°C, a temperature that becomes lethal to cells (Plesofsky-Vig & Brambl, 1985). Normal protein synthesis is reduced at 45°C and heat shock protein synthesis is at a sharp maximum at this temperature. A lethal temperature for the germinating conidia is at 48°C to 50°C, but if the 30°C cells are incubated at the intermediate temperature of 45°C before transfer to 50°C, about 65% of them survive, compared to about 1% without exposure to the adaptive heat shock temperature.

Addition of cycloheximide during the 45°C incubation blocks the induction of heat shock protein synthesis and the induced protection, implying that the synthesis of the heat shock proteins is essential for cell survival, although not necessarily sufficient.

Accumulation of transcripts of heat shock genes

We employed specific cDNA probes to measure accumulation of RNAs for both heat shock genes and normal genes during heat shock and recovery at high temperature (Plesofsky-Vig & Brambl, 1987). In one set of experiments we analyzed these transcripts in mycelial cells, where hsp30 and hsp83 transcripts reach a maximum level within 60 min. We established that hsp30 transcripts are not present in normal cells in advance of heat shock, whereas transcripts for hsp83 are present at moderate levels in un-induced cells. Recovery of normal protein synthesis at high temperature depends upon translational regulation and the use of pre-existing mRNAs that are retained (at least partially) in the heat shocked cells. The transcripts for a normal gene, encoding subunit 9 of the mitochondrial ATPase, within 5 min began a sharp decrease to about 10 to 20% of the original level, and it is this residual RNA that supports the recovery of translation capability at high temperature. Transcripts for other normal genes, such as a nuclear subunit of cytochrome c oxidase, only decreased by about 50% during heat shock (unpublished).
The activated conidia of *Neurospora* show an even more rapid response to heat shock, with heat shock transcripts reaching their maximum level by 15 min. Transcripts for hsp83 are conserved in the dormant conidia, whereas those for hsp30 are not. The spores also showed a more rapid increase after 15 min in transcripts for normal genes, such as the ATP9, which increased 4-fold during a subsequent 30-min interval, when the mycelial cells showed no increase. We believe that recovery of activated spores from heat shock depends upon accumulation of new mRNAs for normal proteins, whereas mycelial cells appear to change their translational preference during continued incubation at high temperature, from a discrimination against normal mRNAs to a resumption of their translation into normal cellular proteins.

Hsp30 gene sequence

We isolated a cDNA for hsp30 and then sequenced both the cDNA and the gene for hsp30 (Plesofsky-Vig & Brambl, 1990). Promoter regions of heat shock genes contain variations of a repeated heat shock consensus element (HSE), CTnGAAnnTTCnAG, which is essential for gene transcription, and the *Neurospora* hsp30 gene contains several multiple overlapping HSEs. These are extended by GAA repeated every 10 nucleotides. The mRNA has a long leader with little apparent secondary structure. There are no introns in the hsp30 gene, and it exists as a single copy, according to Southern hybridization analyses and molecular mapping to a single locus.
There are certain features within the predicted protein sequence that tell us what properties the heat shock protein may have. Within hsp30 are two regions that have the potential to form strongly amphiphilic helices. These amphiphilic helices are characterized by a large hydrophilic-hydrophobic gradient across the helix, and this type of structure has been found to interact with membranes and also with other proteins (Johnsson *et al.*, 1988). The helical sequence near the amino terminus of hsp30 has an especially high hydrophobic moment and maximum hydrophobicity, leading us to predict that it might be part of a targeting structure that determines the association of the heat shock protein with mitochondrial membranes. As

explained below, our experiments to delete different segments of this protein led us to focus on this region of the protein as a localization determinant. The region that is α-crystallin-related, near the carboxy end, is one of the more hydrophobic regions within this protein, which is generally hydrophilic. Beginning with this region, we have identified an extended region of homology among a large number of proteins that we have employed for phylogenetic analysis, as described later.

Heat shock proteins in gradient-purified mitochondria

It has been of major interest to us to determine the cellular localization of hsp30, since we expect that this information eventually will lead to an understanding of its function. We purified mitochondria by Percoll gradient centrifugation and by sucrose flotation gradients and found that hsp38, hsp34, and hsp30 were enriched in the mitochondria (Plesofsky-Vig & Brambl, 1985; Plesofsky-Vig & Brambl, 1990). The mitochondria from nonheat-shocked cells lacked these proteins. Other experiments in which we employed autoradiography of the radiolabeled mitochondrial fraction and electron microscopy also supported this association of heat shock proteins with the mitochondria. The electron microscopy shows that the radiolabel is mainly associated with mitochondria or with membrane fragments. There is no evidence for co-sedimenting structures associated with the radiolabeled proteins.

Cells exposed to 45°C were examined by electron microscopy to determine whether the heat shock caused major structural changes in mitochondria or mitochondrial membranes (Habel *et al.*, 1991). These mitochondria in stressed cells appeared to be swollen, but there was no evident change in structural complexity of mitochondria. A striking, incidental change we observed was that glycogen granules rapidly disappeared in those cells exposed to heat shock, and we proposed that this glycogen was converted to trehalose, a protective disaccharide (Habel *et al.*, 1991). This proposal has since received direct experimental support (Neves *et al.*, 1991).

Mitochondria respond bioenergetically to heat shock

We found that mitochondria themselves were participating in the heat shock response. One indication came from measurements of oxygen consumption, where we and others have found that transfer to heat shock conditions caused a dramatic drop (of about 30%) in the rates of cyanide-sensitive respiration. The catalytic activity of the oligomycin-sensitive mitochondrial ATPase decreased by about 25% in *Neurospora* cells exposed to a 45°C heat shock (unpublished). A more graphic illustration of the response of mitochondria came from our recent *in vivo* studies of mitochondria that were loaded with a cationic cyanine dye, 3,3'-diheptyloxacarbocyanine iodide, whose fluorescence reports the energization of the mitochondria and the electrochemical potential across the mitochondrial membranes. Using fluorescence microscopy, we found that the mitochondria of heat shocked cells (45°C) show a sharply decreased fluorescence in comparison to those maintained at 30°C. We believe that heat shock immediately causes the mitochondrial membranes to undergo a partial depolarization. Later, with prolonged heat shock, the mitochondrial energization is restored at high temperature (Habel *et al.*, 1991).

In vitro binding of hsp30 to mitochondria

To analyze the association of hsp30 with mitochondria, we employed *in vitro* binding assays with isolated mitochondria and with hsp30 that was made *in vitro* by coupled transcription and translation. We were especially interested in learning how hsp30 binding might be regulated and what regions of the protein might be important for binding (Plesofsky-Vig & Brambl, 1990). The radiolabeled protein was incubated with mitochondria that we isolated from heat-shocked cells and also with mitochondria from nonheat-shocked cells. We made this comparison between the two types of mitochondria, because experiments showed that the temperature of cell incubation determined the association of hsp30 with mitochondria *in vivo*. That is, after heat-shocked cells were shifted down to normal temperature, we found that significant amounts of hsp30 left the mitochondria within 30 min. We performed the *in vitro* binding assay at 29°C.

In this binding assay, *in vitro* synthesized hsp30 showed a much greater affinity for mitochondria from the heat-shocked cells than for mitochondria from normal cells, which corresponds to the type of association that we found *in vivo*. Succinate and ATP apparently are required for binding of hsp30 to mitochondria *in vitro*. The succinate may be required for energization of the mitochondrial membranes and the ATP may be required as a non-hydrolyzed cofactor in protein interaction with the membranes.

We constructed derivatives of hsp30 that are deleted in specific regions of the peptide to test their role in binding *in vitro* to mitochondria. We eliminated most of the amino-terminal helix of hsp30 to test the importance of this amphiphilic structure for mitochondrial targeting. A comparison of the binding of this mutant protein to heat shock mitochondria and to normal mitochondria indicated that hsp30 binding was not blocked by deleting the potential amphiphilic helix. In fact, the deletion mutant appeared to bind more strongly to mitochondria than the original hsp30, by about 2-fold. However, the mutant protein also bound strongly to mitochondria from the nonheat-shocked cells. The strong binding of the mutant peptide to both types of mitochondria suggests that deletion of the amphiphilic helix somehow reduced or abolished the selectivity of hsp30 for binding only to heat-shock mitochondria. The amphiphilic helix may be important *in vivo* for regulating hsp30 binding.

Another contrast between the two proteins is that the full-length hsp30 had more stringent requirements for binding to heat shock mitochondria than the mutant protein. Performing the incubation at 4°C lowered the binding of hsp30 at least 2-fold, compared with its binding at the standard assay temperature of 29°C, but binding of the mutant protein was not reduced at 4°C compared with 29°C.

We have prepared antibodies to hsp30 through use of a fusion protein that we produced in *E. coli*. The anti-hsp30 IgG precipitated its protein from mitochondrial membranes of radiolabeled heat-shocked cells, but not from cells radiolabeled at 30°C and then transferred to 45°C in the absence of

radiolabel. Pre-immune antiserum did not recognize any protein.
Several types of evidence suggest to us that hsp30 may be binding to the outer mitochondrial membrane of *Neurospora*. *In vivo*, the hsp30 dissociates from the mitochondria as the cells are shifted from 45°C to 30°C. An outer membrane association seemingly would be necessary to account for the reversibility of the binding to the organelle. In addition, the association of hsp30 with mitochondria is susceptible to digestion with trypsin. We expect that the hsp30 binds to a protein that is normally resident on the membrane but which is altered by heat shock to thereby become an effective receptor of the heat shock protein. We are now testing this location using antiserum against a mixture of components of the purified outer membrane of *Neurospora*, which was provided to us by Carmen Mannella (New York State Department of Health; Albany). In our first experiments, we added the antiserum to the *in vitro* binding reaction and we found that it strongly reduced hsp30 binding, while pre-immune serum had no effect on binding. In the same type of experiment, we included antiserum against porin, which is a prominent channel-forming protein complex of the outer membrane. The porin-specific serum had only a small effect on hsp30 binding, which suggests that the principal binding sites for hsp30 lie elsewhere in the membrane than porin.

Cellular distribution of hsp98

In our earlier cell fractionation studies to examine the distribution of the heat shock proteins, we noticed that one of them, hsp98, was enriched in the microsome fraction (unpublished). We recently re-examined this issue, and we confirmed that hsp98 is concentrated in the microsomal fraction, although it is also present elsewhere, such as the microsomal supernatant.
A. Vassilev in our laboratory learned that hsp98 binds strongly to heparin-agarose, which allowed him to purify hsp98 through several types of chromatography. We performed an amino acid sequence analysis of the peptides derived from cyanogen bromide cleavage of the purified protein. For one 28-amino acid sequence (derived from two overlapping peptides), we learned that hsp98 has 75% homology with a region of the ClpB protein of *E. coli* (Gottesman *et al.*, 1990) and 86% homology to a corresponding region of a related protein from *Trypanosoma brucei* (Gottesman *et al.*, 1990). Both of these proteins are closely related to the ClpA protein of *E. coli* that is reported to be a regulatory subunit of an ATP-dependent protease (Gottesman *et al.*, 1990). The sequence that we determined for the *Neurospora* protein has a 71% homology with the corresponding segment of hsp104 of *Saccharomyces* (Parsell *et al.*, 1991).
Because of the apparent enrichment of hsp98 in the microsomes, we next fractionated the microsomes by sucrose density gradient centrifugation. We collected fractions of the gradients that corresponded to the soluble meniscus fraction, the ribosome subunits, the monoribosomes, and the polyribosomes. The polyribosomes responded dramatically to heat shock, with the majority of the polyribosomes converted to monoribosomes upon shift to 45°C, followed by their restoration upon return of the cells to 30°C. An electrophoretic analysis of the fractions from the sucrose gradients showed that the polyribosomes contain hsp98, and its enrichment in the polyribosomes was considerably greater than in the monoribosomes. Nevertheless, this analysis

indicated that polyribosomes contained other heat shock proteins as well, and hsp98 was strongly represented in the soluble meniscus fraction, thus showing that hsp98 is not necessarily targeted specifically to polyribosomes. Transferring the cells to normal growth temperature after 30 min heat shock treatment led to a rapid loss of polyribosome-bound heat shock proteins. This redistribution also occurred more slowly during a continuous incubation of the cells at high temperature. We dissociated the polyribosomes with EDTA and subjected the resulting subunits to electrophoresis. Nearly all the heat shock proteins associated with the polyribosomes, in fact, were associated with the large subunits. We could detect only one heat shock protein (22,000 daltons) that was unique to the small subunits. More importantly, this analysis led to the detection of a new heat shock protein of 28,000 daltons that is found as the most strongly radiolabeled protein in the large subunit. This protein was not a conspicuous heat shock protein in total cell extracts, but it is the dominant heat shock protein in the large ribosomal subunit (Vassilev *et al.*, 1992).

Phylogenetic analysis of proteins related to *Neurospora* hsp30

We have examined the phylogeny of hsp30 of *Neurospora* and those proteins of the same class from a number of other plants, animals, and microorganisms (Plesofsky-Vig *et al.*, 1992b). A number of sequences for these genes have been reported recently, making possible an analysis of these proteins across distant evolutionary relationships. The similarity of the α-crystallin-related proteins near their carboxy terminus is well known, but we were able to extend the alignment considerably among a number of diverse protein sequences. Approximately one-third (30) of the aligned amino acids in these proteins were conserved in 75% of the proteins, and three blocks of consensus sequence were identified. Relationships were established by maximum parsimony and distance matrix analysis of the aligned amino acid sequences. The inferred phylogeny trees indicate that the plant proteins clearly divided into three major groups that were not related by taxonomy: two groups of proteins that originate from a common ancestral plant protein and the nucleus-encoded, chloroplast-localized proteins. The animal proteins, in contrast, branch in accordance with taxonomy. The only non-taxonomic subgrouping, occurring within the vertebrates, is the separation of the α-crystallins from the conventional heat shock proteins. This analysis indicates that the small heat shock proteins of animals, which follow taxonomy, have diverged rapidly and more widely than have the plant proteins, of which one group is especially stable. The slowly evolving plant heat shock proteins presumably group according to function. In the fungi there is wide divergence between *Neurospora* and yeast, with these two ascomycetes having identity in only a third of their amino acids within the region compared. Distance matrix analysis shows that the distance between these two fungi is as great as between the human hsp27 and the *Drosophila* hsp22.

The response of germinating Neurospora ascospores to heat shock

Although the heat shock response is a universal reaction of organisms to the stress of high temperature, some developmental stages, such as the germinating ascospores of *Neurospora tetrasperma*, are an exception. In studies of ascospore germination, we observed, as had other workers many years previously, that these spores developed a rapid respiration immediately after activation, and we found that this mitochondrial respiratory activity depended upon the activities of enzyme systems conserved in the ascospores during dormancy for function at the outset of spore germination (Hill *et al.*, 1992). What we also found was that the spores could not synthesize protein until after 90 min of incubation at 30°C. More sensitive immunochemical assays of individual respiratory enzyme subunits also showed that there was no incorporation of radiolabel into spore protein during the first 90 min of spore incubation, but after this time protein synthesis accelerated rapidly (Hill *et al.*, 1992).

We next asked if these spores could accumulate RNA for specific genes, and we found that they were blocked in RNA accumulation for at least 90 to 120 min of incubation after their activation. In using a probe for a mitochondrial ATPase subunit we observed that new nuclear transcripts were not accumulated appreciably until after 90 min, the point when this specific mRNA is translated (Hill *et al.*, 1992).

Since the ascospores appeared to be blocked transcriptionally, we asked if they were capable of a heat shock response (Plesofsky-Vig *et al.*, 1992a). We found that the ascospores not only were blocked in their molecular response to heat shock but that they also were rapidly killed by high temperatures (60°C) after activation, with fewer than 5% surviving. These cells at the outset of vegetative growth are extremely sensitive to elevated temperature, a paradox, since these ascospores are activated in nature and in the laboratory by brief exposure to high temperature (60°C). By use of one of the heat shock probes, in this case the hsp30, we learned that heat shock gene expression was blocked and that heat shock RNAs did not accumulate during the first 300 min; the ascospores do not acquire the ability to express a full heat shock response until after 300 min of incubation. We believe that this transcriptional blockage could involve a stage-specific methylation of the ascospore DNA, since for at least one heat shock probe the inability to accumulate transcripts is kinetically coincident with removal of cytosine methylation at *Sau*3AI restriction sites in the DNA. The dormant ascospore DNA appears to be unmethylated, but within the first 30 to 60 min of incubation the DNA develops a methylation that is retained through at least 330 min and then disappears (Plesofsky-Vig *et al.*, 1992a). This stage-specific interruption in the cell capacity to express a heat shock response apparently has parallels in other organisms. We have recently learned that the germinating pollen of maize also cannot express heat shock genes, and in these cells accumulation of heat shock transcripts is specifically blocked (Hopf *et al.*, 1992).

Acknowledgments

Research from the authors' laboratory was supported by grants from the US National Institute of General Medical Sciences (GM 19398) and the Competitive Research Grants Office of the USDA.

References

Arrigo A-P, Suhan JP, Welch WJ (1988) Dynamic changes in the structure and intracellular locale of the mammalian low-molecular-weight heat shock proteins. Mol. Cell. Biol. 8:5059-5071.

Bardwell JCA, Craig EA (1984) Major heat shock gene of *Drosophila* and the *Escherichia coli* heat inducible dnaK gene are homologous. Proc. Natl. Acad. Sci. USA 81:848-852.

Bardwell JCA, Craig EA (1987) Eukaryotic M_r 83,000 heat shock protein has a homologue in *Escherichia coli*. Proc. Natl. Acad. Sci. USA 84:5177-5181.

Bond U, Schlesinger MJ (1985) Ubiquitin is a heat shock protein in chicken embryo fibroblasts. Mol. Cell. Biol. 5:949-956.

Chappell TG, Welch WJ, Schlossman DM, Palter KB, Schlesinger MJ, Rothman JE (1986) Uncoating ATPase is a member of the 70 kilodalton family of stress proteins. Cell 45:3-13.

Cheng JY, Hartl F-U, Martin J, Pollock R A, Kalousek F, Neupert W, Hallberg EM, Hallberg RL, Horwich, AL (1989) Mitochondrial heat-shock protein hsp60 is essential for assembly of proteins imported into yeast mitochondria. Nature 337:620-625.

Chou M, Chen Y-M, Lin C-Y (1989) Thermotolerance of isolated mitochondria associated with heat shock proteins. Plant Physiol. 89:617-621.

Craig EA, Kramer J, Shilling J, Werner-Washburne, M, Holmes S, Kosic-Smithers J, Nicolet CM (1989) SSC1, an essential member of the yeast HSP70 multigene family, encodes a mitochondrial protein. Mol. Cell. Biol. 9:3000-3008.

Deshaies RJ, Koch BD, Werner-Washburne M, Craig EA, Schekman R (1988) 70kd stress protein homologues facilitate translocation of secretory and mitochondrial precursor polypeptides. Nature 332:800-805.

Flynn GC, Chappell TG, Rothman JE (1989) Peptide binding and release by proteins implicated as catalysts of protein assembly. Science 245:385-390.

Gottesman S, Squires C, Pichersky E, Carrington M, Hobbs M, Mattick JC, Dalrymple B, Kuramitsu H, Shiroza T, Foster T, Clark WP, Ross B, Squires CL, Maurizi MR (1990) Conservation of the regulatory subunit for the Clp ATP-dependent protease in prokaryotes and eukaryotes. Proc. Natl. Acad. Sci. USA 87:3513-3517.

Habel D, Plesofsky-Vig N, Brambl R (1991) The respiratory response to heat shock in *Neurospora crassa*. FEMS Microbiol. Lett. 81:317-322.

Hemmingsen SM, Woolford C, van der Vies SM, Tilly K, Dennis DT (1988) Homologous plant and bacterial proteins chaperone oligomeric protein assembly. Nature 333:330-334.

Hill EP, Plesofsky-Vig N, Paulson A, Brambl R (1992) Respiration and gene expression in germinating ascospores of *Neurospora tetrasperma*. FEMS Microbiol. Lett 90:111-116.

Hopf N, Plesofsky-Vig N, Brambl R (1992) The heat shock response of pollen and other tissues of maize Plant Mol. Biol. (In press).

Ingolia TD, Craig EA (1982) Four small Drosophila heat shock proteins are related to each other and to mammalian α-crystallin. Proc. Natl. Acad. Sci. USA 79:2360-2364.

Johnsson N, Marriott G, Weber K (1988) p36, the major cytoplasmic substrate of src tyrosine protein kinase, binds to its p11 regulatory subunit via a short amino-terminal amphiphatic helix. EMBO J. 7:2435-2442.

Kloppstech K, Meyer G, Schuster G, Ohad I (1985) Synthesis, transport and localization of a nuclear coded 22-kd heat-shock protein in the chloroplast membranes of peas and *Chlamydomonas reinhardi*. EMBO J. 4:1901-1909.

Kurtz S, Rossi J, Petko L, Lindquist S (1986) An ancient development induction: heat-shock proteins induced in sporulation and oogenesis. Science 231:1154-1157.

Leicht BG, Biessmann H, Palter KB, Bonner JJ (1986) Small heat shock proteins of *Drosophila* associate with the cytoskeleton. Proc. Natl. Acad. Sci. USA 83:90-94.

Lindquist, SL (1986) The heat-shock response. Annu. Rev. Biochem. 55:1151-1191.

Lindquist S, Craig EA (1988) The heat-shock proteins. Annu Rev Genet 22:631-677.

Marshall JS, DeRocher AE, Keegstra K, Vierling E (1990) Identification of heat shock protein hsp70 homologues in chloroplasts. Proc. Natl. Acad. Sci. USA 87:374-378.

Maiti M, Masahiro K, Chakrabarti B (1988) Heat-induced changes in the conformation of the α- and β-crystallins: unique thermal stability of α-crystallin. FEBS Lett. 236:109-114.

Munro S, Pelham HRB (1986) An HSP70-like protein in the ER: identity with the 78kd glucose-regulated protein and immunoglobulin heavy chain binding protein. Cell 46:291-300.

Nerland AH, Mustafa AS, Sweetser D, Godal T, Young RA (1988) A protein antigen of *Mycobacterium leprae* is related to a family of small heat shock proteins. J Bacteriol. 170:5919-5921.

Neves MJ, Jorge JA, Francois JM, Terenzi HF (1991) Effects of heat shock on the level of trehalose and glycogen, and on the induction of thermotolerance in *Neurospora crassa*. FEBS Lett. 283:19-22.

Nover L, Scharf K-D, Neumann D (1983) Formation of cytoplasmic heat shock granules in tomato cell cultures and leaves. Mol. Cell. Biol. 3:1648-1655.

Oppermann H, Levinson W, Bishop JM (1981) A cellular protein that associates with the transforming protein of Rous sarcoma virus is also a heat shock protein. Proc. Natl. Acad. Sci. USA 78:1067-1071.

Ostermann J, Horwich AL, Neupert W, Hartl F-U (1989) Protein folding in mitochondria requires complex formation with hsp60 and ATP hydrolysis. Nature 341:125-130.

Parsell DA, Sanchez Y, Stitzel JD, Lindquist S (1991) Hsp104 is a highly conserved protein with two essential nucleotide-binding sites. Nature 353:270-273.

Pelham HRB (1984) HSP70 accelerates the recovery of nucleolar morphology after heat shock. EMBO J. 3:3095-3100.

Pelham HRB (1986) Speculations on the functions of the major heat shock and glucose regulated proteins. Cell 46:959-961.

Petko L, Lindquist S (1986) Hsp26 is not required for growth at high temperatures, nor for thermotolerance, spore development, or germination. Cell 45:885-894.

Pinelli E, Shapira M (1990) Temperature-induced expression of proteins in *Leishmania mexicana amazonensis*. Eur. J. Biochem. 194:685-691.

Plesofsky-Vig N, Brambl R (1985) Heat shock response of *Neurospora crassa*: protein synthesis and induced thermotolerance. J. Bacteriol. 162:1083-1091.

Plesofsky-Vig N, Brambl R (1987) Two developmental stages of *Neurospora crassa* utilize similar mechanisms for responding to heat shock but contrasting mechanisms for recovery. Mol. Cell. Biol. 7:3041-3048.

Plesofsky-Vig N, Brambl R (1990) Gene sequence and analysis of hsp30, a small heat shock protein of *Neurospora crassa* which associates with mitochondria. J. Biol. Chem. 265:15432-15440.

Plesofsky-Vig N, Paulson, A, Hill EP, Glaser L, Brambl R (1992a) Heat shock gene expression in germinating ascospores of *Neurospora tetrasperma*. FEMS Microbiol. Lett 90:107-112.

Plesofsky-Vig N, Vig J, Brambl R (1992b) Phylogeny of the α-crystallin-related heat shock proteins J. Mol. Evol. (In press).

Riehl RM, Sullivan WP, Vroman BT, Bauer VJ, Pearson GR, Toft DO (1985) Immunological evidence that the nonhormone binding component of avian steroid receptors exists in a wide range of tissues and species. Biochemistry 24: 6586-6591.

Rossi JM, Lindquist S (1989) Intracellular location of hsp26 in yeast cells varies with metabolism. J. Cell. Biol. 108: 425-439.

Sanchez ER, Toft DO, Schlesinger MJ, Pratt WB (1985) Evidence that the 90-kDa phosphoprotein associated with the untransformed L-cell glucocorticoid receptor is a murine heat shock protein. J. Biol. Chem. 260: 12398-12401.

Susek RE, Lindquist SL (1989) hsp26 of *Saccharomyces cerevisiae* is related to the superfamily of small heat shock proteins but is without a demonstrable function. Mol. Cell. Biol. 9: 5265-5271.

Vassilev AO, Plesofsky-Vig N, Brambl R (1992) Isolation, partial amino acid sequence, and cellular distribution of heat shock protein hsp98 from *Neurospora crassa*. submitted.

Vierling E, Nagao RT, DeRocher AE, Harris LM (1988) A heat shock protein localized to chloroplasts is a member of a eukaryotic superfamily of heat shock proteins. EMBO J. 7: 575-581.

Wilhelmsson A, Cuthill S, Denis M, Wikström A-C, Gustafsson J-Å, Poellinger L (1990) The specific DNA binding activity of the dioxin receptor is modulated by the 90 kd heat shock protein. EMBO J. 9: 69-76.

Blue light regulated expression of geranylgeranyl pyrophosphate synthetase (Albino-3) gene in *Neurospora crassa*

G. Macino, S. Baima*, A. Carattoli, G. Morelli*, E. M. Valle
Dept. of Human Biopathology
Section of Cellular Biology
Policlinico Umberto 1°
University of Rome "La Sapienza"
00161 Rome
Italy

Introduction

In the filamentous fungus *Neurospora crassa* several effects are induced by blue light during its life cycle. Carotenoid biosynthesis in mycelia (Harding, Shropshire 1980) and phase shifting and photosuppression of the circadian rhythm of conidiation (Feldman, 1982) are induced by blue light during vegetative growth. When sexual cycle takes place blue light stimulates the protoperitecia formation (Degli Innocenti, Russo 1983) and the positive phototropism of peritecial beaks (Harding, Melles, 1983). In *N.crassa* all these photoresponses are suppressed by mutations in two genes, called white collar (*wc-1*, *wc-2*) (Degli Innocenti, Russo, 1984). Blue light regulates a group of genes involved in the carotenogenesis (*al-1*, *al-2* and *al-3*) (Schmidhauser *et al.*, 1990; Lauter F R , Russo V E A unpublished data; Nelson *et al.*, 1989), in the conidiation (*con-5* and *con-10*) and other genes called *bli*, whose functions are not known yet (*bli-3*, *bli-4*, *bli-7* and *bli-13*) (Sommer *et al.*, 1989; Pandit N N, Russo V E A 1991). Blue light induction of carotenogenesis needs the transcriptional activation of the albino genes coding for the geranylgeranyl pyrophosphate synthetase (*al-3*), phytoene synthetase (*al-2*) and phytoene dehydrogenase (*al-1*). Three albino mutants which are defective in the structural genes necessary for carotenogenesis and described above, have been characterized in *N.crassa*. These mutants have white conidia and mycelia, unlike the orange conidia and mycelia characteristic of wild-type strains (Harding, Turner 1981).
We have cloned the gene coding for the geranylgeranyl pyrophosphate synthetase (Nelson *et al.*, 1989) and its sequence was determined (Carattoli *et al.*, 1991). We have also demostrated that the *al- 3* gene is regulated by blue light: the *al- 3* mRNA is not detected in dark grown mycelia, while its mRNA accumulates transiently in response to a short pulse of blue light. In the

---

* Unità di Nutrizione Sperimentale, Istituto Nazionale della Nutrizione, Via Ardeatina 546, 00178 Roma, Italy.

NATO ASI Series, Vol. H 69
Molecular Biology and its Application to Medical Mycology
Edited by B. Maresca, G. S. Kobayashi and H. Yamaguchi

strains carrying the white collar (wc) mutations, *al-3* gene failed to show any response to blue light (Nelson *et al.*, 1989) demonstrating a role for the *wc* gene products in the light regulation of carotenoid biosynthesis. The minimum light energy fluence necessary to saturate the photoresponse is reached at 130 J/m² (Baima *et al.*, 1991). We determined the kinetic of accumulation of the *al-3* mRNA after a brief illumination (one minute pulse under saturating conditions) of wild type mycelia. We have demonstrated that the *al-3* gene is a very fast regulated gene. The *al-3* mRNA increases rapidly during the first 20 minutes after the light pulse and then its steady state mRNA level decreases until becoming undetectable at a very fast rate. It has been shown that actinomycin D inhibits the accumulation of *al-3* mRNA (Baima *et al.*, 1991). These results strongly indicated that the *al-3* mRNA photoinduction is mainly regulated at the transcriptional level. This is in agreement with previous reports in which light induced carotenoid synthesis was impaired by actinomycin D in *N. crassa* (Subden, Bobowski 1973). Here we present a further characterization of *al-3* gene expression including a kinetic of *al-3* mRNA accumulation under continous light. We also present the *in vivo* characterization of a minimal *al-3* promoter and the *in vitro* analysis of an *al-3* promoter fragment in its interaction with the putative trans-acting factors.

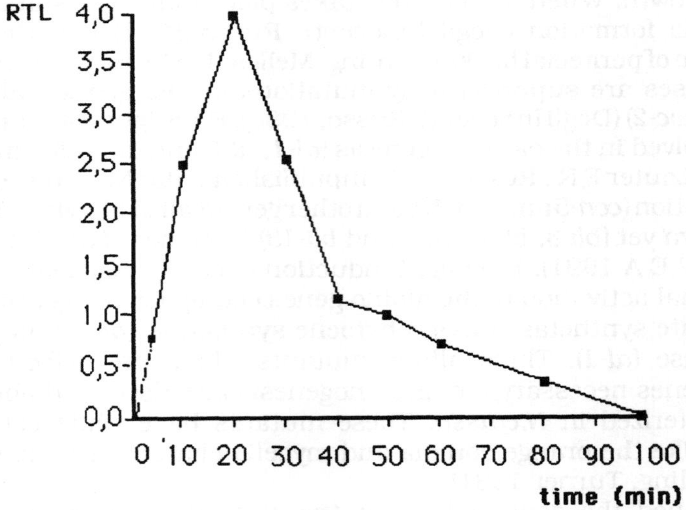

Figure 1. Kinetic of *al-3* mRNA accumulation in mycelia cells exposed to blue light. Dark grown wild type mycelia cells were continously illuminated with an energy fluence rate of 130 J/m² and collected at different times. Filter bound RNA samples were hybridized with the *al-3* specific probe and with the IF2 cDNA probe (IF2 is a light uninducible gene) to normalize the data. The X-ray films from Northern hibridization experiments were analyzed with a 2D LKB Laser Densitomer Ultrascan XL and LKB software. The relative transcript level (RTL) refers to the ratio of the *al-3* transcript to the IF2 transcript.

Characterization of *al-3* gene expression

We analyzed the kinetic of the *al-3* mRNA accumulation under continous saturating light (Figure 1). The *al-3* gene is transiently expressed also under continous light illumination. The *al-3* mRNA accumulation is very fast and reaches a maximum at 20 minutes of light induction. After that time we observed that the *al-3* mRNA decreases at a very fast rate. The *al-3* mRNA becomes undetectable after 100 minutes of continous illumination. This transient pattern of the *al-3* gene expression could be interpreted as a consequence of an adapting response of *N.crassa* to blue light. Although the light stimulus is continously present, the *al-3* mRNA decreases suggesting that the transducting machinery has been adapted to the new light regime. Photosensory adaptation is of great importance in plants (including fungi) that operate in large intensity ranges.
When light-preadapted plants are incubated in the dark, some time elapses before the full sensitivity of the organism to a light induced response is restored (Galland, 1989). In order to test if this dark adaptation is also observed in *N.crassa* we determined the responsiveness of *al-3* gene under light-dark-light incubation periods. We exposed the *N.crassa* mycelia to 20 minutes of saturating light, followed by 60 minutes of incubation in the dark

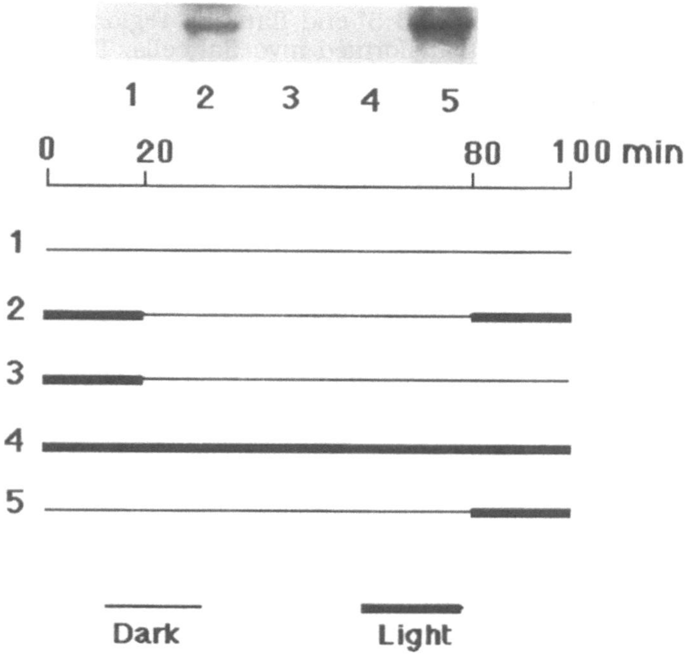

Figure 2. *al-3* mRNA analysis upon light-dark-light periods.
Northern blot analysis were performed as described in Materials and Methods. The experiment's design is shown at the botton. 100 minutes represent the time point of harvesting of all the samples illuminated in various ways. The energy fluence rate used was 130 J/m^2.

and finally by a second period of 20 minutes of saturating light. After these incubations total RNA was extracted and *al-3* mRNA was clearly detected (Fig 2, lane 2). When the first 20 minutes light period was followed by a continous dark incubation, the *al-3* mRNA was undetectable (lane 3). The same result was obtained when the mycelia were subjected to 100 minutes of continous light incubation as expected from the results showed in Fig 1 (lane 4). As controls, lane 5 shows the *al-3* mRNA level after 20 minutes of illumination as described in Fig 1 and lane 1 shows that *al-3* mRNA extracted from dark grown mycelia is undetectable. The difference between the mRNA levels observed in lanes 4 and 2 can be explained by the transiency of *al-3* gene expression under continous light (Fig. 1) and by the recovery of the photoinductive response after 60 minutes of dark incubation. For shorter time of dark incubation the recovery was less effective (data not shown) The difference between the *al-3* mRNA level in lane 2 and lane 5 is probably due to the incomplete recovery of the photoinducibility. These results suggest that the time required to fully recover the sensitivity of the system exceedes 60 minutes. These data indicate that *N.crassa* possesses a photosensory adaptation mechanism. These results are also in agreement with previous studies in light-induced carotenoids accumulation in *N. crassa* (Schrott, 1980).

In vivo characterization of al-3 promoter

We tested the ability of the *al-3* 5' end flanking region to promote light regulated transcription in transformed mycelial cells. The transcriptional fusion to a reporter gene of the *al-3* promoter from -1120 to +40, has demonstrated that this promoter fragment contains the *cis*-elements necessary for the photoinducibility of the *al-3* gene (data not shown). Experiments were carried out in order to test the ability of a short promoter (220bp) to respond to the blue light induction. A 675bp deletion into the *al-3* coding region (pΔ675)

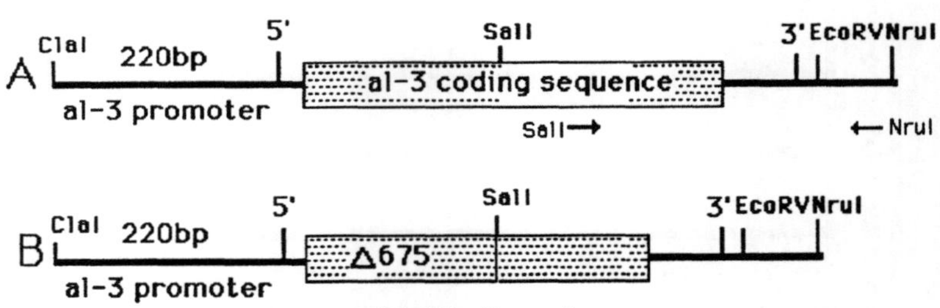

Figure 3. *al-3* gene construction.
Panel A shows the 220 bp 5' end flanking region, the coding sequence and the 3' end flanking region of *al-3* gene. Two oligonucleotides containing SalI or NruI restriction sites, indicated by arrows, were used to amplify by PCR the *al-3* coding sequence starting from 675 bp downstream of the internal SalI site to the indicated 3' end flanking region. The amplified fragment was then ligated in the endogenous SalI and NruI sites. *Panel B* shows the 675 bp deleted *al-3* coding sequence resulting from the above described cloning. The open reading frame obtained is the same of *al-3* coding region without 225 amino acids. A 500 bp DNA probe used in Southern blot hybridizations is indicated.

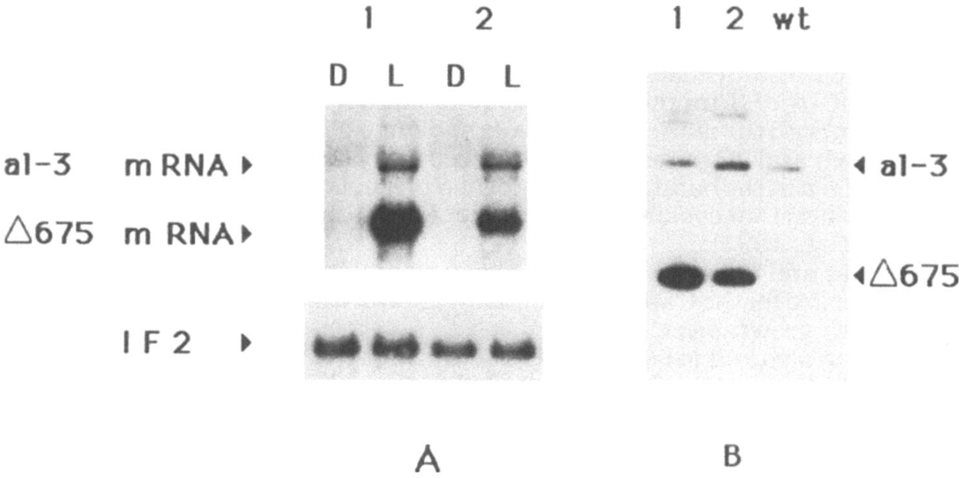

Figure 4. Photoinduction of pΔ675 *al-3* mRNA.
Panel A Northern blot analysis of endogenous and pΔ675 *al-3*, from two independent transformants. Control hybridization with IF2 cDNA, which is not light regulated, is shown on the lower part of the figure. *Panel B* Southern blot analysis of the genomic DNA from the same independent transformants shown in Panel A, restricted with EcoRV and ClaI. The 500 bp DNA probe indicated in Fig 3 was labeled by nick translation. The experimental conditions were as described in Materials and Methods.

was produced to distinguish between the endogenous *al-3* mRNA and the mRNA of the exogenous construct (Fig. 3). Total RNA was extracted from dark-grown and light-induced mycelia of two independent transformants obtained with pΔ675. The Northern blot analysis (Fig. 4A) shows two mRNA bands visible only in lanes containing RNA from light-induced mycelia. The upper band represents the endogenous *al-3* mRNA, while the lower band results from the transcription of the exogenous construct. We can conclude that the Δ675 *al-3* mRNA is transcribed in a light inducible fashion.
Differences in bands intensity are due to the multiple copies of the plasmid pΔ675 introduced by transformation in *N. crassa* cells. The pΔ675 copy number was determined by Southern blot hybridization (Results are shown in Fig. 4B). The lower hybridization band represents the Δ675 *al-3* fragment obtained by digestion of the integrated exogenous DNA while the upper band (see wild type pattern) corresponds to the endogenous *al-3* gene fragment. The ratio between the bands intensity indicates the number of the pΔ675 introduced copies. The faint upper bands could be due to integration events occurring between ClaI and EcoRV sites of pΔ675 plasmid indicated in Fig. 3.
In summary, the *in vivo* results suggest that the minimal *al-3* promoter region (220bp) contains the *cis*-elements necessary for the light induced transcription of the gene.

In vitro characterization of al-3 promoter

In order to test the ability of *al-3* promoter fragment to interact with specific

trans-acting factors we performed an *in vitro* analysis. Gel retardation assays were carried out incubating total protein extracts from dark grown and 20 minutes light-induced mycelia with the minimal *al-3* promoter fragment. The band shift of *al-3* promoter fragment is shown in Fig 5. By competition experiments all the shifted bands were shown to be specific (data not shown). Dark grown and 20 minutes light induced mycelial cell extracts showed the same pattern in the band shift assay of *al-3* promoter DNA. We have previously demonstrated that cycloheximide did not affect the *al-3* gene photoinduction (Baima *et al.*, 1991).
Those data and the gel retardation assay clearly indicate that all the protein factors needed for the transcriptional activation of *al-3* gene are already present in the dark grown mycelial cells.
From the *in vivo* and *in vitro* analysis of the 220bp *al-3* promoter fragment we can conclude that the elements needed for the photoinduction of *al-3* gene are included in the minimal region of the promoter. Experiments are in course to determine the *al-3* promoter regions involved specifically in the interaction with the trans-acting factors.

Strains

The following *Neurospora crassa* wild-type strain was obtained from the Fungal Genetics Stock Center (FGSC; University of Kansas, Kansas City, Kansas): Oak Ridge wild-type 74-OR23A (FGSC no. 987).

Growth and photoinduction conditions.

For photoinduction experiments *N.crassa* growth and light treatment were performed as described in Baima *et al.*, 1991.

RNA extraction and Northern hybridization.

Total RNA was extracted, transfer onto Hybond-N membranes and hybridized as described in Baima *et al.*, 1991.

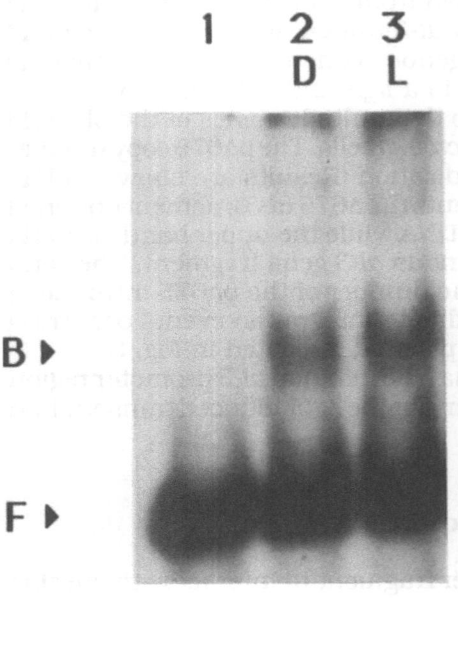

Figure 5. Gel retardation assay of *al-3* promoter.
N.crassa crude extracts were prepared from dark grown (D) and 20 minutes light induced (L) mycelia as described in Materials and Methods. *al-3* promoter fragment used in these experiments does not contain the TATA box at -30. Lane 1 shows the electrophoretic migration of the free *al-3* promoter fragment. Lane 2 and 3 show *al-3* promoter fragment plus D and L extracts respectively.

DNA extraction, Southern Hybridization and nick translation.

N.crassa DNA was prepared as described by Sherman *et al.*, 1978, except that Novozym 234 (Biolabs, Novo Allé, Bagsvaerd, Denmark) was used instead of Zymolyase. Southern blot and nick translation were carried out as described in Maniatis *et al.*, 1982.

N. crassa and Escherichia coli transformations.

Preparation of spheroplasts and transformation with pΔ675 clone were as described by Nelson *et al.*, 1989. Transformation of *E. coli* was carried out as described in Maniatis *et al.*, 1982.

Crude mycelia extracts of N.crassa.

The crude extracts were prepared by grinding the frozen mycelia which were collected by filtration from dark grown or 20 minutes light induced cells with mortar and pestle in 20 mM Hepes pH 7.9, 100 mM KCl, 2 mM EDTA, 10 mM DTT and 1 mM PMSF. After centrifugation at 11000xg for 10 minutes the supernatants were made to 20% glycerol and stored in aliquots at -70°C.

Gel retardation experiments.

The crude extracts (20 µg of proteins) were incubated 10 minutes at room temperature with 2 µg of poly(dI.dC).poly(dI.dC). The *al-3* minimal promoter fragment labeled by random priming kit (Boehringer) (10000cpm) was added and incubated other 20 minutes at room temperature. All other conditions were as described by Schneider *et al.*, 1986.

Acknowledgements

This work was supported by grants from the Ministero Agricoltura e Foreste, Piano Nazionale Tecnologie Applicate alle Piante, and Institute Pasteur-Fondazione Cenci Bolognetti".

References

Baima S, Macino G, Morelli G (1991) Photoregulation of *albino-3* gene in *Neurospora crassa*. Photochem Photobiol B: Biol 11:107-115.

Carattoli A, Romano N, Ballario P, Morelli G, Macino G (1991) The *Neurospora crassa* carotenoid biosynthetic genes (*albino-3*) reveals highly conserved regions among prenyltransferases. J Biol Chem 266:5854-5859.

Degli Innocenti F, Russo VEA (1983) Photoinduction of protoperithecia in *Neurospora crassa* by blue light. Photochem Photobiol 37:49-51.

Degli Innocenti F, Russo VEA (1984) Genetic analysis of blue light-induced responses in *Neurospora crassa*. In Senger H (ed) Blue light effects in biological systems. Springer-Verlag KG, Berlin, p213-219.

Feldman JF (1982) Genetic approaches to cyrcadian clocks. Annu Rev Plant Physiol 33:583-608.

Galland P (1989) Photosensory adaptation in plants. Bot acta 102:11-20.

Harding RW, Shropshire WJr (1980) Photocontrol of carotenoid biosynthesis. Annu Rev Plant Physiol 31:217-238.

Harding RW, Turner RV (1981) Photoregulation of the carotenoid biosynthesis pathway in albino and white collar mutants of *Neurospora crassa*. Plant Physiol 68:745-749.

Harding RW,Melles S (1983) Genetic analysis of phototropism of *Neurospora crassa*

perithecial beaks using white collar and albino mutants. Plant Physiol 72:996-1000.
Maniatis T, Fritsch EF, Sambrook J (1982) Molecular cloning. A Laboratory Manual. Cold Spring Harbor Laboratory, Cold Spring Harbor, New York.
Nelson MA, Morelli G, Carattoli A, Romano N, Macino G (1989) Molecular cloning of a *Neurospora crassa* carotenoid biosynthetic gene (*albino-3*) regulated by blue light and the products of the white collar genes. Mol Cell Biol 9:1271-1276.
Pandit NN, Russo VEA (1991) The bli regulon-a network of blue light inducible genes of *Neurospora crassa*. Neurospora Newsl 38:93-95.
Schmidhauser TJ, Lauter FR, Russo VEA, Yanofsky C (1990) Cloning, sequence, and photoregulation of *al-1*, a carotenoid biosynthetic gene of *Neurospora crassa*.Mol Cell Bioll10:5064-5070.
Schneider R, Garder I, Mueller U,Mertz R, Winnacker EL (1986) A sensitive and rapid gel retardation assay for nuclear factor I and other DNA binding proteins in crude nuclear extracts. Nucleic Acid Res 14:1303-1317.
Schrott EL (1980) Fluence response relationship of carotenogenesis in *Neurospora crassa*. Planta 150:174-179.
Sherman F, Fink GR, Lawrence CW (1978) Methods in yeast genetics. Cold Spring Harbor Laboratory , Cold Spring Harbor, New York.
Sommer GJ, Chambers AA, Eberle J, Lauter FR, Russo VEA (1989) Fast light regulated genes of *Neurospora crassa*. Nucleic Acid Res. 17:5713-5724.
Subden RE, Bobowski G (1973) Evidence for actinomycin D inhibition of transcription of carotenoid loci in *Neurospora*. Experientia 29:965-967.

Expression of genes for the biosynthesis of penicillin

G. Turner, P.E. Browne and A.A. Brakhage
Department of Molecular Biology and Biotechnology
University of Sheffield
P.O. Box 594
Sheffield S10 2UH
U.K.

Introduction

Genetic studies on *Aspergillus nidulans* have been in progress for around 40 years, making it one of the best analysed filamentous fungi. This ascomycete has a well defined sexual cycle which is easy to use for meiotic analysis, and together with a parasexual cycle which facilitates assignment of new mutations to one of the 8 linkage groups, has helped generate a detailed genetic map (Clutterbuck, 1987). The ability of the organism to grow on a simple, defined minimal medium, and the considerable metabolic versatility of the fungus has also led to the characterisation and detailed analysis of many regulated systems (Scazzocchio, this volume). Regulation by nutrient availability, such as nitrogen and carbon source (Davis, Hynes, 1989), and development (Timberlake, this volume) have been studied in detail.
In addition to the ability of this organism to use a wide variety of nutrients, like many other microorganisms, *A. nidulans* is able to produce secondary metabolites. These are defined as compounds which, unlike primary metabolites, have no obvious function in the growth of the organism, and include antibiotics such as ß-lactam and polyketide compounds.
Many reasons have been put forward to explain the reason for secondary metabolite production, including competitive advantage, safety valve for unwanted compounds, involvement in development and sporulation, or simply a metabolic playground or biochemical irrelevance from which new, useful primary metabolic pathways might arise from time to time (Zähner *et al.*, 1983). For the latter, it is further suggested that such wasteful metabolic processes must not be detrimental to the survival of the organism, and might therefore be subject to regulation related to nutrient supply. This might lead to certain patterns of regulation of secondary metabolism such that it does not drain the resources of the organism in unfavourable conditions such as starvation. Evidence for this is sought by examining the conditions required for secondary metabolite production.
Some of the clearest regulation is seen in certain prokaryotes such as *Bacillus* (Marahiel *et al.*, 1987) or *Streptomyces spp.* (Guthrie, Chater, 1990), where the onset of antibiotic production coincides with sporulation or stationary phase. The growth phase is sometimes referred to as the *trophophase*, and the secondary metabolite production phase the *idiophase*. Interest in this question is not purely academic, since commercial antibiotic

NATO ASI Series, Vol. H 69
Molecular Biology and its Application to Medical Mycology
Edited by B. Maresca, G. S. Kobayashi and H. Yamaguchi

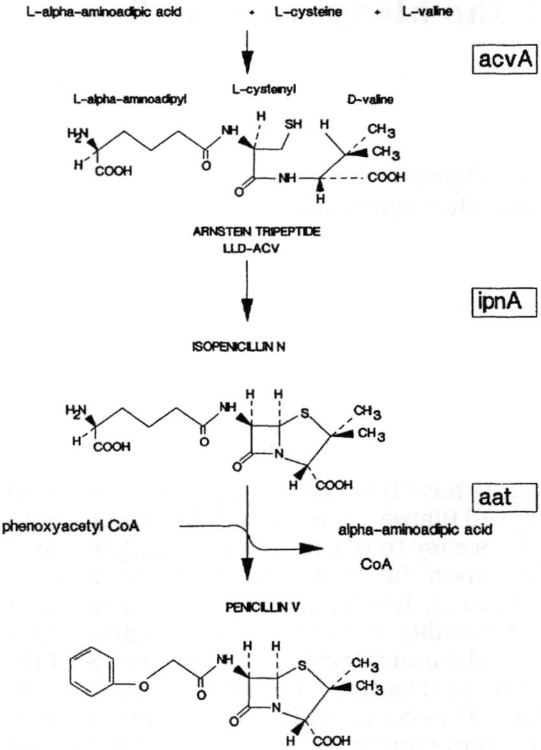

Figure 1. Biosynthesis of penicillin V. The structural genes encoding the enzymes for the 3 steps are given in the boxes.

production has involved many studies designed to achieve fermentation conditions for maximum economic yield. Considerable work has gone into fermentation development for penicillin production in *Penicillium chrysogenum*, complementing parallel strain improvement programmes. However, much of this work is empirical. Since conditions must be optimized for the particular strain in use by a company, and these results are not generally published, studies on the regulation of penicillin production in wild-type strains are limited, and only recently has it become possible to examine gene expression. Some of the parameters have been summarized in reviews (Martin, Demain, 1980) (Martin, Aharonowitz, 1983), where it is acknowledged that the division between the *trophophase* and *idiophase*, may not ßе so clear cut in fungi as in prokaryotes, at least for ß-lactam antibiotic biosynthesis. In addition to possible temporal/developmental regulation of expression, growth rate, carbohydrate, and oxygen have been identified as factors which influence penicillin production. It is generally believed that penicillin production is controlled at the level of enzyme synthesis and stability (Revilla *et al.*, 1986).

Until recently, direct study of the regulation of individual steps in penicillin biosynthesis was rather difficult. It has taken a long time to elucidate the precise details of the biosynthetic pathway, and enzyme assays for some of the steps are still inconvenient.

Fortunately, the recent identification of the genes encoding the 3 steps of penicillin biosynthesis from amino acid precursors (Fig. 1) has altered the situation, and opened new possibilities for investigation of the mode of expression of these genes (Smith *et al.*, 1990a; MacCabe *et al.*, 1990; Queener, 1990). Further, there is evidence for enhanced transcription of the biosynthetic genes in commercial, high-titre strains of *P. chrysogenum*, possibly resulting from mutations in trans-acting regulatory genes (Smith *et al.*, 1989).

Penicillin biosynthetic gene cluster

The organization of the biosynthetic genes of *A. nidulans* is very similar to that of *P. chrysogenum*. Since the latter organism has no sexual cycle, and is not so well characterized genetically, *A. nidulans* provides a unique system for the analysis of regulation of a fungal secondary metabolic pathway. Until recently (Shah *et al.*, 1991)(Gomez-Pardo, Penalva, 1990), almost all of the work on regulation of penicillin production had been carried out with *P. chrysogenum*. The aim of the work described in this article was to examine, in *A. nidulans*, some of the factors claimed to influence penicillin production, and to determine whether any of these factors act at the level of secondary metabolic gene expression, i.e. can genetic regulation of secondary metabolism be detected?
The structural genes are clustered (Fig. 2), and it has been demonstrated that a cosmid carrying this cluster confers the ability to produce penicillin on fungi which cannot produce the antibiotic, if introduced by transformation (Smith *et al.*, 1990b).

Assessing gene expression

While enzyme assays normally offer the best initial approach to study gene regulation, the assays used for the 3 enzymes of interest are not convenient. No sensitive quantitative assay exists for ACV synthetase, and the other two

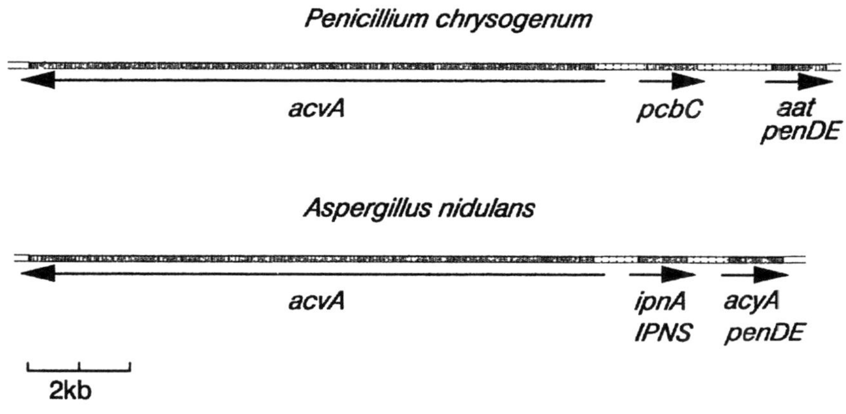

Figure 2. Penicillin biosynthetic gene clusters in *P. chrysogenum* and *A. nidulans*. Arrows show the direction of the transcripts. More than one gene symbol has been used in the literature, and some of these are given.

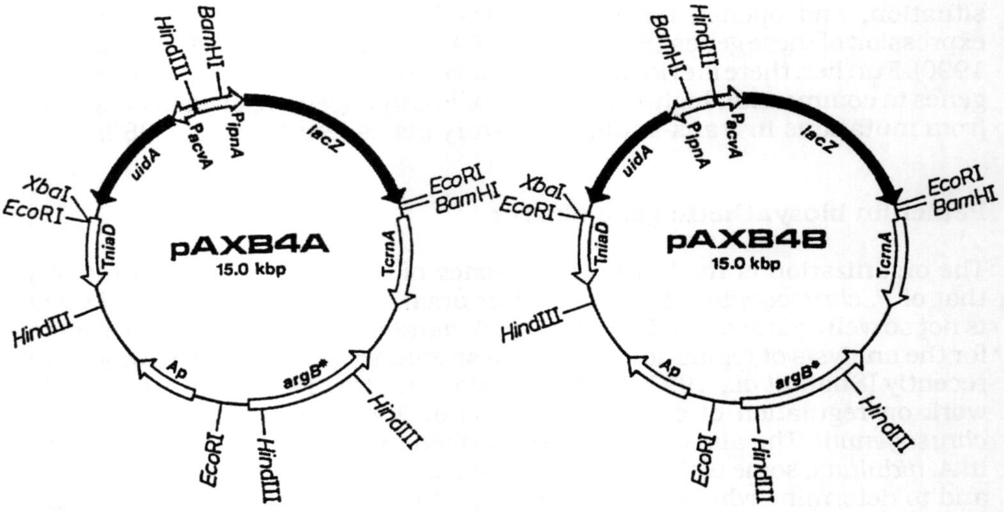

Figure 3. Reporter gene vectors. *lacZ*, β-galactosidase; *uidA*, a-glucuronidase; *TcrnA*, terminator of the *crnA* gene; *TniaD*, terminator of the *niaD* gene; *argB**, mutant allele of the *argB* gene; Ap, ampicillin resistance. The *acvA-ipnA* intergenic region has been inserted in both orientations between the *uidA* and *lacZ* sequences to generate translational fusions.

enzymes require bioassay systems. In view of this difficulty, we have utilized a twin reporter vector to monitor expression of the first 2 genes of the pathway (Fig. 3) (Punt *et al.*, 1991). The reporter enzymes can be quantified by assay of extracts, or detected in indicator plates, the latter also providing a possible approach to the isolation of mutations affecting gene expression.

β-galactosidase is detectable in wild-type *A. nidulans*, but reduced to negligible levels under the conditions used in a strain carrying the *bgaO* mutation (Fantes, Roberts, 1973).β-glucuronidase is absent from the organism. Since the *acvA* and *ipnA* genes are divergently transcribed, and separated by only 872 bp, this short intergenic region should contain the promoter regions for both genes. A fragment containing this region and a short stretch of the 5' coding region of each gene was ligated into the binary vector in both orientations to maintain the open reading frames and generate translational fusion genes (Fig. 3). The vector also carries a mutant *argB* allele such that recombination events between the transforming DNA and the *argB* locus of the recipient are necessary to yield Arg^+ transformants. Since gene conversion and multiple copy transformation events also occur, transformants were screened for single copy integration by hybridization analysis on transformant DNA. Single copy transformants for pAXB4A and pAXB4B were generated to examine the effects of different gene fusions on reporter gene expression. This showed that *acvA* fusions had about 3 to 4-fold lower activity than *ipnA* fusions, probably reflecting a weaker promoter for the former. Although expression has been studied using both fusion vectors, results presented in this article are for pAXB4A.

Choice of fermentation conditions

While a minimal, defined medium would normally provide the best option for studying nutrient effects on gene expression, early work on penicillin development showed that penicillin production was poor in such media, and was greatly enhanced in complex corn steep liquor medium with lactose as the main carbon source (Swartz, 1985). In accordance with this, growth on *Aspergillus* minimal medium leads to a drastic reduction in the level of all biosynthetic gene transcripts compared to fermentation medium (FM) (MacCabe *et al.*, 1990). While this large difference in gene expression might itself be a useful one to follow up, no single medium component has yet been identified which accounts for it. For the present work, we have therefore used FM. When supplemented with phenoxyacetic acid as a side chain precursor, penicillin V, more stable than penicillin G at low pH, is formed (Fig. 1). Since the isopenicillin N synthetase step (Fig. 1) requires oxygen, vigorous aeration is also essential for optimum yield. In order to obtain this, 250ml flasks shaken at 250 rpm and containing only 20ml of culture medium were used for each determination. Penicillin production also decreases considerably at 37°C, so experiments were conducted at the optimal temperature of 26°C (Swartz, 1985).

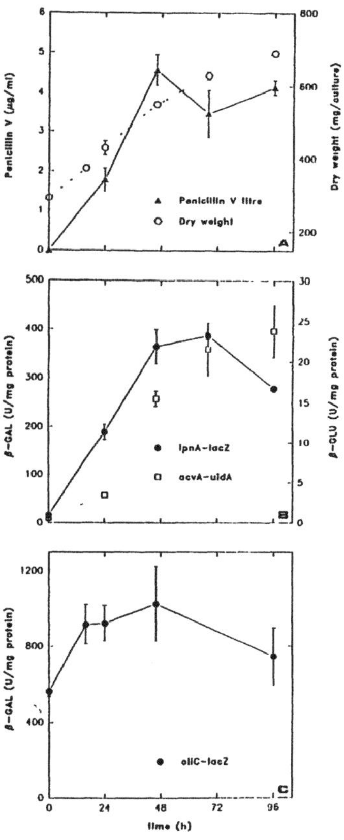

Figure 4. Antibiotic production and reporter gene expression under optimal conditions with 4% (w/v) lactose. Each point is the mean data from 3 20ml cultures, with standard deviations shown.

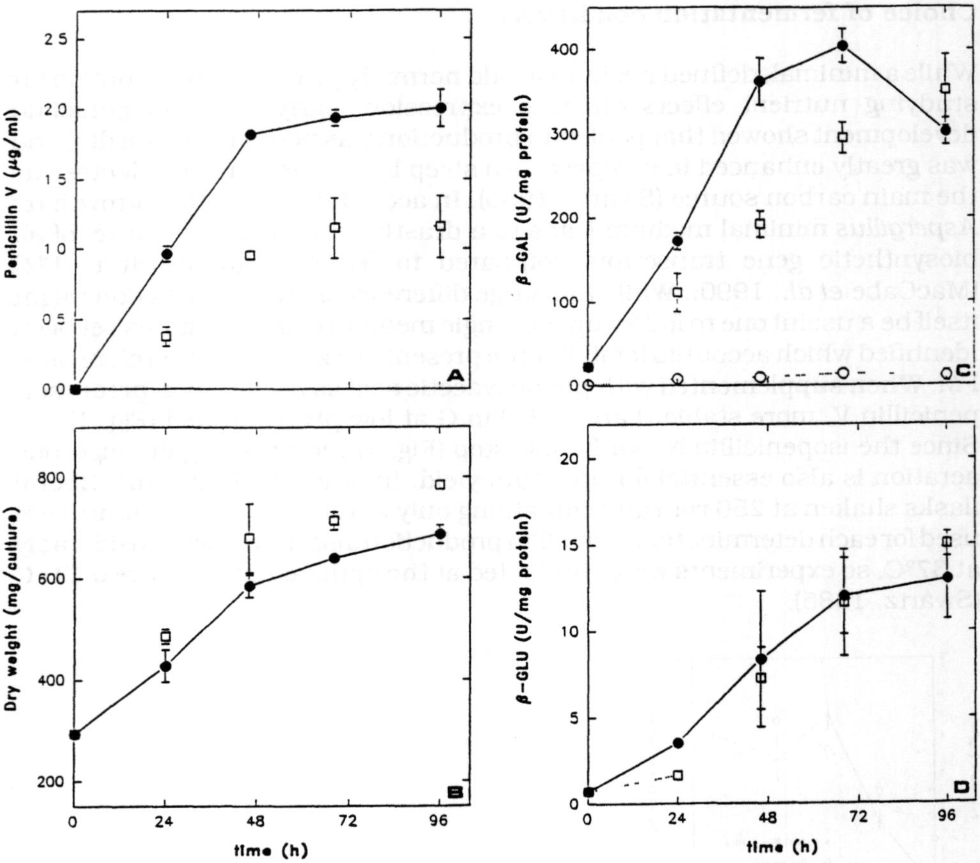

Figure 5. Effect of glucose on penicillin production and reporter gene expression (pAXB4A). Cultures were grown with 4% (w/v) lactose (•) and 4% (w/v) glucose (□).

Finally, conidia of *A. nidulans* were first cultured for 24h in a seed medium of corn steep liquor with 1.5% glucose (Smith *et al.*, 1990a) to provide actively growing, young mycelium as an experimental inoculum. This mycelium was washed in saline prior to inoculation of FM.

Since penicillin yield and reporter enzyme specific activities are prone to variation between experiments, data shown in each figure is from a single experiment, at least 3 flasks were used in each experiment for each time point, and standard deviations within an experiment are shown.

Temporal expression

Time course of penicillin production and gene expression under optimal conditions are shown in Fig. 4. It can be seen that penicillin accumulation occurs mainly during the first 48 hours, while the mycelial mass is still increasing, and is paralleled by a rapid increase in specific activity of each reporter enzyme. Results are also presented for a strain carrying a single copy of the *oliC-lacZ* fusion gene (mitochondrial ATP synthetase, a "housekeeping gene")(Ward, Turner, 1986). In contrast to the *acvA* and *ipnA* fusion genes, the specific activity of the *oliC* fusion, already high in the young mycelium from the seed culture, does not show a great increase during the subsequent fermentation.

Glucose repression

It has been reported that glucose repression of penicillin biosynthesis in *P. chrysogenum* acts at the level of the biosynthetic enzymes (Revilla *et al.*, 1986). Cultures differing only in carbon source (4% lactose or 4% glucose) are shown in Fig. 5. Penicillin yield is decreased by around 50% in the presence of glucose. Although the glucose drops below 1% after 46h hours under the conditions described, maintenance of a high glucose level by batch feeding (data not shown) did not give any further decrease in penicillin yield. Accumulation of the *ipnA-lacZ* fusion product is slowed on glucose, but interestingly, not the *acvA-uidA* fusion product. Experiments with the pAXB4B (reverse orientation) vector showed that this difference was not a result of the reporter gene used.
In contrast, the *oliC-lacZ* fusion shows about 40% higher expression on glucose compared to lactose (data not shown).
What is the mechanism of "glucose repression"? A number of other genes in *A. nidulans* respond to glucose repression, and this repression is abolished by mutations at the *creA*, *creB* and/or *creC* loci (Bailey, Arst, 1975)(Hynes, Kelly, 1977). Although the mechanism of action of these *cre* genes has yet to be determined, glucose repression of glyoxylate cycle enzymes, alcohol dehydrogenase, and acetamidase can be explained in terms of catabolite repression, where glucose is the preferred carbon source. We found that penicillin production in strains carrying *creA*^{d}-1, *creB304* and *creC302* mutations was still repressed by glucose to the same extent as in the wild-type background, suggesting a different mechanism of repression. It remains possible that faster growth rate rather than glucose itself is the cause of the lower penicillin production.

Lysine repression

Lysine repression of penicillin biosynthesis has been reported for *P. chrysogenum*, and it has been suggested that this might operate through feedback inhibition of lysine biosynthesis at one or more steps (Martin, Aharonowitz, 1983) (Jaklitsch, Kubicek, 1990)(Affenzeller *et al.*, 1989), thus depleting the level of α-aminoadipic acid, a precursor of penicillin and lysine (Figs. 1 and 7). Relatively high levels of lysine (0.1M) in the medium have been necessary to obtain such inhibition. It can be seen that such a high level has no apparent effect on growth rate (Fig. 6) but decreases penicillin accumulation in FM with lactose by about 50%. Unlike the results for glucose, both the *acvA* and *ipnA* expression are similarly reduced.

There is some evidence that intracellular α-aminoadipate level regulates penicillin biosynthesis in *P. chrysogenum* (*Honlinger*, Kubicek, 1989), by availability of substrate for ACVS and/or regulation of expression of the biosynthetic enzymes.

In view of the observed effect of lysine in *A. nidulans*, it would be interesting to investigate further the effect of α-aminoadipic acid pool size on expression of the reporter genes. Lysine auxotrophs are available for *A. nidulans* (Clutterbuck, 1987), and have been mapped to 6 unlinked loci. A mutant

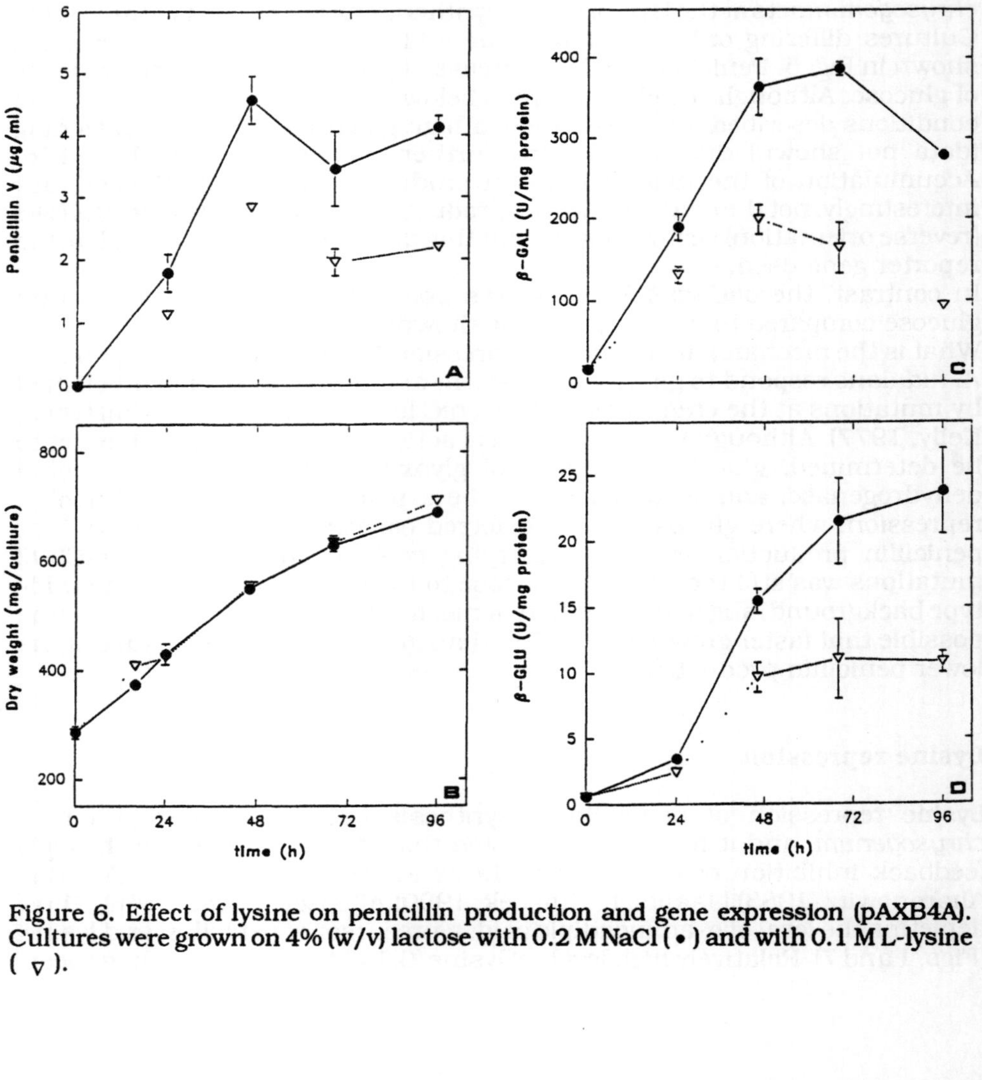

Figure 6. Effect of lysine on penicillin production and gene expression (pAXB4A). Cultures were grown on 4% (w/v) lactose with 0.2 M NaCl (•) and with 0.1 M L-lysine (▽).

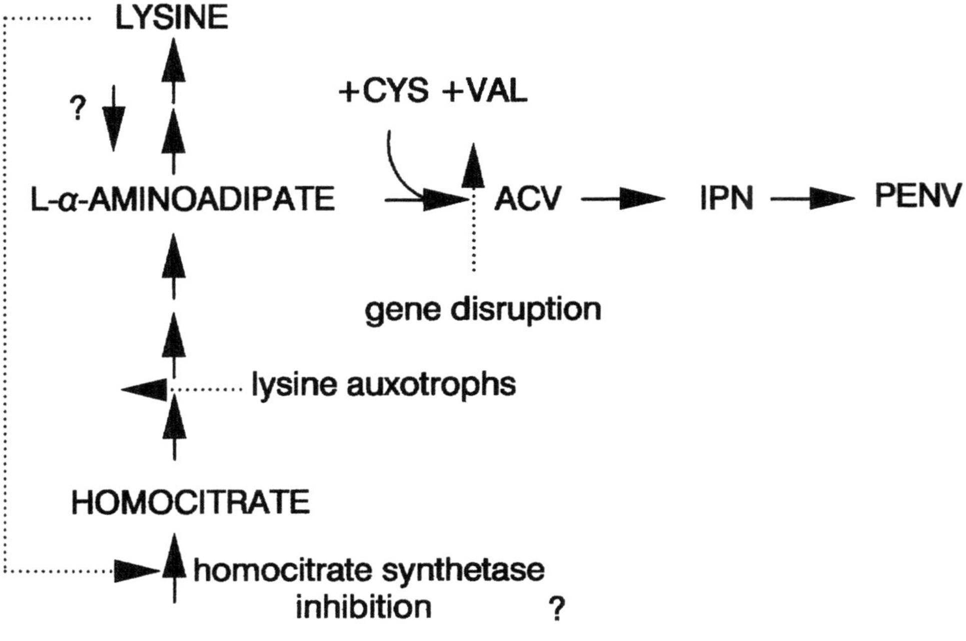

Figure 7. Ways of perturbing the pathway

unable to synthesize α-aminoadipic acid formation might offer an alternative way of controlling the pool size of this precursor. Since no further biochemical characterization of available mutants had been carried out, we tested some of these mutants by feeding with α–aminoadipic acid on minimal medium. Available *lysA*, *lysF* and *lysE* strains could not be fed, but *lysB* and *lysD721* did grow with α-aminoadipic acid. Some permeability barrier was suggested by the requirement for 10mM α-aminoadipic acid in medium lacking lysine. While *lysB* and *lysD721* strains offer potential for further investigation, we also need to ascertain to what extent lysine can be converted to α-aminoadipic acid in this organism. *Streptomyces clavuligerus*, which produces the β-lactam cephamycin, does not possess the same lysine biosynthetic pathway as fungi (Martin, Aharonowitz, 1983). Lysine is made via the diaminopimelic acid pathway in which α–aminoadipate is not an intermediate. Cephamycin biosynthesis requires the enzyme lysine ε-aminotransferase (LAT) to produce α–aminoadipic acid directly from lysine, and it was recently shown that the cephamycin biosynthetic gene cluster of *Streptomyces clavuligerus* conveniently carries the gene encoding this enzyme (Tobin *et al.*, 1991). Do fungi also possess a similar enzyme in addition to the α-aminoadipate pathway? If so, then lysine added to the growth medium would provide α-aminoadipic acid. We are currently investigating these possibilities using the available lysine mutants.

Sequential regulation

A mutant N-2 of *Cephalosporium acremonium* defective in cephalosporin production was reported to show simultaneous loss of IPNS, IPN epimerase and deacetoxycephalosporin C synthetase, while accumulating ACV (Ramos *et al.*, 1986), and was believed to carry a regulatory mutation of some kind. Subsequent studies on this mutant identified a point mutation in the IPNS gene, and demonstrated the presence of an inactive IPNS enzyme (Ramsden *et al.*, 1989). It was postulated that an active IPNS is necessary for expression of subsequent steps in the pathway. Such a regulation could be explained by sequential induction, where IPN is necessary for induction of IPN epimerase and so on.

Is ACV necessary for induction of IPNS? This question can be approached directly in *A. nidulans* using a molecular approach. A gene disruption technique had already been used to aid identification of the *acvA* gene region (Smith *et al.*, 1990a), and a similar disruption is shown in Fig. 8. Such a disruption totally abolishes penicillin formation, without damaging the gene sequences encoding IPNS and AAT. Clearly, such a strain should lack the ACV tripeptide, and it can be asked whether this intermediate is needed for IPNS synthesis.

The disrupted *acvA* shown in Fig. 8 was introduced into a strain carrying pAXB4A via a sexual cross, and suitable progeny identified and checked by hybridization. The effect of the disruption on the expression of the reporter genes is shown in Fig. 9, and it can be seen that ACV is not needed for expression of the IPNS fusion gene.

IPNS enzyme activity was also measured in wild-type and disrupted strains, and is shown in Fig. 10. It can be seen that enzyme activity is present in both

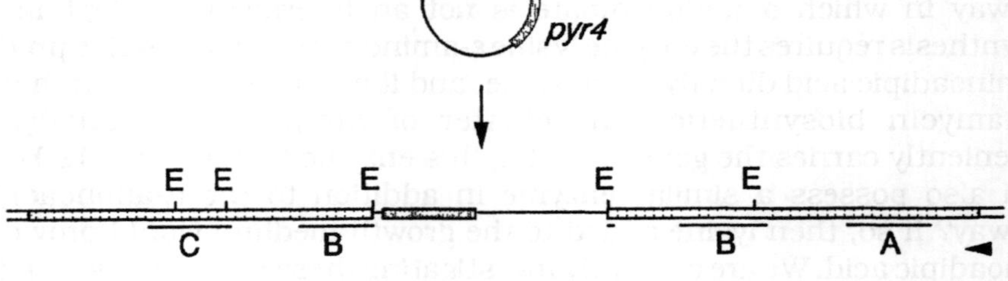

Figure 8. Disruption of *acvA* by homologous transformation. pXEP3 carries an EcoRI fragment internal to the *acvA* gene, and the *pyr4* gene of *N. crassa* as a selectable marker. A, B and C are related domains identified in the *acvA* sequence (Smith *et al.* 1990c).

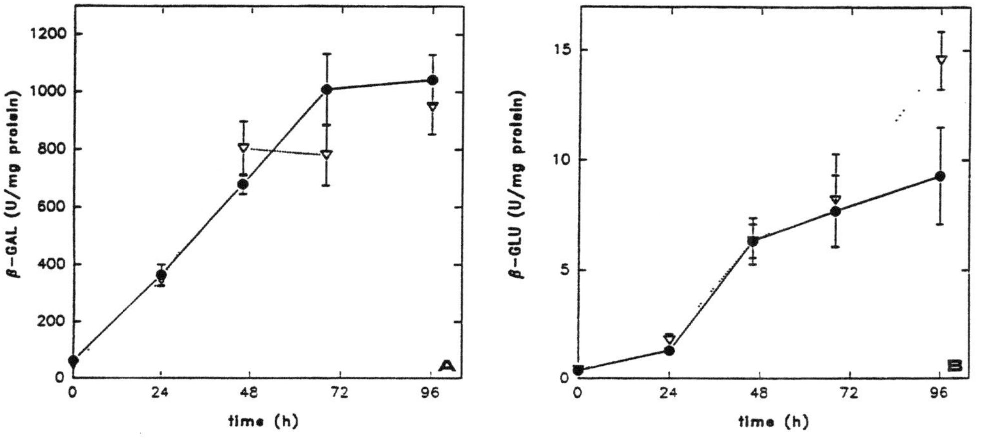

Figure 9. Effect of the *acvA* disruption on reporter gene expression (pAXB4A). Cultures were grown on 4% (w/v) lactose. Control (▿), acvA disrupted strain (●).

strains. IPNS activity was very low after 48 hours in the wild-type strain compared to the disrupted strain, and may reflect a difference in enzyme stability. We are currently using anti-IPNS antibody to examine the amount of IPNS protein in the two strains.

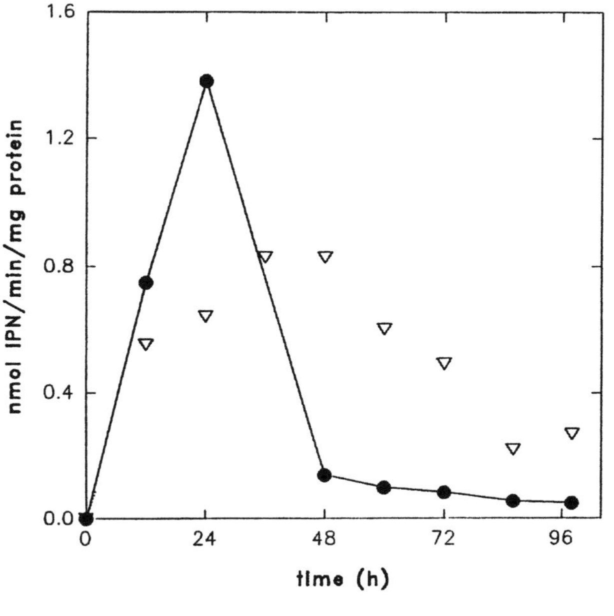

Figure 10. Effect of the *acvA* disruption on IPNS enzyme activity in mycelial extracts. Control (●), *acvA* disrupted strain (▿).

Discussion

Since our main aim in this work was to examine nutritional factors which might regulate gene expression, we have used young mycelium as our inoculum, and have not yet investigated gene expression events in the seed medium with different carbon sources. Nevertheless, it is interesting to note that the specific activity of the *ipnA-lacZ* and *acvA-uidA* gene products continues to increase for longer than *oliC-lacZ* product, suggesting some difference in the timing of expression of primary and secondary metabolic genes.

It is clear that both glucose and lysine repression of penicillin biosynthesis occur to some extent at the level of secondary metabolic gene regulation in *A. nidulans*, though the extent of regulation is limited, and the mechanisms as yet unknown.

In addition, it has been demonstrated that ACV is not required for expression of the *ipnA* gene.

Another reason for the use of reporter genes was to facilitate the isolation of mutants altered in regulation of β-lactam biosynthesis by using X-GAL or X-GLU indicators in plates. However, the limited degree of repression which can be achieved makes this approach difficult. Even on minimal medium containing glucose and indicators, expression of both *ipnA*- and *acvA*-fusion genes results in blue colonies. At present, we are therefore focussing on the lysine repression, where several avenues of investigation are open, in particular the possibility of controlling the α-aminoadipate pool. In addition, we have found that lysine repression of fusion gene expression in minimal medium indicator plates is sufficient to obtain white colonies, which opens up the possibility of isolating regulatory mutants.

Acknowledgments

This work was supported by a grant to A.A.B. from the Deutsche Forschungsgemeinschaft (Br 1130-1, 1130-2), by the University of Sheffield (grant RSG 197) (A.A.B. and G.T.) and by an SERC CASE studentship with SmithKline Beecham Pharmaceuticals (P.E.B.).

References

Affenzeller K, Jaklitsch WM, Honlinger C, Kubicek CP (1989) Lysine biosynthesis in *Penicillium chrysogenum* is regulated by feedback inhibition of α-aminoadipate reductase. FEMS Microbiol. Lett. 58:293-298.

Bailey C, Arst HNJr (1975) Carbon catabolite repression in *Aspergillus nidulans*. Eur. J. Biochem. 51:573-577.

Clutterbuck AJ (1987) *Aspergillus nidulans* (n=8). In: O'Brien SJ (ed) Genetic maps. A compilation of linkage and restriction maps of genetically studied organisms. Cold Spring Harbor Laboratory, Cold Spring Harbor, p 325.

Davis MA, Hynes MJ (1989) Regulatory genes in *Aspergillus nidulans*. Trends Genet. 5:14-19.

Fantes PA, Roberts CF (1973) β-galactosidase activity and lactose utilization in *Aspergillus nidulans*. J. Gen. Microbiol. 77:471-486.

Gomez-Pardo E, Penalva MA (1990) The upstream region of the IPNS gene determines expression during secondary metabolism in *Aspergillus nidulans*. Gene 89:109-115.

Guthrie EP, Chater KF (1990) The level of a transcript required for production of a *Streptomyces coelicolor* antibiotic is conditionally dependent on a tRNA gene. J. Bacteriol. 172:6189-6193.

Hynes MJ, Kelly JM (1977) Pleiotropic mutants of *Aspergillus nidulans* altered in carbon metabolism. Mol. Gen. Genet. 150:193-204.

Honlinger C, Kubicek CP (1989) Regulation of δ-(L-α-aminoadipyl)-L-cysteinyl-D-valine and isopenicillin N biosynthesis in *Penicillium chrysogenum* by the α-aminoadipate pool size. FEMS Microbiol. Lett. 65:71-76.

Jaklitsch WM, Kubicek CP (1990) Homocitrate synthase from *Penicillium chrysogenum*. Localization, purification of the cytosolic isoenzyme, and sensitivity to lysine. Biochem. J. 269:247-253.

MacCabe AP, Riach MBR, Unkles SE, Kinghorn JR (1990) The *Aspergillus nidulans npeA* locus consists of three contiguous genes required for penicillin biosynthesis. EMBO J. 9:279-287.

Marahiel MA, Zuber P, Czekay G, Losick R (1987) Identification of the promoter for a peptide antibiotic biosynthesis gene from *Bacillus brevis* and its regulation in *Bacillus subtilis*. J. Bacteriol. 169:2215-2222.

Martin JF, Aharonowitz Y (1983) Regulation of biosynthesis of β-lactam antibiotics. In: Demain AL, Solomon NA (eds) Antibiotics. Containing the β-lactam structure I. Springer Verlag, Berlin, p 229.

Martin JF, Demain AL (1980) Control of antibiotic biosynthesis. Microbiol. Rev. 44:230-251.

Punt PJ, Greaves PA, Kuyvenhoven A, Van Deutekom JCT, Kinghorn JR, Pouwels PH, Van den Hondel CAMJJ (1991) A twin-reporter vector for simultaneous analysis of expression signals of divergently transcribed, contiguous genes in filamentous fungi. Gene 104:119-122.

Queener SW (1990) Molecular biology of penicillin and cephalosporin biosynthesis. Antimicrob. Agents Chemother. 34:943-948.

Ramos FR, Lopez-Nieto MJ, Martin JF (1986) Coordinate increase of isopenicillin N synthetase, isopenicillin N epimerase and deacetoxycephalosporin C synthetase in a high cephalosporin-producing mutant of *Acremonium chrysogenum* and simultaneous loss of the three enzymes in a non-producing mutant. FEMS Microbiol. Lett. 35:123-127.

Ramsden M, McQuade BA, Saunders K, Turner MK, Harford S (1989) Characterization of a loss-of-function mutation in the isopenicillin N synthetase gene of *Acremonium chrysogenum*. Gene 55:267-273.

Revilla G, Ramos FR, Lopez-Nieto MJ, Alvarez E, Martin JF (1986) Glucose represses formation of δ-(L-α–aminoadipyl)-L-cysteinyl-D-valine and isopenicillin N synthase but not penicillin acyltransferase in *Penicillium chrysogenum*. J. Bacteriol. 168:947-952.

Shah AJ, Tilburn J, Adlard MW, Arst HN Jr (1991) pH regulation of penicillin production in *Aspergillus nidulans*. FEMS Microbiol. Lett. 77:209-212.

Smith DJ, Bull JH, Edwards J, Turner G (1989) Amplification of the isopenicillin N synthetase gene in a strain of *Penicillium chrysogenum* producing high levels of penicillin. Mol. Gen. Genet. 216:492-497.

Smith DJ, Burnham MKR, Bull JH, Hodgson JE, Ward JM, Browne P, Brown J, Barton B, Earl AJ, Turner G (1990a) β-Lactam antibiotic biosynthetic genes have been conserved in clusters in prokaryotes and eukaryotes. EMBO J. 9:741-747.

Smith DJ, Burnham MKR, Edwards J, Earl AJ, Turner G (1990b) Cloning and heterologous expression of the penicillin biosynthetic gene cluster from *Penicillium chrysogenum*. Biotechnol. 8:39-41.

Smith DJ, Earl AJ, Turner G (1990c) The multifunctional peptide synthetase performing the first step of penicillin biosynthesis is a 421073 dalton protein similar to *Bacillus brevis* peptide antibiotic synthetases. EMBO J. 9:2743-2750.

Swartz RW (1985) Penicillins. In: Blanch HW, Drew S, Wang DIC (eds) Comprehensive Biotechnology. The principles, applications and regulations of biotechnology in industry, agriculture and medicine. Vol.3. The practice of biotechnology: Current commodity products. Pergamon Press, Oxford, p 7.

Tobin MB, Kovacevic S, Madduri K, Hoskins JA, Skatrud PL, Vining LC, Stuttard C, Miller JR (1991) Localization of the lysine ε-aminotransferase (lat) and δ-(L-α-aminoadipyl)-L-cysteinyl-D-valine synthetase (pcbAB) genes from *Streptomyces clavuligerus* and production of lysine ε-aminotransferase activity in *Escherichia coli*. J. Bacteriol. 173:6223-6229.

Ward M, Turner G (1986) The ATP synthase subunit 9 gene of *Aspergillus nidulans*. Mol. Gen. Genet. 205:331-338.

Zähner H, Anke H, Anke T (1983) Evolution and secondary pathways. In: Bennett JW, Ciegler A (eds) Secondary metabolism and differentiation in fungi. Marcel Dekker Inc., New York & Basel, p 153.

New Tools and Prospectives for Medical Mycology

Imaging of the yeast killer phenomenon

L. Polonelli, S. Conti, M. Gerloni, W. Magliani,
G. Morace*, J. C. Cailliez
Institute of Microbiology
University of Parma
Via Gramsci, 14
43100 Parma
Italy

Idiotypes, or antigens on the inside (Jerne NK, 1982), are peptide determinants expressed for the antigen on the variable domains of specific immunoglobulin receptors in limphocytes as well as for viruses, toxins or hormones in other cells. The potential of antiidiotypic antibodies as vaccines for immuno-intervention in microbial infections emerged in the concept that the idiotope of anti-antibodies, being the internal image of the antigenic epitope, might induce primarily an antibody-like response. Ideally, antiidiotype-based vaccines might present several attractive advantages on conventional antigens such as unlimited availability, non infectious nature and ontogenic restriction of the host. Anaphylactic reactions, restricted antigenic response repertoire and high titer of long lasting response should be considered, on the contrary, as possible limits. Conventional vaccines based on live etiologic agents, such as attenuated viruses, may induce reversion to a virulent form or, because of their contents of genomic materials, transformation of host cells. Toxic side effects might occur during immunization with whole microorganisms for the prophylaxis of bacterial, fungal and protozoan diseases. Recombinant vaccination by synthetic peptides may suffer of high costs, poor immunogenicity, side effects of expression vectors (Finberg R, 1981) and be uneffective in case of carbohydrate nature of the immunogenic epitope of the antigen. Theoretically, antiidiotypic antibodies should act as non infectious, non toxic, easy to purify, administer and engineer vaccines. The stimulation of a restricted immune response by antiidiotypic vaccines should avoid damaging immunoreactions to the host, although the responsiveness directed to only a few epitopes of a microorganism is more susceptible to antigenic drift or genetic restriction. Challenges of developing an effective vaccine to human imunodeficiency virus or *Trypanosoma rhodesiense* by antiidiotype immunity are represented by the definition of an antigenic region of the viral envelope without high mutation rates and the genetic restriction provided by the strain dependence of resistance to the protozoan (Sacks DL, 1983). Internal-imaged-based vaccines displaying the same restriction pattern as the immunogenic epitope of the antigen should induce immunity across species barriers and overcome limitations related to genetic restriction which might occur because of the lack of cross reactive antiidiotypes. Responsiveness

* Istituto di Microbiologia, Università Cattolica del Sacro Cuore, Roma, Italy

NATO ASI Series, Vol. H 69
Molecular Biology and its Application to Medical Mycology
Edited by B. Maresca, G. S. Kobayashi and H. Yamaguchi

of the immune system to internal image antiidiotypes may be regulated in an antigen-like MHC restricted fashion and differ significantly in comparison with the one to cross reactive antiidiotypes which may directly interact with lymphocyte clones, thus perturbing the immune response through the idiotypic network and inducing silent antigen-binding clones (Bona C, 1981; Francotte M, 1984) which might provide immunity in some conditions of immune incompetence such as neonatal infections (Gaulton GN, 1986a; Stein K, 1984). Regulation of the strenght and duration of the immune response to antiidiotypic antibodies might be indeed more variable than with the antigens. The prolonged humoral and cellular immune response which is usually obtained by immunization with live viruses is also obtainable by using antiidiotypes (Gaulton GN, 1986a; Kennedy RC, 1986) in contrast with the rapid decline of the immune response after antigen clearance by administration of attenuated viruses or peptides (Finberg R, 1981). The feasibility of antiidiotypic antibodies as vaccines might be particularly important in the immunization against infectious agents other than viruses such as protozoan and other microorganisms for which the growth of some life cycle stages is difficult to obtain or carbohydrate epitopes are protective antigens (Hiernaux JR, 1988; Sacks DL, 1982; Sacks DL, 1985). The surprisingly successfull demonstration of humoral and cellular antiidiotypic responses in both experimental and clinical studies to schistosomiasis and Chagas' disease is suggestive for something in common with these infections (Colley DG, 1989a; Colley DG, 1989b; Lima MSG, 1986; Olds GR, 1985; Sher A, 1989). Although the etiologic agents are not related and have different effects on their hosts, still the microorganisms are both responsible for chronic endemic infections, an opportunity which might allow the development of extensive idiotypic interactions. Chronicity and endemicity may result in active infections during pregnancy yielding most of future patients and influencing by idiotypic regulation the expression of immunological responses. The regulatory impact is apparent in different chronic conditions (autoimmune and neoplastic diseases, allogenic transplants) (Bona CA, 1987) and particularly evident in some parasitic infections such as schistosomiasis (Colley DG, 1989b; Kresina TS, 1986). In clinical schistosomiasis mansonii and Chagas' disease, idiotypic-antiidiotypic interactions correlate with immunoregulatory functions and clinical disease states and the role of antiidiotypic responses in the immunopathology is suggested (Colley DG, 1990; Doughty BL, 1989; Montesano MA, 1989). Possibility exists that idiotypic-antiidiotypic responses may be involved in the expression of protective immune responses in parasitic infections either in chronic or intrauterine infections. Cord blood mononuclear cells of newborns from mothers affected with Chagas' disease or schistosomiasis are reactive with the idiotypes of their mother's antiparasitic antibodies (Eloi-Santos SM, 1989). The correlation of idiotypic-antiidiotypic interactions with the clinical forms of various parasitic diseases should be well considered for the development of vaccines destined to populations of endemic areas.

Theoretically, the main feature of antiidiotypic antibodies, internal image of the antigen, is epitope mimicry. In case that the epitope has a relevant biological function by binding or attachment to specific receptors of another molecule or cell, antiidiotypic antibodies may mimic that interaction (Sege K, 1978). Idiotypic and antiidiotypic mimicry of receptor functions and ligand signal delivery have been reported in many biological systems and several laboratories have adopted the approach of idiotypic-antiidiotypic interactions for the study of ligand-receptor binding (Gaulton GN, 1986b).

In medical mycology, opportunistic fungi represent an emerging problem to immunocompromised hosts because of their characteristics of environmental ubiquity and endogenicity in man. The prophylaxis and therapy of the mycoses have been longly hampered by the relative understanding of the host immunoresponse to fungal infections and structural similarities between the eukaryotic fungal and animal cells. New strategies, based on bioaction modalities and cellular targets effectively used by microorganisms in their natural competition, might be applied to prevent and control the diseases caused by fungal agents.

Killer toxins secreted by particular yeast isolates may exert a lethal activity against sensitive strains characterized by the occurrence of complementary cell wall receptors and lack of specific immunity systems for the yeast killer toxins (Tipper DJ, 1984).

The genetic determinants which are responsible for toxin production have been characterized in several yeast killer systems: double-stranded RNA virus-like particles in *Saccharomyces cerevisiae* and *Ustilago maydis* (Fujimura T, 1988), couples of linear DNA plasmids in *Kluyveromyces lactis* (Gunge N, 1981), chromosomal genes in *Pichia anomala* (Starmer WT, 1987). Different bioaction modalities have been reported according to the different yeast killer systems: alteration of the permeability of the plasma membrane for *S. cerevisiae* and *U. maydis*, inhibition of the cyclic AMP for *K. lactis* and arrest of β-glucan synthesis in the cell wall of sensitive yeast isolates for *P. mrakii* killer toxin (Wickner RB, 1985).

In the K1 system of *S. cerevisiae*, the first product of the dsRNA is a preprotoxin whose cytoplasmic process of cleavage into 4 subunits (α, β, γ and δ) is mediated by nuclear genes. Subunit δ is a leader protein implicated in the evolution process of the mature killer toxin which is secreted as α and β subunits bound by disulphuric links. Subunit β specifically recognizes the superficial receptors of sensitive yeast cells constituted by (1,6)β-D-glucan for the formation of which the activity of some nuclear genes is also essential. The binding between the subunit β of the exogenous k1 killer toxin and the specific cell wall receptor of the sensitive yeast cell liberates the α subunit which interacts with a plasma membrane receptor also regulated by a specific nuclear gene and activates the killing process.

In the killer toxin producing cells, the α subunit is likely to be responsible for the phenomenon of self-immunity by interacting during the cytoplasmic maturation process with the same plasma membrane receptor and making it unavailable to the activity of the α subunit of the exogenous mature killer toxin (Douglas CM, 1988). *In vitro* site-directed mutagenesis carried out by synthetic oligonucleotides on the gene encoding for the k1 killer toxin permitted to identify a specific region of the α subunit which appear to be involved in the phenomenon of self-immunity. The substitution of the asparagine residue located in position 94 included in the specific region may determine the loss of immunity. Specific mutations have induced *S. cerevisiae* killer strains to express a sensitive phenotype to the activity of their own killer toxin, thus behaving as suicidal mutants (Magliani W, 1990a; Magliani W, 1990b).

The yeast killer phenomenon is usually restricted within strains of the same or congeneric species. Killer toxins produced by isolates of *P. anomala* have shown, on the contrary, a lethal activity against many taxonomically unrelated eukaryotic and prokaryotic microorganisms (Polonelli L, 1986a). The wide spectrum of antimicrobial activity is presumptive for the occurrence of common superficial receptors for *P. anomala* killer toxin in evolutionary divergent microorganisms and the potential use, indeed, of those toxins as antibiotics. A *P. anomala* killer toxin, KT 25F, selected for the widest spectrum of antimicrobial activity proved to be therapeutic in the topical treatment of pityriasis versicolor-like lesions of superficially infected laboratory animals (Polonelli L, 1986b). Experimental intravenous injections, however, proved to be toxic and antigenic when applied to the treatment of systemic candidosis. The recognized *in vitro* lability of the KT 25F molecule at physiological values of temperature and pH, moreover, have furtherly precluded the possibility of administering KT 25F as systemic antibiotic. It was conceivable that some immunological derivatives of yeast killer toxins might maintain their antibiotic activity and preserving by undesired side effects.

Theoretically, the steric interaction between the biologically functional epitope of yeast killer toxin and its specific cell wall receptor on sensitive yeasts may be imagined by the complementarity of the idiotype of a yeast killer toxin neutralizing monoclonal antibody and its antiidiotype. A monoclonal antibody (MAb KT4) produced against KT 25F proved to neutralize its *in vitro* killing activity against a sensitive strain of *Candida albicans* (CDC B385) (Polonelli L, 1987). MAb KT4 showed a differential reactivity by indirect immunofluorescence and double immunodiffusion with killer toxin produced by yeast isolates belonging to different species of the genus *Pichia*, excluding the recognition of the metabolic products of other yeast killer systems (Polonelli L, 1989a; Polonelli L, 1991a). The use of MAb KT4 by colloidal gold immunoelectromicroscopy permitted the ultrastructural visualization of the reacting epitope in the process of intracellular toxinogenesis and secretion which apparently occurred through specialized although undefined areas of the cell wall of *P. anomala* 25F (Cailliez JC, 1992). The use of MAb KT4 as immunogen elicited in rabbits the production of antiidiotypic antibodies (K-antiIds) which proved to be cross reacting with KT 25F by competing for the binding site (idiotype) of MAb KT4 (Polonelli L, 1988).

K-antiIds were used as immunological probes to visualize, by indirect immunofluorescence, the presumptive cell wall receptors specific for killer toxin on sensitive yeasts (Polonelli L, 1990). The demonstration of superficial receptors on the killer toxin producing cells themselves was presumptive for a specific self-immunity system in *P. anomala*. On the contrary, the lack of reactivity of K-antiIds with *in vitro* cultured mammalian cells by immunofluorescence testified the potential specificity of the target for killer toxins in yeast cells (Polonelli L., 1991c). Rabbit K-antiIds purified by affinity chromatography against the complementary immobilized MAb KT4 displayed *in vitro* a killer activity against *C. albicans* CDC B385 recognized to be sensitive to the activity of KT 25F. Affinity cromatography purified K-antiIds, moreover, proved to act as antibiotics also against cells of *P. anomala* 25F thus presuming that the physiological immunity systems for self killer toxin were uneffective to yeast killer toxin-like K-antiIds. "Antibiobodies" has been the term proposed for antiidiotypic antibodies exerting a receptor mediated antibiotic activity (Polonelli L, 1991b). K-antiIds elicited by vaccination with MAb KT4 in syngeneic mice exerted a significative immmunoprotection against lethal inocula of *C. albicans* CDC B385 administered intravenously (idiotypic vaccination). Immunoprotection was expressed as increasing time of survival of the vaccinated animals comparing with the controls and was related to the K-antiIds titer *in vivo* detectable in the sera of the immunized animals by competitive inhibition immuno-enzymatic immunosorbent assay with KT 25F for MAb KT4. Mouse affinity chromatography purified K-antiIds proved to be killing, *in vitro*, *C. albicans* CDC B385 cells used as infecting inoculum (Polonelli L, 1989b). Although in the idiotypic network it might be difficult to state precise identities since any idiotope is likely to act as a paratope, it is claimed that idiotypic vaccination differs from antiidiotypic vaccination in that antiidiotypic antibodies are the effect and not the cause of the immunoresponse and, then, they act as antibiotics rather than immunoglobulins against the etiologic agent. The most serious problem with K-antiidiotypic therapy or idiotypic vaccination might be the repeated administration of heterologous antibodies. Even the development of human hybridomas for use in man will not escape potentially serious allogenic, allergic and immuno-complex reactions when multiple injections are used. A possible avenue is that molecular dissection of recombinant yeast killer toxins and receptor-like idiotypes may allow the development of synthetic peptides that mimic complementarity.

Killer strains belonging to the genus *Pichia* appear to be peculiar, in that nuclear and not plasmid-borne genetic determinants of toxin production have been suggested. The analysis of the killer phenotype of the meiotic tetrads derived from the mating of killer and non killer parental strains has apparently confirmed this hypothesis (Starmer WT, 1987). No RNA or DNA plasmids, moreover, have been detected in any of a large number of *Pichia* isolates investigated at this purpose (Gunge N, 1988, personal communication). In order to identify and clone killer toxin encoding genes, a *P. anomala* cDNA expression library in the bacteriophage lambda gt11 was constructed and the immunoscreening performed by using MAb KT4. Poli A+ mRNA was isolated from *P. anomala* 25F cells grown under conditions favoring the maximum production of yeast killer toxin. Oligo (dT) cellulose purified mRNA was converted to an RNA-cDNA hybrid by reverse transcriptase using oligo dT priming and the hybrid was converted into double stranded blunt-ended cDNA using RNase H in combination with DNA polymerase I, *Escherichia coli* DNA ligase and T4 DNA polymerase (Gubler N, 1983). Eco RI adapters with internal Not I sites where added and the linker-adapted, phosphorylated cDNA was ligated to Eco RI cut lambda gt11 vector. This lambda gt11 cDNA expression library was immunoscreened by the plaque-lift assay (Maniatis T, 1989). Phages were plated onto *E. coli* host strain Y1090 at a density of $5x10^6$ plaques per 90 mm plate. After incubation at 42°C for 3 hours, nitrocellulose filters, presoaked in 10 mM IPTG, were placed onto the plates followed by further incubation at 37°C for 4 hours. Filters were removed, washed and probed with ammonium sulfate precipitated MAb KT4 ascitic fluid (120 mg/ml) diluted 1/10,000 in buffered solution and incubated in a second antibody solution (biotinylated anti-mouse IgG). Biotin moieties of the bound antibodies were detected by avidin-conjugated horseradish peroxidase followed by immunostaining using 4-chloro-1-naphtol as the substrate. The recombinant phages in plaques exhibiting a color signal were purified to homogeneity by repeated serial dilution and retested on *E. coli* Y1090. Phage DNA was prepared by standard procedures (Maniatis T, 1989) and analyzed by restriction

digestion and agarose gel electrophoresis. The cDNA inserts were excised by Eco RI digestion and subcloned into the Eco RI site of M13 mp 18 and M13 mp 19 DNAs. Sequencing is going to be accomplished by the ^{35}S-dideoxy chain termination procedure (Sanger FG, 1977; Yannisch-Perron C, 1985). SDS-PAGE and Western blotting analysis of the fusion proteins produced by the cells containing the recombinant DNA, using MAb KT4, will be used to compare the data with the previous preliminary findings on the crude secreted toxin. Cloning and sequencing of *P. anomala* killer toxin-encoding, and possibly receptor synthesizing, genes could enable to understand the nature, structure and mechanism of action of the yeast killer toxin and possible relationships with the physiologycal host immunity system and the receptor mediated resistance to the toxin activity.
The production of small synthetic peptides of the yeast killer toxin still maintaining the killer activity and avoiding undesired side effects to the host as well as the artificial synthesis of the idiotype representing the internal image of killer toxin receptors might envisage a new way for the treatment and prophylaxis of *Candida* infections.

Acknowledgments

This work was supported by the National Research Council (CNR) Target Projects "Biotechnology and Bioinstrumentation", "Prevention and Control of Disease Factors" Subproject "Etiology of Infectious Diseases" (contract number 91.00121.41/115.15897) and by Ministero della Sanità-Istituto Superiore di Sanità Progetto A.I.D.S. (1990) - Roma - Italia, and it was developed inside the European Group for *Candida* Cell Biology and Pathogenicity.

References

Bona C, Heber-Katz E, Paul WE (1981) Idiotype-anti-idiotype regulation. I. Immunization with a levan-binding myeloma protein leads to the appearence of autoanti-(antiId) antibodies and to activation of silent clones. J. Exp. Med. 153:951-967.

Bona CA (1987) Idiotypes and lymphocytes. Academic Press, Inc., New York.

Cailliez JC, Gerloni M, Morace G, Conti S, Cantelli C, Polonelli L (1992) Ultrastructural immunodetection of a *Pichia anomala* killer toxin: a preliminary study. Biol. Cell, in press.

Colley DG, Goes AM, Doughty BL, Correa-Oliveira R, Parra JC, Lairmore KM, Montesano MA, Rocha RS, Gazzinelli G (1989a) Antiidiotypic T cells and factors in murine and human schistosomiasis. In: Kaplan JG, Green DR, Bleackley RC (eds) The Cellular Basis of Immune Modulation. Alan R. Liss, Inc., New York, p 367-378.

Colley DG, Montesano MA, Eloi-Santos SM, Powell MR, Correa-Oliveira R, Rocha RS, Gazzinelli G (1989b) Idiotype networks in schistosomiasis. In: McAdam KPWJ (ed) New Strategies in Parasitology. Churchill Livingstone, Edinburgh, p 179-190.

Colley DG, Gazzinelli RT, Brener Z, Montesano MA, Eloi-Santos SM, Correa-Oliveira R, Parra JC, Lima MS, Gazzinelli G (1990) Idiotypes and anti-idiotypic T lymphocytes in chronic endemic parasitic diseases. In: UytdeHaag F, Urbain J,

Osterhaus A (eds) Idiotypes in Biology and Medicine. Elsevier Biomedical Press, Amsterdam.

Doughty BL, Goes AM, Parra JC, Rocha R, Cone JC, Colley DG, Gazzinelli G (1989) Anti-idiotypic T cells in human schistosomiasis. Immunol. Invest. 18:373-388.

Douglas CM, Sturley SL, Bostian KA (1988) Role of protein processing, intracellular trafficking and endocytosis in production of and immunity to yeast killer toxin. Eur. J. Epidemiol. 4:400-408.

Eloi-Santos SM, Novato-Silva E, Maselli VM, Gazzinelli G, Colley DG, Correa-Oliveira R (1989) Idiotypic sensitization in utero of children born to mothers with schistosomiasis or Chagas' disease. J. Clin. Invest. 84:1028-1031.

Finberg R, Benacerraf B (1981) Induction, control and consequences of virus specific cytotoxic T cells. Immunol. Rev. 58:157-180.

Francotte M, Urbain J (1984) Induction of anti-tobacco mosaic virus antibodies in mice by rabbit antiidiotypic antibodies. J. Exp. Med. 160:1485-1494.

Fujimura T, Esteban R, Wickner RB (1988) Double-stranded RNA replication in *Saccharomyces cerevisiae*: genetic control and *in vitro* replication. In: Koltin Y, Leibowitz M (eds) Viruses of Fungi and Simple Eukaryotes. Marcel Dekker, Inc., New York, p 91-115.

Gaulton GN, Sharpe AH, Chang DW, Fields BN, Greene MI (1986a) Syngeneic monoclonal internal image anti-idiotopes as prophylactic vaccines. J. Immunol. 137:2930-2936.

Gaulton GN, Greene MI (1986b) Idiotypic mimickry of biological receptors. Annu. Rev. Immunol. 4:253-280.

Gubler N, Hoffman BJ (1983) A simple and very effective method for generating cDNA libraries. Gene 25:263-269.

Gunge N, Tamaru A, Ozawa F, Sakaguchi K. Isolation and characterization of linear deoxyribonucleic acid plasmids from *Kluyveromyces lactis* and the plasmid associated killer character. J. Bacteriol. (1981) 145:382-390.

Hiernaux JR (1988) Idiotypic vaccines and infectious diseases. Infect. Immun. 56:1407-1413.

Jerne NK (1982) Idiotypes - Antigens on the Inside, Editions 'Roche', Basel, p 238-239.

Kennedy RC, Eichberg JW, Lanford RE, Dreesman GR (1986) Anti-idiotypic antibody vaccine for type B viral hepatitis in chimpanzees. Science 232:220-223.

Kresina TF, Olds GR (1986) Concomitant cellular and humoral expression of a regulatory cross-reactive idiotype in acute *Schistosoma japonicum* infection. Infect. Immun. 53:90-94.

Lima MS, Gazzinelli G, Nacimento E, Cavalho-Parra J, Montesano MA, Colley DG (1986) Immune responses during human schistosomiasis mansoni: evidence for antiidiotypic T lymphocyte responsiveness. J. Clin. Invest. 78:983-988.

Magliani W, Douglas CM, Bostian KA (1990a) Clonazione e sequenziamento di mutanti del gene per la tossina killer K1 di *Saccharomyces cerevisiae*. In: Atti XXII Congresso Nazionale della Società Italiana di Microbiologia, Palermo 26-30 Novembre 1990, p 86.

Magliani W (1990b) Mutagenesi diretta sul gene per la tossina killer K1 di *Saccharomyces cerevisiae* mediante utilizzazione di oligonucleotidi sintetici. In: Oligonucleotidi sintetici in Biologia Molecolare e Biotecnologia. Ed. Roberto Gambari e Claudio Nastruzzi, Ferrara, p 25-30.

Maniatis T, Fritsch EF, Sambrook J (1989) Molecular cloning. 2nd edn, Cold Spring Harbor Laboratory, Cold Spring Harbor, NY.

Montesano MA, Lima MS, Correa-Oliveira R, Gazzinelli G, Colley DG (1989) Immune responses during human schistosomiasis mansoni. XVI. Idiotypic differences in antibody preparations from patients with different clinical forms of infection. J. Immunol. 142:2501-2506.

Olds GR, Kresina TF (1985) Network interactions in *Schistosoma japonicum* infection. Identification and characterization of serologically distinct immunoregulatory auto-antiidiotypic antibody population. J. Clin. Invest. 76:2338-2347.

Polonelli L, Morace G (1986a) Reevaluation of the yeast killer phenomenon. J. Clin. Microbiol. 24:866-869.

Polonelli L, Lorenzini R, De Bernardis F, Morace G (1986b) Potential therapeutic effect of yeast killer toxin. Mycopathologia 96:103-107.

Polonelli L, Morace G (1987) Production and characterization of yeast killer toxin monoclonal antibodies. J. Clin. Microbiol. 25:460-462.

Polonelli L, Morace G (1988) Yeast killer toxin-like anti-idiotypic antibodies. J. Clin. Microbiol. 26:602-604.

Polonelli L, Manzara S, Conti S, Dettori G, Morace G (1989a) Serological study of yeast killer toxins by monoclonal antibodies. Mycopathologia 108:211-215.

Polonelli L, Lorenzini R, Gerloni M, Conti S, Campani L, Morace G (1989b) Yeast killer toxin-like anti-idiotypic antibodies protecting mice against candidosis. In: Proceedings 3rd Symposium Topics in Mycology, Mycoses in AIDS Patients, Paris, November 20-23, 1989, P17.

Polonelli L, Fanti F, Conti S, Campani L, Gerloni M, Castagnola M, Morace G, Chezzi C (1990) Detection by immunofluorescent anti-idiotypic antibodies of yeast killer toxin cell wall receptors of *Candida albicans*. J. Immunol. Methods 132:205-209.

Polonelli L, Conti S, Campani L, Gerloni M, Morace G, Chezzi C (1991a) Differential toxinogenesis in the genus *Pichia* detected by an anti-yeast killer toxin monoclonal antibody. Antonie van Leeuwenhoek 59:139-145.

Polonelli L, Conti S, Gerloni M, Campani L, Castagnola M, Morace G, Chezzi C (1991b) "Antibiobodies": antibiotic-like anti-idiotypic antibodies. J. Med. Vet. Mycol. 29:235-242.

Polonelli L, Morace G, Conti S, Gerloni M, Magliani W, Chezzi C (1991c) Interfaces of the yeast killer phenomenon. Crit. Rev. Microbiol. 18:47-87.

Polonelli L, Morace G, Conti S, Gerloni M, Menozzi MG, Cailliez JC (1991) Conceptions et perspectives du phenomene "killer" chez les levures. Journal de Mycologie Medicale, 1:284-295.

Sacks DL, Esser KM, Sher A (1982) Immunization of mice against African trypanosomiasis using anti-idiotypic antibodies. J. Exp. Med. 155:1108-1119.

Sacks DL, Sher A (1983) Evidence that anti-idiotype induced immunity to experimental African trypanosomiasis is genetically restricted and requires recognition of combining site-related idiotopes. J. Immunol. 131:1511-1515.

Sacks DL, Kirchoff LV, Hieny S, Sher A (1985) Molecular mimicry of a carbohydrate epitope on a major surface glycoprotein of *Trypanosoma cruzi* by using anti-idiotypic antibodies. J. Immunol. 135:4155-4159.

Sanger FG, Michlen S, Coulson AR (1977) DNA sequencing with chain-terminating inhibitors. Proc. Natl. Acad. Sci. USA 70:5463-5467.

Sege K, Peterson PA (1978) Use of anti-idiotypic antibodies as cell-surface receptor probes. Proc. Natl. Acad. Sci. USA 75:2443-2447.

Sher A, Colley DG (1989) Immunoparasitology. In: Paul WE (ed) Fundamental Immunology, 2nd ed., Raven Press, New York, p 957-983.

Starmer WT, Ganter PF, Aberdeen V, Lachance M-A, Phaff HJ (1987) The ecological role of killer yeasts in natural communities of yeasts. Can. J. Microbiol. 33:783-796.

Stein K, Soderstrom T (1984) Neonatal administration of idiotype primes for protection against *Escherichia coli* K13 infection in mice. J. Exp. Med. 160:1001-1011.

Tipper DJ, Bostian KA (1984) Double-stranded ribonucleic acid killer system in yeasts. Microbiol. Rev. 48:125-156.

Wickner RB (1985) Killer yeasts. In: McGinnis MR (ed) Current Topics in Medical Mycology. Springer-Verlag, New York, p 286-312.

Yannisch-Perron C, Vieira JC, Messing JR (1985) Improved M13 phage cloning vectors and host strains: Nucleotide sequences of the M13 mp18 and pUC 19 vectors. Gene 33:109-119.

cDNA cloning of *Candida albicans* aspartic proteinase and its diagnostic application

S. Y. Murayama
Department of Bacteriology
Teikyo University School of Medicine
2-11-1, Kaga, Itabashi-ku
Tokyo 173
Japan

Introduction

The expanding of immunocompromised individuals has focused attention on the importance of *Candida* species as an opportunistic fungal pathogen. *C. albicans* is a major infectious fungal agent in humans, invading by attachment to the host cells via cell-surface adhesive molecules. Once attached, *C. albicans* cells proliferate and release extracellular aspartic proteinase, which aids them to invade the integument of the host. Few extracellular enzymes have been described as virulence factors, but only extracellular an aspartic acid proteinase has received intense attention. There seems to be a relationship between the production of aspartic proteinase by *C. albicans* and its virulence (7, 9, 13), but the actual role of proteinase as the virulence factor remains unclear. Yamamoto *et al.* (20) reported that all strains of *C. albicans* tested secrete proteinase extracellularly, and they sequenced the first 23 amino acids of the N-terminal of the proteinase from their *C. albicans* No. 114. We have undertaken the cDNA cloning of this enzyme using the polymerase chain reaction (PCR) with mixed primers constructed from the information published by that group. This enzyme has very recently sequenced from the *C. albicans* strain ATCC10231 (6). We also show the results of comparison of our sequence with theirs and other related enzymes.

cDNA cloning of aspartic proteinase secreted by *C. albicans*

C. albicans serotype A, No.114 strain obtained from the Research Center for Medical Mycology, Teikyo University School of Medicine, was studied. The procedure of Sripati and Warner (15) was used to prepare total RNA from *C. albicans* cells, and the poly(A)$^+$ RNA was isolated by Oligotex-dT30 beads (Takara Shuzo Co., Ltd., Kyoto, Japan). Double-stranded cDNA was synthesized with the cDNA Synthesis System Plus kit (Amersham, Buckinghamshire, UK) according to the manufacturer's protocol, except that an M4-oligo(dT) primer was used instead of oligo(dT)$_{12-18}$.
To obtain a portion of the cDNA clone of the aspartic proteinase secreted by *C. albicans* using PCR, 5' mixed primers, CANP-1 and CANP-2, and a 5' mixed probe, CANP-3, were constructed according to the N-terminal amino acid

NATO ASI Series, Vol. H 69
Molecular Biology and its Application to Medical Mycology
Edited by B. Maresca, G. S. Kobayashi and H. Yamaguchi

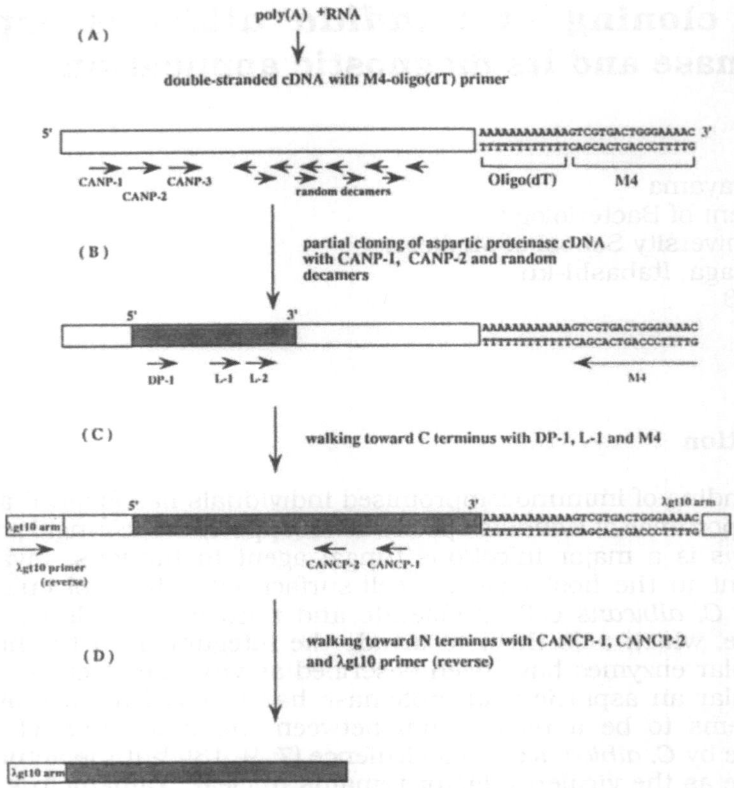

Figure 1. Strategy to obtain the aspartic proteinase cDNA from *C. albicans*. (A) Poly(A)+ RNA preparation followed by double-stranded cDNA synthesis with M4-oligo(dT) as the primer. (B) Nested PCR amplification of the cDNA with mixed oligonucleotide upstream primers and random decamers. (C) Nested PCR amplification of the cDNA to walk toward the C terminus with specific upstream primers we constructed and an M4 downstream primer. (D) Nested PCR amplification of the cDNA to walk toward the N terminus with specific downstream primers we constructed and λgt10 (reverse) upstream primer. In this reaction only, double-stranded cDNA ligated with the *Eco*RI-digested phage vector λgt10 was used as the template. Open box, unknown region; shaded box, region obtained.

sequence (20). The specific primers DP-1, L-1, CANCP-1, and CANCP-2 and the specific probes L-2 and DP-1 were constructed from earlier findings on nucleotide sequences in the portion of the cDNA fragments obtained for the purpose of walking toward the N- or C-terminus.
The sequences of the primers and probes were as follows:

5' mixed primers

```
CANP-1: 5'-CAAGCGGTGCCGGTGAC-3' (17mer)
                G  A  A  A  A
                T  T  T  T
                C  C  C  C
```

```
CANP-2: 5'-CATAATGAACAAGTGACGTA-3' (20mer)
             C  C  G  G  A  A
                         T  T
                         C  C
```

Mixed probe

```
CANP-3: 5'-GCGGCGGATATTACAGTGGG-3' (20mer)
             A  A  C  C  G  A
             T  T     A  C  T
             C  C        T  C
```

Specific primers (sense primers)

```
DP - 1: 5' - GCTGCTGATATTACCGTTGG - 3' (20mer)
L  - 1: 5' - GGATCTTCATCTCAAGGTAC - 3' (20mer)
```

Specific primers (antisense primers)

```
CANPCP-1: 5'-ACTACCACCGGCTTCATTGG-3' (20mer)
CANPCP-2: 5'-TAACCAACCCCTAAAATACC-3' (20mer)
```

*

Specific probes

```
L-1: (above)
L-2: 5'-TTGGATTTGGTGGTGTTTCG-3' (20mer)
```

M13 primer M4, λgt10 primer (reverse), and RV primer were products of Takara Shuzo.

The M4-oligo(dT) primer (5'-GTTTTCCCAGTCACGACTTTTTTTTTT TTTTTTTTT-3') was made with an Applied Biosystems DNA synthesizer (380B; Foster City, CA).

The general strategy by which aspartic proteinase cDNA was obtained from *C. albicans* is shown in Figure 1 (10). PCR was performed in a DNA Thermal Cycler (Perkin Elmer-Cetus, Norwalk, CT) with a GeneAmp DNA amplification kit (Perkin Elmer-Cetus). The amplification profile was 94°C for 0.5 min, 55°C for 2 min, and 72°C for 1 min. Each round of amplification was done for 35 cycles in a total volume of 100 μl.

For the first round, the 5' mixed primer CANP-1 and the random decamers were used with the double-stranded cDNA. In a second round, 1μl of this reaction mixture was reamplified with the 5' mixed primer CANP-2 and the random decamers. In the first nested PCR, the amplification product, designated CA621, gave a discrete band of about 300 base pairs (Figure 1). We designed specific primers DP-1 and L-1 and specific probe L-2 from CA621 to walk toward the C terminus. Next, double-stranded cDNA was used as the template with the specific primer DP-1 and M13 primer M4 to walk toward the C terminus. As mentioned above, the second round was performed with specific primer L-1 and M13 primer M4. In the second nested PCR, the desired clone, designated CA722, was amplified; it included the C terminus. Next, we constructed the specific primers CANCP-1 and CANCP-2 from CA722 to walk toward the N terminus. Finally, to walk toward the N terminus, double-stranded cDNA ligated with the *Eco*-RI-digested phage vector λgt10 was used as the template with specific primer CANCP-1 and λgt10 primer (reverse). The second round was done with the specific primer CANCP-2 and

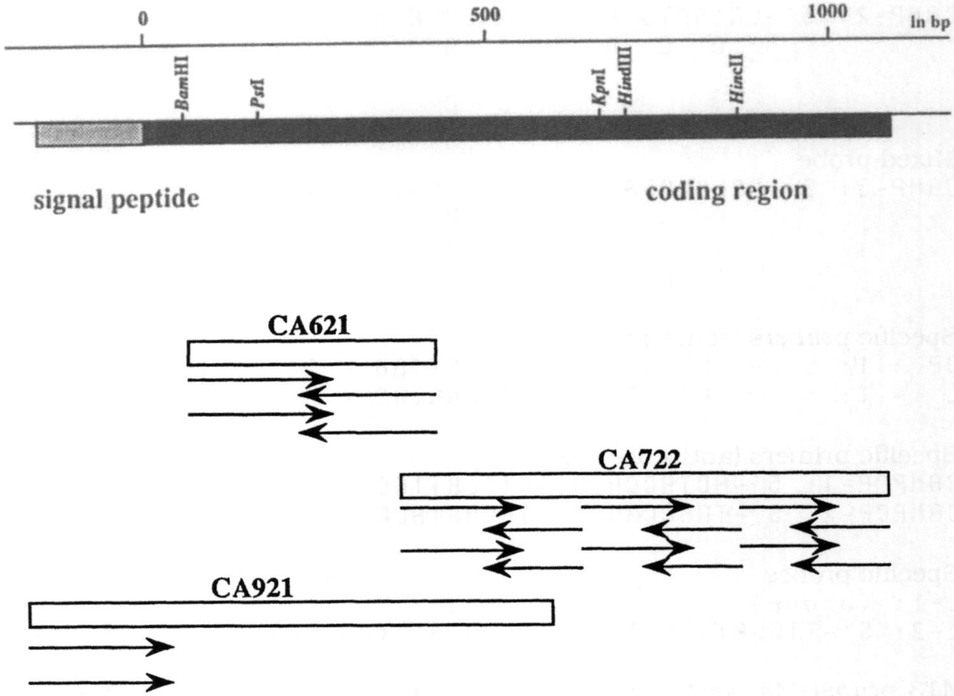

Figure 2. Restriction map of cDNA clones of aspartic proteinase secreted by *C. albicans* and the sequencing strategy. The coding region is indicated by the closed box and the signal peptide region is indicated by the shaded box. Coordinates are given as the nucleotide base number (bp) starting from the 5'-terminal of mature proteinase. Native numbers are used for the signal peptide regions. Open boxes under the restriction map show the region covered by the three clones CA621, CA722, and CA921. Arrows below each clone show the direction of sequencing and the region sequenced

λgt10 primer (reverse). In the final nested PCR, the cDNA clone containing the N terminus, designated CA921, was obtained. PCR fragments were identified by their size by agarose gel electrophoresis and confirmed by Southern blotting with use of the probe internal to the 3' or 5' PCR primer; CANP-3, specific probe L-2, or specific probe DP-1 (data not shown). Using the same procedure we could get the clone harboring the total length of the enzyme.

In the isolation of cDNA clones that encode the aspartic proteinase secreted by *C. albicans*, important steps include the use of mixed primers corresponding to the N-terminus amino acid sequences as the 5' primer, and the use of random decamers as the 3' primer. The first nested PCRs were done with mixed primers and random decamers. Examination of the clone CA621 that we obtained showed that the random decamers had not been made use of in the PCR, but that mixed primers acted as both upstream and downstream primers (data not shown). It seems that degenerate primers accidentally (and fortunately) could be used as both 5' and 3' primers, specifically amplifying cDNA. Still, we saw that random primers were useful in PCR.

```
 -56  M   F   L   K   N   I   F   I   G   L   A   I   A   L   L   V   D   A   T   P   T   -36
   1 ATG TTT TTA AAG AAT ATT TTC ATT GGC CTT GCT ATT GCT TTA TTA GTC GAT GCT ACT CCA ACA   63

 -35  T   T   K   R   S   A   G   F   V   A   L   D   F   S   V   V   K   T   P   K   A   -15
  64 ACA ACC AAA AGA TCA GCT GGT TTC GTT GCT TTA GAT TTC AGT GTT GTG AAA ACT CCT AAA GCA  126

 -14  S   P   V   T   N   G   Q   E   G   K   T   S   K   R   Q   A   V   P   V   T   L     7
 127 TCC CCA GTT ACT AAT GGC CAA GAA GGT AAA ACT TCC AAA AGA CAA GCT GTC CCA GTG ACT TTA  189

   8  H   N   E   Q   V   T   Y   A   A   D   I   T   V   G   S   N   N   Q   K   N   L    28
 190 CAC AAT GAA CAA GTC ACT TAT GCT GCT GAT ATT ACC GTT GGA TCC AAT AAT CAA AAA CTT AAT  252

  29  V   I   V   D   T   G   S   S   D   L   W   V   P   D   V   N   V   D   C   Q   V    49
 253 GTT ATT GTT GAT ACT GGA TCA TCA GAT TTA TGG GTT CCT GAT GTT AAT GTT GAT TGT CAA GTC  314

  50  T   Y   S   D   Q   T   A   D   F   C   K   Q   K   G   T   Y   D   P   S   G   S    70
 315 ACT TAT AGT GAT CAA ACT GCA GAT TTC TGT AAA CAA AAG GGG ACA TAT GAT CCA AGT GGT TCA  377

  71  S   A   S   Q   D   L   N   T   P   F   K   I   G   Y   G   D   G   S   S   S   Q    91
 378 TCA GCT TCA CAA GAT TTG AAT ACT CCA TTC AAA ATT GGT TAT GGT GAT GGA TCT TCA TCT CAA  440

  92  G   T   L   Y   K   D   T   V   G   F   G   G   V   S   I   K   N   Q   V   L   A   112
 441 GGT ACT TTA TAT AAG GAT ACC GTT GGA TTT GGT GGT GTT TCG ATT AAA AAT CAA GTT TTA GCT  503

 113  D   V   D   S   T   S   I   D   Q   G   I   L   G   V   G   Y   K   T   N   E   A   133
 504 GAT GTT GAT TCT ACT TCA ATT GAT CAA GGT ATT TTA GGG GTT GGT TAT AAA ACC AAT GAA GCC  566

 134  G   G   S   Y   D   N   V   P   V   T   L   K   K   Q   G   V   I   A   K   N   A   154
 567 GGT GGT AGT TAT GAT AAT GTC CCT GTC ACT TTA AAA AAA CAA GGA GTC ATT GCT AAG AAT GCT  629

 155  Y   S   L   I   L   I   L   Q   M   L   P   R   D   K   S   F   S   V   G   L   I   175
 630 TAT TCA CTT ATC TTA ATT CTC CAG ATG CTG CCA CGG GAC AAA TCA TTT TCG GTG GGG TTG ATA  692

 176  M   L   N   I   V   G   S   L   I   A   L   P   V   T   S   D   R   E   L   R   I   196
 693 ATG CTA AAT ATA GTG GGT TCA TTA ATT GCA TTA CCA GTT ACT TCT GAT CGT GAA TTA AGA ATT  755

 197  S   L   G   S   V   E   V   S   G   K   T   I   N   T   D   N   V   D   V   L   L   217
 756 AGT TTG GGT TCA GTT GAA GTT TCT GGT AAA ACC ATC AAT ACT GAT AAT GTC GAT GTT CTT TTG  818

 218  D   S   G   T   T   I   T   Y   L   Q   Q   D   L   A   D   Q   I   I   K   A   F   238
 819 GAT TCA GGT ACC ACC ATT ACT TAT TTG CAA CAA GAT CTT GCT GAT CAA ATC ATT AAA GCT TTC  881

 239  N   G   K   L   T   Q   D   S   N   G   N   S   F   Y   E   V   D   C   N   L   S   259
 882 AAT GGT AAA TTA ACT CAA GAT TCC AAT GGT AAT TCA TTC TAT GAA GTT GAT TGT AAT TTG TCA  944

 260  G   D   V   V   F   N   F   S   K   N   A   K   I   S   V   P   A   S   E   F   A   280
 945 GGG GAT GTT GTA TTC AAT TTT AGT AAA AAT GCT AAA ATT TCC GTT CCA GCT TCC GAA TTT GCT 1007

 281  A   S   L   Q   G   D   D   G   Q   P   Y   D   K   C   Q   L   L   F   D   V   N   301
1008 GCT TCT TTA CAA GGT GAT GAT GGT CAA CCA TAT GAT AAA TGT CAA TTA CTT TTC GAT GTT AAT 1070

 302  D   A   N   I   L   G   D   N   F   L   R   S   L   I   L   F   M   I   W   M   I   322
1071 GAT GCT AAC ATT CTT GGT GAT AAC TTT TTG AGA TCA CTT ATA TTG TTT ATG ATT TGG ATG ATA 1133

 323  M   K   F   L   W   L   K   S   N   I   L   L   F   Q   Y   F   S   L   T  stop      341
1134 ATG AAA TTT CTT TGG CTC AAG TCA AAT ATA CTT CTC TTC CAG TAT TTC TCC TTG ACC TAA     1193

1194 GATGAAGGGGTGAGATAAAGTTGAAATATTAAAATATTAGTTCTTGATTAGTTTTTACTTACTTGAAAGGAGTGGCTTTTTTT 1278
1279 TTATAGTTTGATAACTTTTTTTGCTTTCTTTCAAGTTTTTTTTATATTTTGTTTTGTTTTGTAAAAAAAAAAAAAAAAAAA   1354
```

Figure 3. Nucleotide sequence of cloned cDNA and primary structure deduced for aspartic proteinase. The nucleotide numbers and the amino acid residue numbers are shown in the right and left margins. The amino acid residue numbering starts at the NH_2 terminus of mature aspartic proteinase (shown by the vertical arrow). The putative signal peptidase recognition site is underlined (11). *Conserved active-site aspartic acid residue (17).

Nucleotide sequence and predicted polypeptide sequences

The amplified DNA fragments were given blunt ends and subcloned into pUC18 (Takara Shuzo). The cloned plasmids were sequenced with M13 primer M4 and RV primer (Takara Shuzo) by the dideoxy method (14). Chain termination reactions were done with the 7-Deaza sequencing kit (Takara Shuzo). The sequencing strategy used for a 1358-bp region is shown in Figure 2, and the complete nucleotide sequence is shown in Figure 3. The sequence included 167 bp of the 3' flanking sequences in addition to 1191 bp of coding DNA. To judge from the open reading frame, a mature enzyme is 341 amino acids long and has a molecular weight of 37,000. Its amino acid content and codon usage were shown in Table 1. The codon utilization patterns are confirmed the report of Brown *et al.*, (3).

Comparison of the primary structure of the *C. albican* aspartic proteinase with that of other aspartyl proteinase

Sequence homology of this aspartic proteinase with other aspartyl proteinases was expected since the tertiary structures of the aspartic proteinase were homologous(2, 5, 16). The amino acid sequence of the aspartic proteinase secreted by *C. albicans* was aligned with the sequences of *Saccharomyces cerevisiae* proteinase A (1) and vacuolar aspartyl proteinase of *C. albicans* (8; Figure 4), allowing deletions, gaps, and insertions for maximum homology. Mature aspartic proteinase secreted by *C. albicans* had only 17.6% and 15.3% homology with the proteinase A and aspartyl proteinase, respectively. Thus, the *C. albicans* aspartic proteinase had little homology with aspartyl proteinase. This may be due to a difference between the secretive type and the vacuolar type. All such proteinases have some aspartic acid residues at the active site, as exemplified by pepsin and cathepsin D (4, 17). Two active aspartic acid sites, Asp-32 and Asp-218, and the amino acids near those sites were conserved in the *C. albicans* aspartic proteinase. The aspartyl active sites in this aspartic proteinase, the residues 32 and 218 are equivalent to residues 121 and 306 in proteinase A, or 109 and 294 in vacuolar aspartyl proteinase.

```
CA-S    1  QAVPVTLHNEQVTYAADITVGSNNQKLNVIVDTGSSDLWV   40
CA-V   90  KYDAPLTNYLNAQ•FTE•EI•TPG•PFK••L•••••N•••  129
YP-A   79  HD••L•-NYLNAQ•YTD••L•TPP•NFK••L•••••N•••  117

CA-S   41  PDVNVDCQVTYSDQTADFCKQKGTYDPSGSSASQDLNTPF   80
CA-V  130  •SQDCTSLACFLHAKY•H-DASS••KAN••EF•IQYGSGS  168
YP-A  118  •--SNE•--------GSLA•FLHSK••HEA••SYKANG•E•  148

CA-S   81  KIGYG-DGSSSQGTLYKDTVGFGGVSIKNQVLA----DVD  115
CA-V  169  ME••HIQDVLTI•D•VIPGQD•AE-ATSEPG••-----FA  202
YP-A  149  A•Q••-T•-•LE•YISQ••LSI•DLT•PK•DF•EATSEPG  186

CA-S  116  ST-SIDQ--GILGVGYKTNEAGGSYDNVPVTLKKQGVIAK  152
CA-V  203  FG-KF•---••••LA•D•ISVNHIVPPIYNAI-N•ALLE•  237
YP-A  187  L•FAFGKFD••••L••D•ISVDKVVPPFYNAIQQDLLDE•  226

CA-S  153  N-AYSL-IL-ILQMLPRDKSFSV--GLIMLNIVGSLIALP  187
CA-V  238  P-QFGF-Y•GSTDKDENDGGLATFG•YDASLFQ•KITW••  275
YP-A  227  RF•FY•GDT-SKDTENGGEATFG--•IDESKFK•DITW••  263

CA-S  188  VTSDRELRISLGSVEVSGKTINTDNVDVLLDSGTTITYLQ  227
CA-V  276  IRRKDYWEV•FEGIGLGDEYAELHKTGAAI•T••SLIT•P  315
YP-A  264  •RRKAYWEVKFEGIGLGDEYAELESHGAAI•T••SLIT•P  303

CA-S  228  QDLADQIIKAFNGKLTQDSNGNSFYEVDC---NLSG---D  261
CA-V  316  SS••E-••N•----•IGATKSWSGQ•Q•••AKRDSLP---•  348
YP-A  304  SG••EM•NAEIGA•--KGWT•Q--•TL•----•TRDNLP•  336

CA-S  262  VVFNFSKNAKISVPASEFAASLQGDDGQPYDKCQLLFDVN  301
CA-V  349  LTLT•A-GYNFTLTPYDY-YEVS•SCISVFTPMDFPQPIG  386
YP-A  337  LI•••-NGYNFTIGPYDYTLEVS•SCISAITPMDFPEP•G  375

CA-S  302  DANILGDNFLRSLILFMIWMIMKFLWLKSNILLFQYFSLT  341
CA-V  387  •LA•V••A•••KYYSIYDLDKNAVGLAPSKVVLFSVYECV  426
YP-A  376  PLA•V••A•••KYYSIYDLGNNAVGLAKAI            405
```

Figure 4. Alignment of amino acid sequences of aspartic proteinase secreted by *C. albicans* (CA-S), yeast proteinase A (YP-A), and *C. albicans* vacuolar aspartyl proteinase (CA-V). The dots in the lower two sequences indicate identity with the *C. albicans* proteinase sequence. The dash indicates insertions of gaps at positions at which there are no homologous residues. Two underlines show conserved active-site aspartic acid residues.

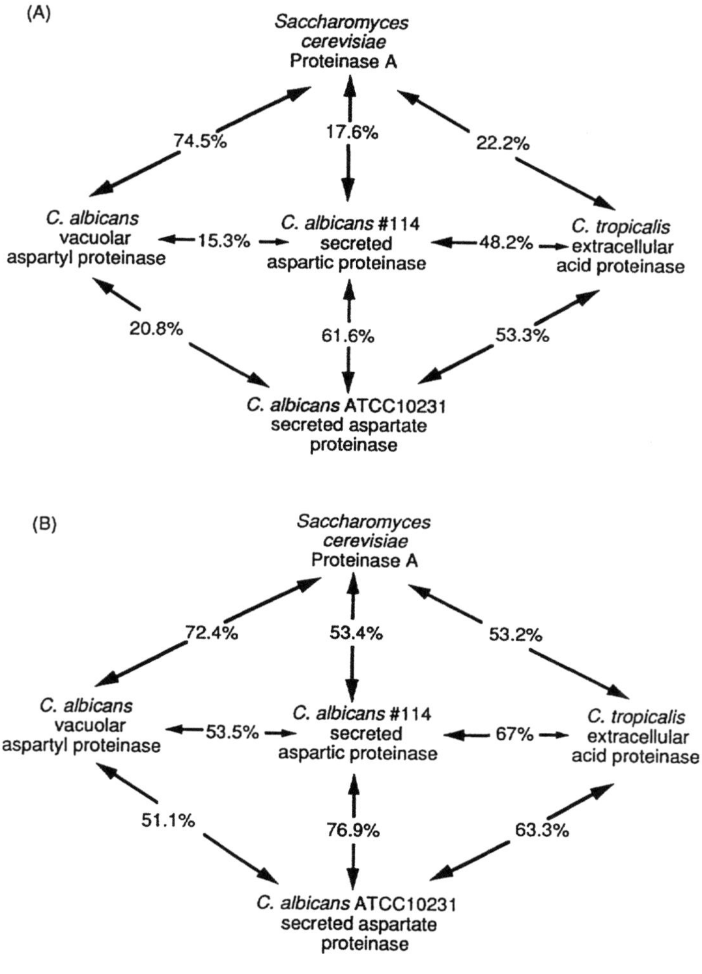

Figure 5. Similarities among the *C. albicans* aspartic proteinase and related proteinase. Similarities at the amino acid level (A) and at the DNA sequence level (B) were shown.

Hube *et al.* very recently reported the DNA sequence of secretory aspartate proteinase of another *C. albicans* strain, ATCC10231 (6). We compared their sequence with our cloned sequence, and the result showed 76.9% DNA sequence homology and 66% amino acid homology (Figure 5). On these points the DNA cloned by Hube *et al.* shows a strong similarity to cDNA described here, but the difference between them seems to be relatively great as usual enzyme secreted by the same species. The difference between the extracellular acid protease gene of *C. tropicalis* (18) and those of *C. albicans* shows almost the same percentage as the difference between ours and Hube's. Hube *et al.* cloned the gene based on the amino acid sequences of 21 N-terminal amino acids of the purified enzyme of the strain CBS2730. And they screened a genomic bank of DNA from *C. albicans* ATCC10231. The N terminal amino acid residue of CBS2730 was reported as tryptophane by

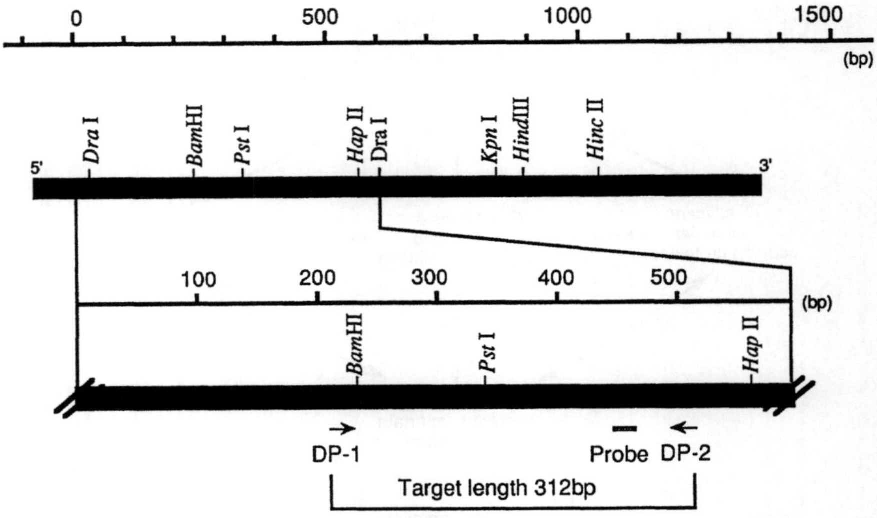

Figure 6. Restriction map of cDNA clone of aspartic proteinase secreted by *C. albicans* No. 114 and the location of the primers and probe. The coding region is indicated by the closed box and the signal peptide region is indicated by the shaded box. Coordinates are given as the number of the nucleotide base pair (bp) starting from the 5'-terminal of mature proteinase. Negative numbers are used for the signal peptide regions. Arrows show the direction and location of the primers.

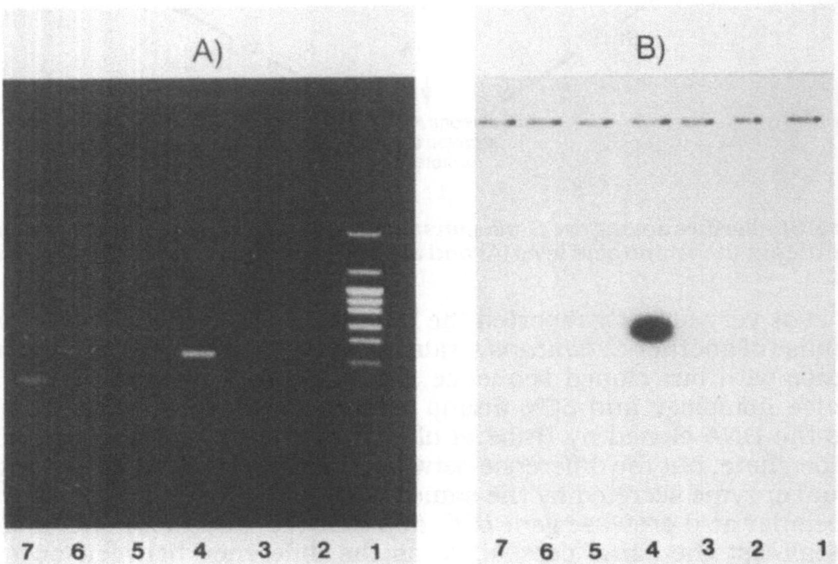

Figure 7. Electrophoresis on 1% agarose gel stained with ethidium bromide (A) and Southern blot hybridization (B) of DNA amplified by the PCR with the primers DP-1 and DP-2 and the radiolabelled probe SP-1. Lanes: 1, Human placenta; 2, *S. cerevisiae*; 3, carp liver, 4, *C. albicans*; 5, *E. coli*, 6, mouse Balb/c 3T3; 7, size markers.

their coworker Rüchel (12), but that of ATCC10231 was glutamine translated from the DNA sequence data. They didn't explain the difference. Therefor ATCC10231, CBS2730 and No.114 seems to have a different secretory aspartic proteinase, respectively. It is interesting that *C. albicans* has such differing secretory aspartate proteinases. Possibly yet another secretory aspartate proteinase of another *C. albicans* strain will be found in the future.

Diagnostic application

In spite of its clinical significance, no effective method for diagnosis has been established for candidiasis, one of the serious opportunistic infections, which tends to have increased in recent years because of the intensive medical treatment now being practiced. A diagnostic reagent currently commercially available is the one for determining anti-*Candida* mannan antibody in serum using mannan antigen, one of the cell wall components of pathogenic *Candida* yeast. However, this reagent often shows a significantly high level even in healthy individuals. Another diagnostic reagent is the one for determining *Candida* antigen in patient's serum using anti-*Candida* whole cell disruption. But the sensitivity of this method is not so high.
There are also methods for biochemically saying metabolites of *Candida* or components of the *Candida* cells. However, these methods too sometimes display a significantly high level even patients who do not suffer pathogenically from candidiasis. It has thus been desired to establish a method for diagnosis specific to candidiasis.
Figure 6 shows the specific primers, DP-1 and DP-2 for PCR on the base of sequence data and tried to detect the target site, 312 bp of *C. albicans* aspartic proteinase gene (19). Using these primers for PCR, only DNA from *C. albicans* was detected and DNAs from mouse Balb/c 3T3, *Escherichia coli* K-12, carp liver, *Saccharomyces cerevisiae*, and human were not (Figure 7A). Confirmatory Southern hybridization was done (Figure 7B). A strong hybridization signal was shown only in *C. albicans*. These specific PCR primers were applied to another *Candida* species (data not shown). Amplified DNAs were detected only from *C. glabrata*, *C. krusei*, *C. kefyr*, *C. lipolytica*, *C. tropicalis*, and *C. parapsilosis* were not detected. These results suggest that the target gene that encodes extracellular aspartic proteinase is specific to *C. albicans*. We have not yet tried the PCR method with patient's serum, but the sensitivity of PCR is very high. Therefor these probe is desired to be useful in diagnosis.
Even if the secreted, inducible acid proteinase of *C. albicans* is a putatative virulence factor of the fungus, its actual function is uncertain. Its precise role would be clear by disruption of this gene based on the information of the sequence.

References

Ammerer G, Hunter CP, Rothman JH, Saari GC, Valis LA, Stereng TH (1986) PEP4 gene of *Saccharomyces cerevisiae* encodes proteinase A, a vacuolar enzyme required for processing of vacuolar precursors. Mol. Cell. Biol. 6:2490-2499.

Andreeva NS, Zdanov AS, Gustchina AE, Federov AA (1984) Structure of ethanol-inhibited porcine pepsin at 2-Å resolution and binding of the methyl ester of phenylalanyl-diiodotyrosine to the enzyme. J. Biol. Chem. 259:11353-11365.

Brown AJP, Bertram G, Feldmann PJF, Peggie MW, Swoboda RK (1991) Codon utilization in the pathogenic yeast, *Candida albicans*. Nucl. Acids. Res. 19:4298.

Faust PL, Kornfeld S, Chirgwin JM (1985) Cloning and sequence analysis of cDNA for human cathepsin D. Proc. Natl. Acad. Sci. USA 82:4910-4914.

Hsu I-N, Delbaere LTJ, James MNG, Hofmann T (1977) Penicillopepsin from *Penicillium janthinellum* crystal structure at 2.8 Å and sequence homology with porcine pepsin. Nature 266:140-145.

Hube B, Turver CJ, Odds FC, Eiffert H, Boilnois GJ, Küchel H, Rüchel R (1991) Sequence of the *Candida albicans* gene encoding the secretory aspartate proteinase. J. Med. Vet. Mycol. 29:129-132.

Kwon-Chung KJ, Lehman D, Good C, Magee PT (1985) Genetic evidence for role of extracellular proteinase in virulence of *Candida albicans*. Infect. Immun. 49:571-575.

Lott TJ, Page LS, Boiron P, Bensen J, Reiss E (1989) Nucleotide sequence of the *Candida albicans* aspartyl proteinase gene. Nucl. Acids Res. 17:1779.

MacDonald F, Odds FC. (1983) Virulence for mice of a proteinase-secreting strain of *Candida albicans* and a proteinase-deficient mutant. J. Gen. Microbiol. 129:431-438.

Mukai H, Takeda O, Asada K, Kato I, Murayama SY, Yamaguchi H (1992) cDNA cloning of an aspartic proteinase secreted by *Candida albicans*. Microbial Immunol. in press.

Perlman D, Halvorson HO (1983) A putative signal peptidase recognition site and sequence in eukaryotic and prokaryotic signal peptides. J. Mol. Biol. 167:391-409.

Rüchel R (1981) Properties of a purified proteinase from the yeast *Candida albicans*. Biochim. Biophys. Acta 659:99-103.

Rüchel R (1983) On the role of proteinases from *Candida albicans* in the pathogenesis of acronecrosis. Zentralbl. Bakteriol. Hyg. Abt. I Orig. A255:524-536.

Sanger F (1981) Determination of nucleotide sequences in DNA. Science 214: 1205-1210.

Sripati CE, Warner JR (1978) Isolation, characterization, and translation of mRNA from yeast. In Prescott DM(ed), Methods in Cell Biology, vol 12. Academic Press, New York, p 61-81.

Subramanian E, Swan IDA, Liu M, Davies DR, Jenkins JA, Tickel IJ, Blundell TL (1977) Homology among acid proteases: comparison of crystal structures at 3 Å resolution of acid proteases from *Rhizopus chinensis* and *Endothia parasitica*. Proc. Natl. Acad. Sci. USA 74:556-559.

Tang J (1979) Evolution in the structure and function of carboxyl proteases. Mol. Cell. Biochem. 26:93-109.

Togni G, Sanglard D, Falchetto R, Monod M (1991) Isolation and nucleotide sequence of the extracellular acid protease gene (ACP) from the yeast *Candida tropicalis*. FEBS Letters 286:181-185.

Takeda O, Mukai H, Takesako K, Sono E, Asada K, Kato I, Murayama SY, Yamaguchi H (1992) Specific detection of *Candida albicans* strains by amplification of its aspartic proteinase gene using PCR. Microbiol. Immunol. in press.

Yamamoto T, Nohara K, Uchida K, Yamaguchi K (1992) Microbiol. Immunol. 36, in press.

Molecular biotyping of *Candida* in experimental and clinical studies

A. Cassone, F. De Bernardis, E. Pontieri[1], G. Carruba[1]
Laboratory of Bacteriology and Medical Mycology,
Istituto Superiore di Sanità,
Viale Regina Elena, 299
00161 Rome
Italy

Summary

Yeasts of the genus *Candida* contain a number of opportunistic pathogens with increasing importance in several human diseases, e.g. the acquired immunedeficiency syndrome (AIDS). As virulence and epidemics-associated properties of these yeasts are clearly strain-dependent, it is essential to define methods for unequivocal identification at sub-species level (strain, biotype, clone). Methods based on physical and/or genetic analysis of both nuclear and mitochondrial DNA offer a great potential for a consistent, highly discriminatory and clinically useful *Candida* biotyping. The paper offers examples of how the analysis of the restriction length polymorphism of whole cell DNA and the electrophoretic karyotype in pulse-field gel electrophoresis lead to useful pattern of biotype identification in clinical settings and sub-species classification for *Candida albicans* and *Candida parapsilosis*, two important pathogens of the group.

Introduction

Rationale for biotyping Candida *species*

Candida species are ubiquitous human opportunistic pathogens causing local, self-limited disease or deep-organ invasion, depending most on the kind of predisposing condition, but also on the actual pathogenicity of the species or strain involved (Odds, 1979; Cassone, 1990).

For instance, oral and vaginal trush range from acute, moderate attack in otherwise healthy subjects to a recurrent, often intractable disease in AIDS patients. On the other hand, the strains of *C. albicans* and other *Candida* spp. isolated from overt disease, are endowed with greater experimental pathogenicity and possession of virulence factors (De Bernardis *et al.*, 1990; Agatensi *et al.*, 1991). Neutropenia due to conditional chemotherapy in bone-marrow transplanted, leukemic patient constitutes a hallmark in the predisposition to invasive candidiasis. Nonetheless, the mortality risk is higher for *C. tropicalis*-

---

[1] Department of Experimental Medicine, University of L'Aquila, L'Aquila, Italy.

NATO ASI Series, Vol. H 69
Molecular Biology and its Application to Medical Mycology
Edited by B. Maresca, G. S. Kobayashi and H. Yamaguchi

Table 1.
Epidemiological assessment of strain variations in *Candida*: some important questions that may be answered

To what extent different individuals are colonized by different biotypes?
Are these biotypes distinct in different colonization sites (different body sites)?
Is the infection from endogenous or exogenous source?
What is the natural history of biotypes during infection and antibiotic therapy?
Are selected biotypes more prone to infect (virulent clones)?
Is that particular candidiasis due to infection, reinfection or relapse?
How frequent is patient to patient or hospital staff (nurse?) to patient spread of *Candida* in nosocomial infections?

invaded subjects (Bodey & Fainstein, 1985).
Epidemiologically, it is of utmost importance that *Candida spp.*, in particular, *C. albicans*, the most pathogenic of all *Candida* species, is a human commensal, easily colonizing most healthy individuals in their gastrointestinal tract and, albeit less prevalent, skin and genital areas (Meunier, 1989). In addition to that, virulent strains can be transmitted from person to person, either sexually or by other means (Bodey, 1984). Therefore, the infection by *Candida* may be either endogenous or exogenous, and the infecting strain in endogenous infection may spread from an adjacent or distant body site (Bodey, 1984). Thus, it may well happen that a predisposed subject is infected by an exogenous strain of *C. albicans* while being colonized, even in multiple body sites, by other strains of the same fungus. Moreover, depending on the specific virulence of the strain, some patients in a given neutropenic setting will develop candidiasis while others, equally if not more compromized and heavily colonized, will not (Wade & Schimpff, 1985). Under these conditions, exact definition of source and route of transmission is essential to the control and prevention of disease, also considering that diagnostic and therapeutic interventions are to be specifically directed against the strain specifically involved in that disease. For all of this, unambiguous identification of *Candida* at subspecies level (strain, biotype) is a prerequisite. Table 1 enlists important epidemiological and pathogenetic questions that could be answered by effective delineating a biotype profile in *Candida* spp.
Recurrences of oro-oesophageal *Candida* infection have become a dramatic problem in AIDS patients, because of the profound immunosuppression and likelyhood to develop antimycotic-resistance (Fox *et al.*, 1989). Here too, a distinction beetween failure of eradicating the infectious strain and reinfection with a new strain is of paramount importance in devising appropriate therapy.

Methods for biotyping

Based on above considerations, epidemiologically useful *Candida* typing methods have been sought for. The main ones are listed in Table 2, where advantages and disadvantages of each proposed method are also considered. In particular, the biotyping method of Odds and Abbott has been the most widely used amongst the phenotypic methods (Odds and Abbott, 1980). By assessing growth on a number of standardized media, it allows detection of 512 potential biotypes of *C. albicans*, of which about 200 have so far been indeed encountered (Odds *et al.*, 1983). Although not yet widely applied, the killer system developed by Polonelli and coworkers has permitted a highly discriminatory and reproducible strain typing in the few studies performed, even among non *albicans* species (Polonelli *et al*, 1983). It is based on the susceptibility-resistance of a given strain of *Candida* to a killer toxin secreted by a panel of other suitably selected fungi. Empirically, it most resembles phage-typing systems widely used in bacteriology. All other phenotypic methods have not gained an appreciable consensus.
Although phenotypic methods have given valuable information on a few outbreaks of nosocomial *Candida* infections (Burnie *et al.*, 1985) and epidemiology of *Candida* vaginitis, (O'Connor and Sobel, 1986) their drawbacks,

Table 2.
Methods of *Candida* biotyping

Method	For	Against
A. Phenotypic 1. Colonial Morphology	Inexpensive, methodologically simple.	Laborious and subjective assessment, morphology switching.
2. Serology	Simple and rapid.	Only two strains distinguished (serotype A and B in *C. albicans*).
3. Biochemistry[a] Physiology	Discriminatory and simple. Already widely used.	Some characters are instable.
4. Killer[b] Phenotype	Precise and highly reproducible.	Large application still lacking, then significance of results to be confirmed.
B. Genotypic[c] 1. DNA fingerprinting 2. Electrophoretic Karyotype	Highly discriminative, unambiguous attribution.	Expensive, long-term stability of the documented pattern sometimes questioned.

a) Ref. Odds & Abbott, 1980; b) Ref. Polonelli *et al.*, 1983; c) Refs. Scherer & Stevens, 1988; Riggsby, 1990; Carruba *et al.*, 1991.

mainly consisting in a limited resolving power, and lack of standardization, hence ambiguous interpretation of some data have limited their widespread use and fostered the application of genotypic methods.

Genotypic biotyping

If a strain or a biotype "really" exists, its characters are determined genetically. Thus, why not determining directly the genetic difference among the strains? Theoretically, this genetic approach should be able to delineate each strain of a given species as unique, being highly discriminatory and unambiguous, with high sensitivity and specificity (Scherer and Stevens, 1988; Riggsby, 1990). Encouraged by preliminary results, DNA-typing methods are now gaining ever increasing popularity, and methodological refinements with computerized analysis of the data will undoubtledly lead to a widespread clinical laboratory use, as has happened in other microbiological fields. Elementary genetic characteristics of *C. albicans* are given in Table 3. Essentially two methods have so far been employed: the determination of restriction length polymorphism of total or mitochondrial DNA, with a moderately repetitive DNA probe, and the determination of chromosome-sized bands by pulse-field gel electrophoresis (electrophoretic karyotype) (Riggsby, 1990; Mathews & Burnie, 1989). The polymorphisms observed by both techniques in *C. albicans* and other *Candida* spp are extremely abundant within a trait of common phenotypic "species" character, documenting a continuous development of genetic diversity. As emphasized by Fox *et al.* (1989), this development may explain the differences observed in DNA-biotypes from different patients (or different body sites). The rate of rearrangement of newly generated sequences appears to be sufficiently slow to permit the identification of a conserved pattern during a clinical outbreak or even a prolonged, chronic disease (see also below).

Results and discussion

We have recently adopted both DNA fingerprinting and the analysis of the electrophoretic karyotype for definition of biotype-related questions in clinical and experimental *Candida* infection. We will report here a synthesis of these

Table 3.
Some elementary genetic characteristics of *Candida albicans*, medically the most important *Candida* species[a].

DNA content	About 40 fg/cell
DNA base composition	34.3 to 37.3% (G+C)
Ploidy	Diploid
Karyotype[b]	7-9 distinct chromosomes[b]
Sex	Asexual

a) See Ref. Riggsby, 1990.
b) From pulse-field gel electrophoresis.

results (Venditti *et al.*, 1992; Carruba *et al.*, 1991; De Bernardis *et al.*, 1992).

Uniqueness of C. albicans biotype in a rare, particular, invasive candidiasis in a surgical patient

This was an unusual case of a 21 year-old man, previously colectomized for diffuse polyposis, who developed an unusual syndrome of catheter-related, right-sided endocarditis, accompained by endophtalmitis and serious, extensive folliculitis. *C. albicans* serotype A was isolated from catheter, blood, skin and fistolous discharge, and because of the rarity of this clinical picture, we wondered whether one or different biotypes could be aetiologically involved in such diverse manifestations of the disease as candidemia and folliculitis. Both DNA fingerprinting (EcoRI and HinfI endonucleases) and the electrophoretic karyotipe, using transverse alternating gel field electrophoresis (TAFE; Beckman) were performed on all isolates, and at different time intervals after isolation and repeated sub-culturing (five subcultures at monthly intervals in Sabouraud dextrose agar). Fig. 1 shows the TAFE profiles of chromosome-sized bands of the isolates from catheter tip, blood and skin.
It is clear that all isolates had the same karyotype, within a pattern typical of *C. albicans* chromosomes analyzed by TAFE. The examination of whole cell DNA fingerprinting with the two endonucleases also demonstrated an identical pattern (data not shown). Thus, it has been assumed that, most likely, one single strain of *C. albicans* was aetiologically involved in all the diverse manifestations of the invasive candidiasis.

Demonstration of unique biotype in a "non-germinative" strain of C. albicans causing "pseudohyphal" development in rat vaginal infection

Recently, our group has been studying the vaginopathic potential and the intravaginal morphology of non-germinative mutants of *C. albicans* in a well-suited model of experimental vaginitis in oophorectomized, estradiol-treated rats (De Bernardis *et al.*, 1989). The goal of these studies are to compare germ-tube forming parental strains with their non-germinative derivatives, as germ-tube formation has been claimed to be a decisive factor in the ability to colonize and infect human vagina (Sobel *et al.*, 1984). We observed that an agerminative mutant of *C. albicans* (strain CA-2) was as vaginopathic for the rat as its parental counterpart 3153 strain. Cytologically, however, the CA-2 strain, which was unable to form germ-tubes in any previously studied *in vivo* or *in vitro* model, proved to form coarse filaments in the vagina, which were ultimately not distinguishable from the hyphal filaments formed by the germ-tube positive 3153 strain. Apart from showing a rather peculiar, unique morphological development of the mutant during intravaginal growth, these experiments could strongly confirm the importance of the filamentous growth for vaginopathy, provided that the identity of the filamentous CA-2 strain isolated during intravaginal growth with the initial challenger could be demonstrated. Thus, we applied both DNA fingerprinting and electrophoretic karyotype determination by CHEF or TAFE to the cultured CA-2 strain and to all isolates from the vaginal fluid during the infection. All the distinct *Candida* isolates from the vaginal fluids (at least 10 colonies examined for each isolation on days 1, 7, 14 and 28 during infection) had the same DNA-biotype, similar

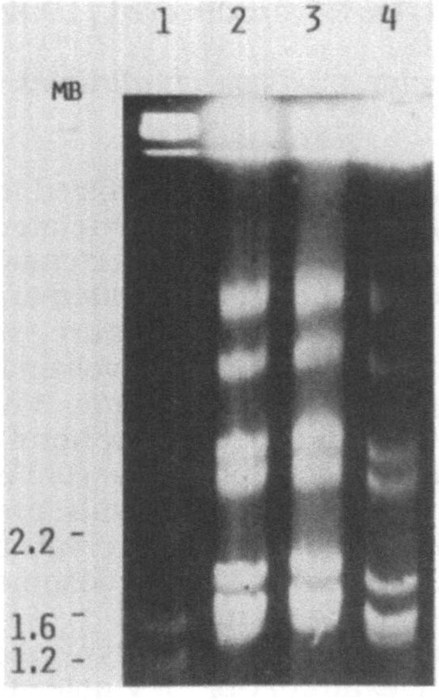

Figure 1. TAFE analysis of isolates of *C. albicans* from different body sites 2, catheter tip; 3, blood; 4, skin of the patient whose clinical case is described in the text. Lane 1 shows part of the *S. cerevisiae* chromosomes as standards. For technical details, see Venditti *et al.*, (1992).

to that of the initial culture for rat challenge. Figure 2 gives an example of the results obtained.

Biotyping of Candida parapsilosis

C. parapsilosis is an increasingly important cause of fungemia in neutropenic or otherwise-severely debilitated subjects undergoing parenteral nutrition (De Bernardis *et al.* 1989). Endocarditis may occur as severe complication of candidemia. Moreover, we have recently proved that *C. parapsilosis* is an

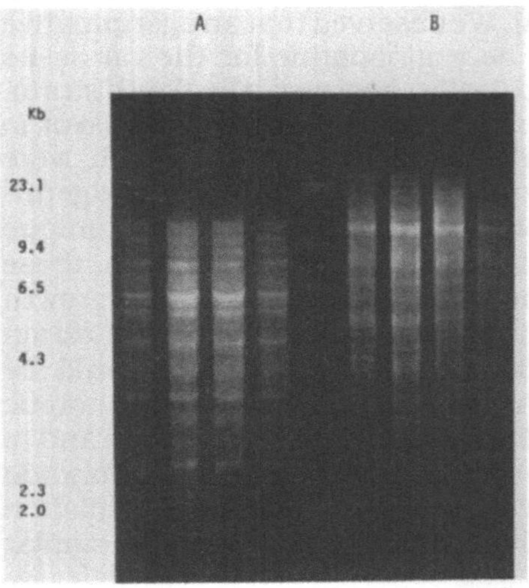

Figure 2. DNA fingerprinting of rat vaginal isolates of CA-2 cells (days 1, 7, 14, 21), lanes 1-4, respectively, during the experimental vaginal infection in oophorectomized, estradiol-treated rats (see text). Panel A, EcoRI; panel B, Hind III-digested DNA. Molecular weight standards (Hind III digest of lambda DNA) are also indicated (De Bernardis *et al.*, submitted).

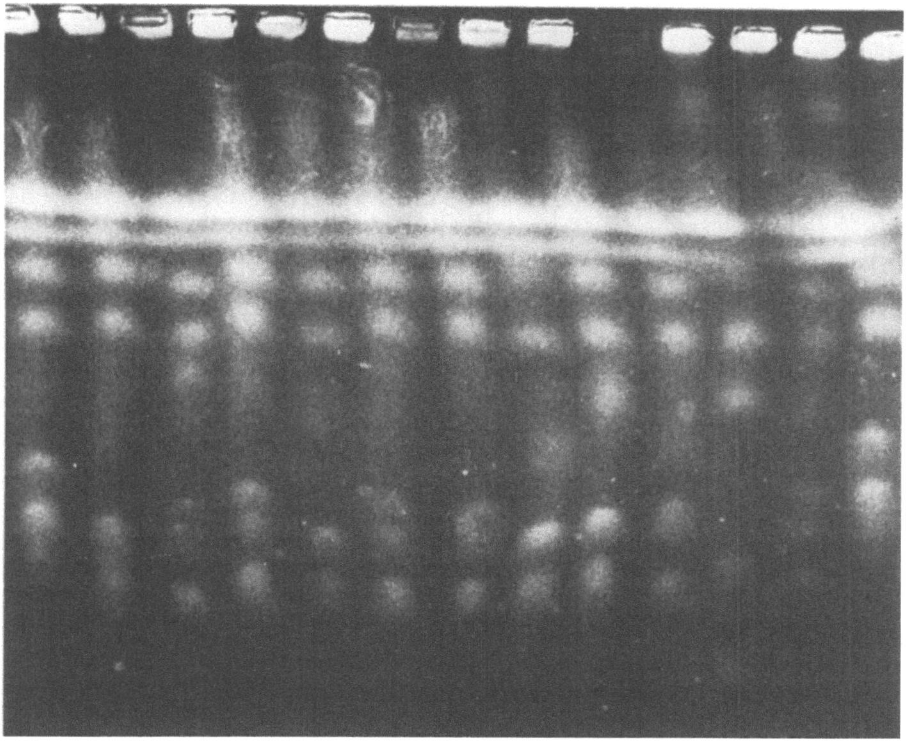

Figure 3. CHEF pattern of chromosome-sized DNA-bands from 13 different *C. parapsilosis* isolates from the blood of candidemic subjects. For technical details, see Carruba *et al.*, 1991.

aethiological agent of a vaginitis clinically indistinguishable from that caused by *C. albicans* or *Torulopsis glabrata* (Agatensi *et al.*, 1991; De Bernardis *et al.* 1989). There are, however, marked differences in experimental pathogenicity and secretion of virulence enzymes by the different isolates of *C. parapsilosis*, even when they are absolutely identical from the point of view of classic taxonomy. In particular, all fresh clinical isolates of this fungus (either from vaginitis or candidemia) were pathogenic for cyclophosphamide-neutropenic mice, whereas the environmental isolates were not (De Bernardis *et al.* 1989). Overall, the above distinctions suggest the existence of different biotypes of *C. parapsilosis*, the disclosure of which could be epidemiologically and clinically important.

We addressed this problem by determining the electrophoretic karyotype, by different technical approaches, in some thirty different isolates of *C. parapsilosis*, from different source and different clinical conditions (Carruba *et al.*, 1991). The investigation performed with CHEF or RFGE or TAFE gave an optimal resolution, and constancy of electrophoretic patterns for different sub-cultures of the same isolates. Apparently, the isolates could be assigned to at least 7 different karyotype classes based on number of chromosome in the gel. Figure 3 shows representative examples of the heterogeneity of karyotype

Table 4.
Class subdivision for *C. parapsilosis* as determined by karyotype analysis from different pulse-field gel techniques

N° OF PUTATIVE CHROMOSOME INTEGRATED						
Laboratory code	Source of isolation	Chef	RFGE	TAFE	Class proposed	Integrated N° of Chromosome
ISS-4879	SOIL	8 - 7	6 - 7	8	I	9
ISS-4884	SOIL					
ISS-4888	SOIL					
SA - 25	VAGINITIS					
SA - 17	VAGINITIS	7	6 - 7	8	II	9
SA - 19	VAGINITIS	6 - 7	5 - 6	7	III	8
SA - 23	VAGINITIS					
SA - 36	VAGINITIS					
SA - 38	CARRIAGE					
SA - 42	VAGINITIS					
SA - 28	CARRIAGE	7 - 8	6 - 7	9 - 8	IV	9
SA - 197	CARRIAGE	6 - 7	5	8	V	8
HEM - 1	CANDIDEMIA ENDOCARDITIS	7 - 8	6 - 7	8	VI	9
HEM - 3	CANDIDEMIA					
HEM - 4	CANDIDEMIA					
HEM - 2	CANDIDEMIA	8 - 7	6 - 7	8 - 9	VII	10

pattern within the isolates from hematological setting (candidemia and/or endocarditis), whereas Table 4 recapitulates class attribution for some isolates. In general, karyotype variation amongst isolates occurred in the lower Mr portion of the gel, where chromosomes of 0.7 to 1-3 Kb were located. The minimum estimates of chromosome number obtained by combining the results from both CHEF, TAFE and RFGE range from 6 to 9 for *C. parapsilosis*, figures not too distinct from chromosomes number (8, or 9) in *C. albicans* (Magee & Magee, 1987).
The data in Table 4 also demonstrate that, although the pathogenic potential and source of isolation, were somewhat shared among the different isolates, their distribution was not random at all. In particular, none of the twelve hematological isolates belonged to a class containing the vaginopathic isolates or those from vaginal carriage, suggesting some form of correlation between some biotypes and propensity for a given disease. Moreover, it is worthwile noting that the environmental isolates, though relatively few, did not split into different classes.

Conclusions

The determination of DNA-biotype of *Candida* spp. in clinical settings may prove invaluable for a full assessment of the source of infection and the virulence of the responsible strain, both of paramount importance in the epidemiological control of the disease. It may also greatly help in defining particular questions of strain identity in the laboratory, particularly during experimental infections in animals which either carry *Candida* spp as normal commensal or can be easily cross-contaminated by *Candida*. Both the analysis of DNA restriction pattern, and the electrophoretic kariotype have already shown to be a valid marker of strain identity and constancy, at least for the time required to monitor this identity in the clinical situation.

Acknowledgements

The authors' work described in this review has been supported by grants from National Research Council, Projects F.A.T.M.A. and "Biotecnologie" (Units A. Cassone and G. Carruba) (Contract N° 9200210.41).
Thanks are due to Mrs A. M. Marella for help in the preparation of the manuscript.

References

Agatensi L, Franchi F, Mondello F, Bevilacqua RL, Ceddia T, De Bernardis F, Cassone A (1991) Vaginopathic and proteolytic *Candida* species in outpatients attending a gynaecology clinic. J. Clin. Pathol. 44:826-830.

Bodey GP (1984) Candidiasis in cancer patients. Am. J. Med. 77(4D):13-19.

Bodey GP, Fainstein V (1985) Systemic candidiasis. In: Bodey GP, Fainstein V, eds. Candidiasis. New York: Raven Press, 135-168.

Burnie JP, Odds FC, Lee W, Webster C, Williams JD (1985) Outbreak of systemic *Candida albicans* in intensive care unit caused by cross infection. Brit. Med. J. 290:746-748.

Carruba G, Pontieri E, De Bernardis F, Martino P, Cassone A (1991) DNA fingerprinting and electrophoretic karyotype of environmental and clinical isolates of *Candida parapsilosis*. J. Clin. Microbiol. 29:916-922.

Cassone A (1990) Cell wall of *Candida albicans*, its function and its impact on the host. Curr. Top. Med. Mycol. 3:249-314.

De Bernardis F, Agatensi L, Ross IK, Emerson GW, Lorenzini R, Sullivan PA, Cassone A (1989) Evidence for a role for secreted aspartate proteinase of *Candida albicans* in vulvovaginal candidiasis. J. Infect. Dis. 161:1276-1283.

De Bernardis F, Lorenzini R, Verticchio R, Agatensi L, Cassone A (1989) Isolation, acid proteinase secretion, and experimental pathogenicity of *Candida parapsilosis* from outpatients with vaginitis. J. Clin. Microbiol. 27:2598-2603.

De Bernardis F, Agatensi L, Ross IK, Emerson GW, Lorenzini R, Sullivan A, Cassone A (1990) Evidence for a role for secreted aspartate proteinase of *Candida albicans* in vulvovaginal candidosis. J. Infect. Dis. 161:1276-1283.

De Bernardis F, Adriani D, Lorenzini R, Carruba G, Cassone A (1992) Association of filamentous growth with elevated vaginopathic potential in a "non-germinative" mutant of *Candida albicans* experimentally infecting rat vagina. Submitted.

Fox BC, Mobley HLT, Wade JC (1989) The use of a DNA probe for epidemiological studies of candidiasis in immunocompromised hosts. J. Infect. Dis. 159: 488-493.

Magee B, Magee PT (1987) Electrophoretic karyotypes and chromosome numbers in *Candida species*. J. Gen. Microbiol. 133:425-430.
Mathews R, Burnie JP (1989) Assessment of DNA fingerprinting for rapid identification of outbreaks of systemic candidiasis. Brit. Med. J. 298:354-357.
Meunier F (1989) Candidiasis. Eur. J. Clin. Microbiol. Infect. Dis. 8(5):438-447.
O'Connor MI, Sobel JD (1986) Epidemiology of recurrent vulvovaginal candidiasis: Identification and strain differentiation of *Candida albicans*. J. Infect. Dis. 154(2):358-363.
Odds F C (1979) *Candida* and candidosis. Leicester University Press, Leicester, England. 51-97.
Odds FC, Abbott AB (1980) A simple system for the presumptive identification of *Candida albicans* differentiation of strains within the species. J. Med. Vet. Mycol. 18:301-317.
Odds FC, Abbott AB (1983) Modification and extension of tests for differentiation of *Candida* species and strains. J. Med. Vet. Mycol. 21:79-81.
Polonelli L, Archibusacci C, Sestito M (1983) Killer system: a simple methof for differentiating strains of *Candida albicans*. J. Clin. Microbiol. 17:774-780.
Riggsby WS (1990) The genetics of *Candida*. CRC Press.
Scherer S, Stevens DA (1988) A *Candida albicans* dispersed, repeated gene family and its epidemiologic applications. Proc. Natl. Acad. Sci. 156:1452-1456.
Sobel JD, Muller G, Buckley H (1984) Critical role of germ tube formation in the pathogenesis of candidal vaginitis. Infect. Immun. 44:576-580.
Venditti M, De Bernardis F, Micozzi A, Pontieri E, Chirletti P, Cassone A, Martino P (1992) Fluconazole treatment of catheter-related right-sided endocarditis caused by *Candida albicans* and associated with endophthalmitis and folliculitis. Clin. Infect. Dis. 14:422-426.
Wade JC, Schimpff SC (1985) Epidemiology and prevention of *Candida* infections. in Bodey GP, Fainstein, V (eds). Candidiasis. Raven Press.

Homoserine dehydrogenase as a selective target molecule for antifungal action

Hideyo Yamaguchi
Research Center for Medical Mycology
Teikyo University School of Medicine
359 Otsuka, Hachioji,
Tokyo 192-03
Japan

Introduction

Remarkable progress has been made recently in antifungal chemotherapy with the advent of imidazole-or triazole-containing antifungal drugs. However, their clinical usefulness appears to be limited by insufficiency of the selective toxicity. This is probably ascribable to the fact that both fungal and human cells share the common target molecule for attack of azoles, cytochrome P-450, although the affinity of recently developed triazoles such as itraconazole and fluconazole with this enzyme from fungal cells is much higher than that from host cells (Vanden Bossche *et al.*, 1989). It should be those enzymes which are of vital importance for fungal growth but absent in human cells that provides the potential target for higher selectivity of an antifungal action. In the last decade, our laboratory has been involved, in collaboration with several other research groups, in the search for and preclinical evaluation of new antifungal compounds with a favorable profile of activity and safety. In the course of rather empirical screening of antifungal antibiotics with microbial culture filtrates, we discovered a couple of novel ones considered to be promising candidates or leads for clinically useful drugs. Among them is an antibiotic with aspartate-related structure, (s)2-amino-4-oxo-5-hydroxypentanoic acid (RI-331) (Fig. 1).

$$\begin{array}{ccccccccc} HO - CH_2 & - & C & - & CH_2 & - & CH & - & COOH \\ & & \| & & & & | & & \\ & & O & & & & NH_2 & & \end{array}$$

Figure 1. Structure of (s) 2-amino-4-oxo-5-hydroxypentanoic acid (RI-331).

Intensive studies in our laboratory demonstrate its higher selective toxicity and unique mode of antifungal action than has ever before been reported for any antifungal compound; the target molecule of RI-331 attack is homoserine dehydrogenase as reviewed in this paper. This suggests a novel potential target for selective inhibition of fungal growth and a new approach to a more rational screening of promising antifungal compounds.

NATO ASI Series, Vol. H 69
Molecular Biology and its Application to Medical Mycology
Edited by B. Maresca, G. S. Kobayashi and H. Yamaguchi

In vitro and *in vivo* activities and toxicology of RI-331

Although the *in vitro* antifungal activity of RI-331 does not look particularly potent, it substantially inhibits the growth of *Candida albicans* as well as several other medically important yeast species (Yamaguchi *et al.*, 1988). When the susceptibility testing was performed by a microbroth dilution technique on glucose-supplemented Yeast Nitrogen Base, RI-331 was demonstrated to be more or less active against most of the yeast strains tested; complete inhibition was achieved by RI-331 at concentrations up to 40μg/ml for *C. albicans*, *C. tropicalis*, *C. parapsilosis*, *C. kefyr* and *C. glabrata* as well as for *Saccharomyces cerevisiae*. *C. krusei*, *C. guilliermondii* and *Cryptococcus neoformans* were less susceptible and aspergilli almost insusceptible to the antibiotic. It did not inhibit any Gram-positive and -negative bacteria tested. *In vivo* antifungal activity of RI-331 was demonstrated by experiments in which mice were intravenously infected with a fatal dose of *C. albicans* (2). With this model of systemic candidiasis, twice-a-day oral treatment with RI-331 at dosages of 25 to 100 mg/kg starting on the day of infection and continuing for 14 consecutive days increased dose-dependently the survival rate of infected mice as compared with a placebo (saline) treatment (100% survival versus 0% with 100 mg/kg).

The preliminary studies on the toxicology of RI-331 showed that the lethal dose, orally or intraperitoneally, of the antibiotic for 50% of the male ICR mice was more than 5,000 mg/kg, and that there was no drug-related macroscopic or microscopic pathological finding or behavioral change in male Wistar rats dosed up to 500 mg/kg once daily for 14 consecutive days (Yamaguchi *et al.*,

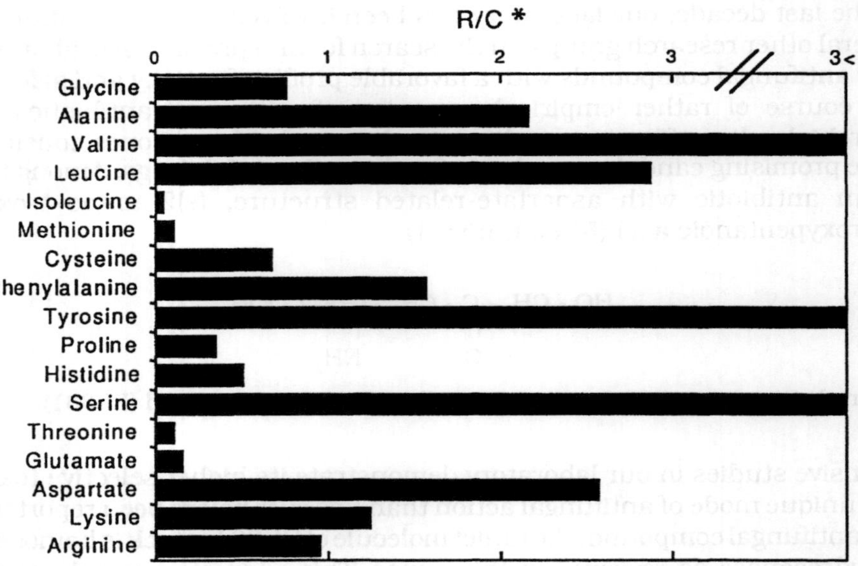

Figure 2. RI-331-induced alteration of composition of amino acid pools in *S. cerevisiae* cells. * represents the amount of intracellular amino acid in RI-331 (15μg/ml)-treated cells relative to that in untreated cells.

1988). It is, therefore, suggested that RI-331 is well tolerated by experimental animals.
This favorable profile of activity and safety of RI-331 tempted us to clarify the biochemical and molecular basis of its antifungal action which confers the antibiotic such a high selective toxicity. For this purpose, experiments were conducted to study the mode of action of the antibiotic using a susceptible wild-type strain, along with several different mutant strains pertaining certain amino acid metabolisms, of *S. cerevisiae*.

Effect of RI-331 on protein synthesis and related cellular metabolism in yeasts

Considering the amino acid-related structure of RI-331, we first compared the effect of this antibiotic on protein synthesis in growing *S. cerevisiae* D273-11A (*ade1, his1*) cells, with the effect on RNA and DNA synthesis, by determination of the radioactivity taken up from the medium containing the specific precursors. In the presence of 150 µg/ml of RI-331, incorporation of [^{14}C] asparagine or [^{14}C] glutamine into cellular proteins was inhibited by 90% or more, while incorporation of [^{3}H] adenine into RNA or DNA was inhibited by approx. 30 to 40% (Yamaki *et al.*, 1988; Yamaguchi *et al.*, 1990). These results strongly suggest that RI-331 preferentially affects protein synthesis in growing cells of susceptible organisms. However, native messenger-directed polypeptide synthesis by cell-free extracts from yeast was completely refractory to the action of RI-331 at a concentration as high as 1,500 µg/ml, indicating that the antibiotic does not directly interfere with the protein-

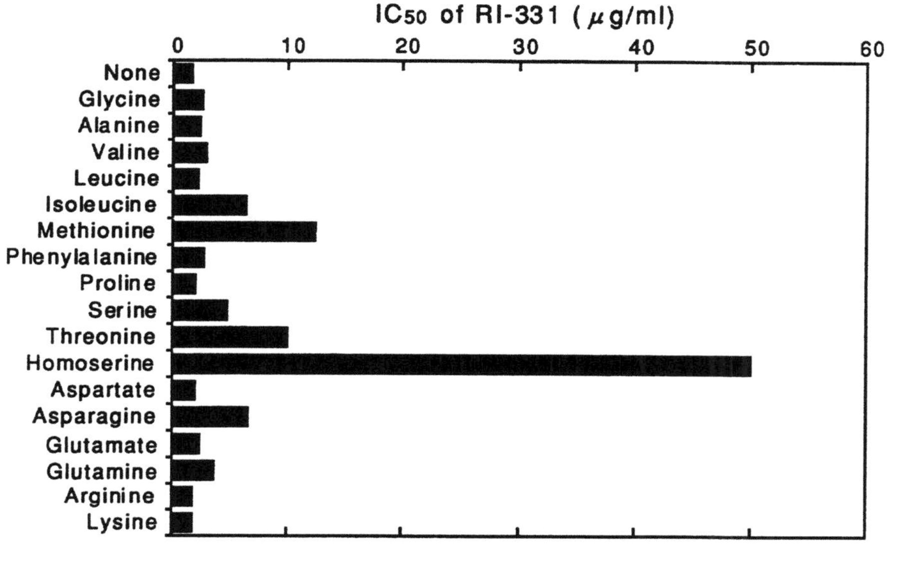

Figure 3. The effect of exogenous amino acids on the growth-inhibitory activity of RI-331 against *S. cerevisiae* cultures.
IC_{50} values were determined turbidimetrically with cultures grown in basal medium without supplementation of any amino acid.

synthesizing machinery itself (Yamaki *et al.*, 1988; Yamaguchi *et al.*, 1990). It was also demonstrated in this study that RI-331 taken up by growing yeast cells from the ambient medium is mostly recovered from the cytosol without any biochemical modification and none is detected in the protein fraction (Yamaguchi *et al.*, 1990).

These results led us to postulate that RI-331 induces exhaustion of some amino acid(s) in the intracellular pool. This possibility was supported by the following two studies. First, the free amino acid fraction was prepared from yeast cells previously incubated for 60 minutes with or without 15 µg/ml of RI-331 and then analyzed chromatographically for the relative amount of each amino acid recovered to compare its level in the intracellular pool of RI-331-treated cells with that of untreated control cells. As shown in Fig. 2, the antibiotic produced a remarkable change in the size and composition of the amino acid pool (Yamaki *et al.*, 1988; Yamaguchi *et al.*, 1990). It was characterized by a profound decrease in the level of isoleucine, methionine, threonine and glutamate, suggesting that the pool of these four amino acids is preferentially exhausted after the antibiotic treatment.

In the second study, various amino acids including most of those mentioned above were examined for their effect on the growth-inhibitory activity of RI-331 against *S. cerevisiae*. The results obtained from this reversal experiment, which was performed in an amino acid-free medium (Yeast Carbon Base plus 1% glucose) with supplementation of 200 µg/ml of testing compound, are illustrated in Fig. 3 (Yamaki *et al.*, 1988; Yamaguchi *et al.*, 1990). The antagonistic activity was expressed by the increase in the value of IC_{50} of RI-331 as compared with that measured in the basal medium. Of 18 tested

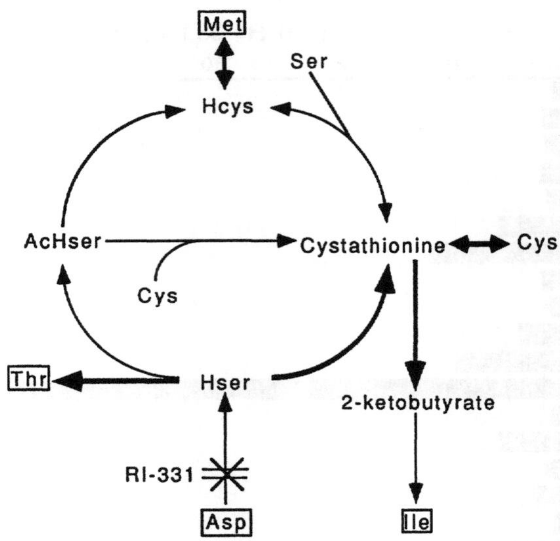

Figure 4. Biochemical pathway involved in metabolism of the aspartate family of amino acids in prototrophic microorganisms including yeasts and assumed target site of RI-331 action.
Abbreviations: AcHser, acetylhomoserine; Asp, aspartate; Cys, cysteine; Hser, homoserine; Ile, isoleucine; Met, methionine; Ser, serine; Thr, threonine.

samples of amino acid or related metabolite, homoserine was most potently antagonistic; it induced a 29-fold increase in the IC_{50} value.
Although to lesser extents, substantial antagonistic action was also exerted by isoleucine, methionine and threonine but was not by glutamate or any other amino acids. It is of interest to note that, as shown in Fig. 4, isoleucine, methionine and threonine are known to share the biosynthetic pathway commonly existing in diverse prototrophic microorganisms which starts from aspartate and proceeds via homoserine and, for this reason, these amino acids are referred as the aspartate family. Considering the most potent antagonistic action of homoserine, it was strongly suggested that the most probable site of attack of RI-331 is on a metabolic step involved in biosynthesis of homoserine (Fig. 4).

Selective inhibition by RI-331 of yeast homoserine dehydrogenase activity

To confirm the validity of this speculation and to identify the target molecule for RI-331 action on the biochemical pathway involved in conversion of aspartate to homoserine, all of the enzymes responsible for the progress of this metabolic process were prepared from yeast to test their susceptibility to RI-331. As shown in Fig.5, there are three steps of biochemical reaction on this pathway sequentially catalyzed by aspartate kinase, aspartate semialdehyde dehydrogenase and homoserine dehydrogenase. Preparation of each of these three enzymes was carried out using two different auxotrophic or block mutant strains of *S. cerevisiae* to avoid any cross-contamination of enzymic activity: both aspartate kinase and homoserine dehydrogenase were purified from a *hom2* mutant strain S2614C lacking aspartate semialdehyde

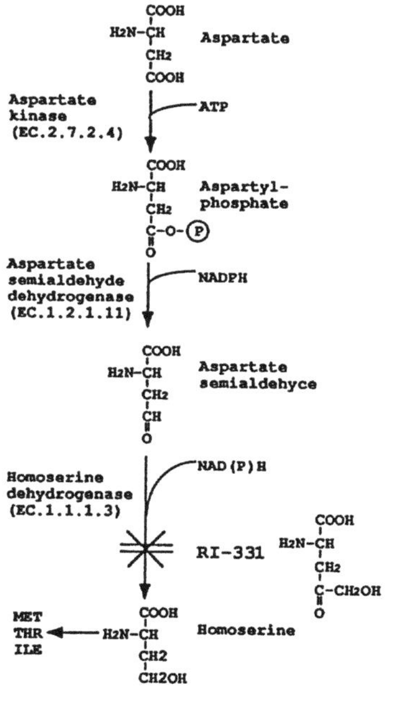

Figure 5. The pathway converting aspartate to homoserine in yeasts and the target site of RI-331 action.

dehydrogenase, and aspartate semialdehyde dehydrogenase was prepared from a *hom6* mutant strain STX25-2A lacking homoserine dehydrogenase. The results of experiments demonstrated that only homoserine dehydrogenase activity was inhibited by RI-331, while the other two enzymic activities were insusceptible (Yamaki *et al.*, 1990).

The enzymological aspect of RI-331 inhibition of homoserine dehydrogenase was studied in more detail. The enzymic activity in the forward reaction was determined by the rate of NADPH dehydrogenation that is dependent on exogenous substrate aspartate semialdehyde (ASA). RI-331 appreciably inhibited this forward reaction and the extent of inhibition increased with an increasing concentration of NADP relative to that of NADPH added to the reaction mixture (Yamaki *et al.*, 1990). Since NADP is not required for the forward reaction by homoserine dehydrogenase, such a high NADP-dependence of the RI-331 action suggests the possibility that NADP may form a complex with the enzyme that is more highly susceptible to the antibiotic. Figure 6A shows kinetics with Lineweaver-Burk plots of RI-331 inhibition of the enzymic activity in the forward reaction. The inhibition was a mixed competitive and noncompetitive one for substrate ASA, with a Ki value of 2 mM versus a Km value of 0.05 mM in the presence of 0.2 mM of NADP. It is also demonstrated that the reaction is inhibited by ASA itself if the concentration of this substrate is higher than 0.04 mM. Therefore, it appears

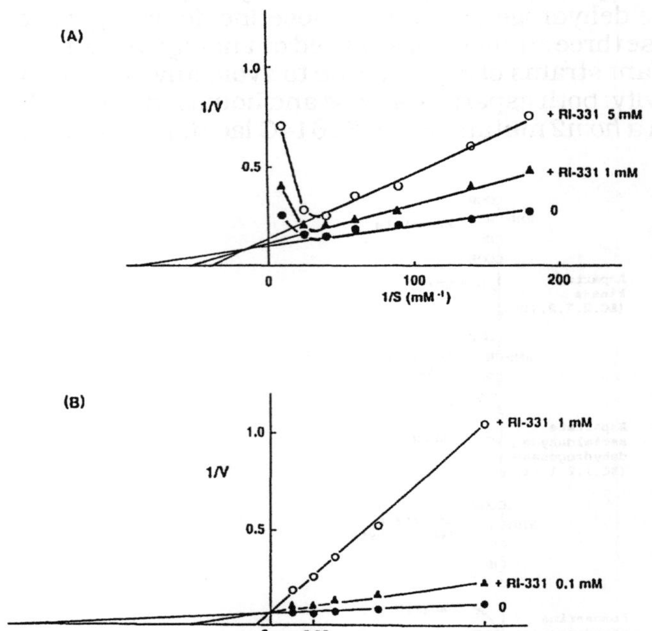

Figure 6. Lineweaver-Burk plots with respect to substrate of RI-331 inhibition of homoserine dehydrogenase in the forward reaction (A) and the reverse reaction (B). A: substrate, aspartate semialdehyde (with 0.2 mM NADP); 1/v, $mmol^{-1} \cdot 30$ min·mg. B: substrate, homoserine; 1/v, $\mu mol^{-1} \cdot 10$ min·mg

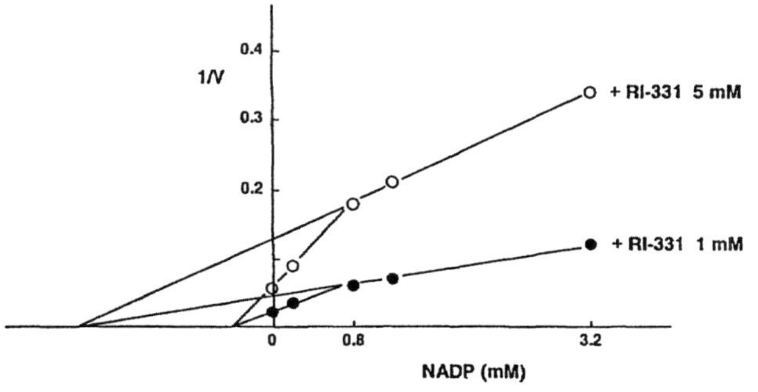

Figure 7. Dixon-plot with respect to NADP of RI-331 inhibition of homoserine dehydrogenase in the forward reaction. Initial velocity of the reaction was plotted versus varying concentrations of NADP in the presence of 1 and 5 mM of RI-331.

likely that RI-331 inhibition of the enzymic activity in the forward reaction is partly due to a high level of accumulation of ASA in the presence of the antibiotic. As given by Dixon-plot analysis (Fig. 7), RI-331 inhibition of this forward reaction is noncompetitive with respect to NADP, and the association constant for NADP in the presence of the antibiotic reveals two different values, 0.4 and 2.0 mM (Yamaki *et al.*, 1992). This would mean that the presumed enzyme-NADP complex would interact with RI-331 to form two types of enzyme-NADP-antibiotic complexes which differ in their metabolic stability.

As shown in Figure 6B, the reverse reaction catalyzed by homoserine dehydrogenase was inhibited by RI-331 to a greater extent that the forward

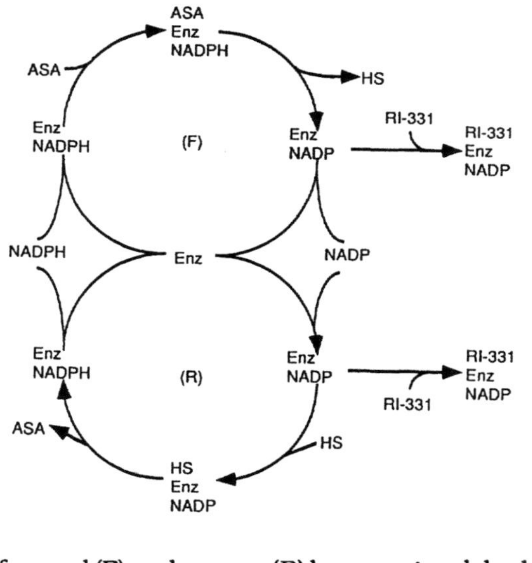

Figure 8. Scheme of forward (F) and reverse (R) homoserine dehydrogenase reactions, with assumed interactions of RI-331 with enzyme-substrate complexes. Abbreviations: ASA, aspartate semialdehyde; Enz, homoserine dehydrogenase; HS, homoserine.

reaction. The inhibition was competitive with respect to substrate homoserine, with a Ki value of 0.025 mM versus a Km value of 17 mM (Yamaki *et al.*, 1992). This and the preceding data indicate that, in the presence of 0.2 mM of NADP (almost equivalent to a physiological concentration of this coenzyme in yeast cells), the enzymic activity for the reverse reaction is at least 10 times more susceptible to RI-331 than that for the forward reaction. Moreover, when the enzyme was preincubated with NADP before starting the reaction by addition of an appropriate substrate, an increase in the extent of RI-331 inhibition of the enzymic activity was attained only for the reverse reaction.
Based on all of these experimental data, we speculated as illustrated in Fig. 8 that the enzyme-NADP complex is formed first and then interacts with RI-331, eventually leading to a formation of enzyme-NADP-antibiotic complex with the resulting inactivation of the enzymic activity. The binding of NADP to homoserine dehydrogenase could induce a conformational change of the enzyme molecule which facilitates its interaction with the antibiotic.

Perspective of RI-331 in antifugal chemotherapy

The high selectivity of RI-331, in terms of profile of activity and safety, can be accounted for by the specific target molecule of its attack, homoserine dehydrogenase. This enzyme which is involved in biosynthesis of the aspartate family amino acids is present in fungal cells but not in host human or animal cells. This implies that fungal homoserine dehydrogenase should be chosen as a target of selective antifungal action. However, so far as RI-331 is concerned, its clinical usefulness may be hampered by a relatively low antifungal activity and narrow spectrum. Tsuchiya recently demonstrated that homoserine dehydrogenase prepared from *Escherichia coli* is highly susceptible to RI-331, despite the antibiotic being virtually without effect on growth of this Gram-negative rod (Tsuchiya T, personal communication). In addition, our preliminary experiments provided data showing that RI-331 is effective in inhibiting conversion of aspartate to homoserine by subcellular extracts prepared from aspergilli and zygomycetes. Presumably the insusceptibility of bacteria or filamentous fungi to the growth-inhibitory activity of RI-331 may be due to difficulty in the access of the antibiotic to the target molecule existing in these microbial cells. Our current interest is a chemical modification of RI-331 so that it gives derivatives with increased influx to fungal cells.
Knowledge of mode of action, particularly that on the molecular level, should enable existing or newly discovered antifungal agents to be exploited more effectively than in the past. Moreover, if novel target molecules like homoserine dehydrogenase for the selective inhibition of fungal growth were to be found, they would open up increasing avenues in the search for and development of antifungal drugs of high clinical usefulness.

References

Vanden Bossche H, Marichal P, Gorrens J, Coene M-C, Willemsens G, Bellens D, Roels I, Moereels H, Janssen PAJ (1989) Biochemical approaches to selective antifungal activity. Focus on azole antifungals. Mycoses 32 (Suppl 1): 35-52.

Yamaguchi H, Uchida K, Hiratani T, Nagate T, Watanabe N, Omura S (1988) RI-331, a new antifungal antibiotic. Ann. N.Y. Acad. Sci. 544: 188-190.

Yamaki H, Yamaguchi M, Nishimura T, Shinoda T, Yamaguchi H (1988) Unique mechanism of action of an antifungal antibiotic RI-331. Drugs Expl. Clin. Res. 14: 467-472.

Yamaguchi M, Yamaki H, Shinoda T, Tago Y, Suzuki H, Nishimura T, Yamaguchi H (1990) The mode of antifungal action of (s) 2-amino-4-oxo-5-hydroxypentanoic acid, RI-331 J. Antibiotics 43: 411-416.

Yamaki H, Yamaguchi M, Imamura H, Suzuki H, Nishimura T, Saito H, Yamaguchi H (1990) The mechanism of antifungal action of (s)2-amino-4-oxo-5-hydroxypentanoic acid, RI-331 ; The inhibition of homoserine dehydrogenase in *Saccharomyces cerevisiae* (1990) Biochem. Biophys. Res. Commun. 168: 837-843.

Yamaki H, Yamaguchi M, Tsuruo T, Yamaguchi H (1992) Mechanism of action of an antifungal antibiotic, RI-331 (s) 2-amino-4-oxo-5-hydroxypentanoic acid; Kinetics of inactivation of homoserine dehydrogenase from *Saccharomyces cerevisiae*. J. Antibiotics 45: 750-755.

Molecular mechanisms of antifungal activity and fungal resistence: focus on inhibitors of ergosterol biosynthesis

H. Vanden Bossche, P. Marichal and H. Moereels[1]
Departments of Comparative Biochemistry and
[1]Theoretical Medicinal Chemistry
Janssen Research Foundation
Turnhoutseweg, 30
B-2340 Beerse
Belgium

Introduction

Metabolic differences between hosts and their fungal pathogens provide scientists, interested in the development of antifungal agents, with selective targets for chemotherapeutic attack. There are many ways to select targets (for a review see Kerridge and Vanden Bossche 1990). A survey can be made of targets that are absent from host cells. The best examples are enzymes involved in the synthesis and hydrolysis of chitin, a structural polysaccharide absent from mammals. An other example is the β-1,3-glucan synthase responsible for the synthesis of the fungal cell wall β-glucan. Such targets are not only involved in the synthesis of cell wall components, they are also present in pathways of intermediary metabolism. An example is the *S*-adenosyl-L-methionine-Δ^{24}-sterol methyltransferase. This mitochondrial enzyme is involved in the 24-alkylation of lanosterol or zymosterol, a step not present in mammalian cells.
The target may be present in both host and fungal cells but is inaccessible in the host cell. An example of a drug interacting with such a target is the antibiotic griseofulvin. Griseofulvin causes major growth abnormalities in sensitive fungal cells, probably by interacting with tubulin (Langcake *et al.*, 1983). It interacts with tubulins from a variety of sources including human cells but it enters sensitive cells only. Indeed, dermathophytes possess a prolonged energy-dependent transport system for griseofulvin, in insensitive cells this is replaced by a short energy-independent transport system (Polak, 1990).
Another possibility to avoid toxicity to the host cells is to transport into the fungal cell a drug that is metabolised to an active compound by an enzyme that is absent or has low activity in the host. The only well studied example is 5-fluorocytosine (5-FC). 5-FC is taken up by a cytosine permease and inside the fungal cell rapidly deaminated to 5-fluorouracil (5-FU) a well-known cytotoxic agent. Selectivity results from the fact that mammalian cells lack the cytosine deaminase responsible for the conversion of 5-FC into 5-FU (Polak, 1990). This knowledge was gathered from mode of action studies. Indeed, the study of the mechanism of action of antifungal agents may

NATO ASI Series, Vol. H 69
Molecular Biology and its Application to Medical Mycology
Edited by B. Maresca, G. S. Kobayashi and H. Yamaguchi

highlight differences between fungi and their host and thus improve our knowledge of the biochemical pathways in both fungus and host.
Studies on the mode of action of amphotericin B and other polyene macrolide antibiotics have emphasised the differences between the plasma membranes of mammals and their fungal pathogens. The selectivity of the polyenes is partly determined by the nature of the sterols present in the membranes; instead of cholesterol the 24-alkylated sterol ergosterol is the main sterol in most pathogenic fungi. Differences in the enzymes involved in the biosynthesis of ergosterol and cholesterol have been elucidated by the use of ergosterol biosynthesis inhibitors (EBI). EBIs constitute the most important group of compounds developed for the control of fungal diseases.

Ergosterol biosynthesis inhibitors

Currently, representatives of three classes of EBI antifungals are available:
- squalene epoxidase inhibitors
- 14α-demethylase inhibitors
- Δ^{14}-reductase and Δ^{8-7}-isomerase inhibitors

A simplified pathway of sterolsynthesis in fungal and mammalian cells is shown in Fig. 1.

Squalene epoxidase inhibitors

Squalene epoxidase (squalene monooxygenase) catalyses the 2,3-oxidation of squalene, an essential step for the subsequent cyclization to lanosterol. The squalene epoxidases from both mammalian and yeast sources are membrane bound, require oxygen , FAD and a reduced pyridine nucleotide. They are not inhibited by carbon monoxide and other cytochrome P450 inhibitors (Ryder, 1990, 1991).
The rat liver squalene epoxidase has been studied in detail. The purified enzyme is a single polypeptide with a molecular weight of 51 kd. The enzymatic reaction requires next to oxygen, NADPH, components of the cytosol and a flavoprotein which is identical with the NADPH-cytochrome P450 reductase (Ono *et al.*, 1977, 1982). The requirement of cytosolic components could be satisfied by phosphatidylglycerol or an other anionic phospholipid and a supernatant protein factor which appeared to facilitate transfer of squalene between membranes and its uptake into the endoplasmic reticulum (Ferguson and Bloch 1977, Ryder, 1990).
The study of the squalene epoxidase in *Saccharomyces cerevisiae, Candida albicans* and *C. parapsilosis* was triggered by the fact that the allylamine naftifine, used for topical therapy of superficial mycoses, was found to inhibit ergosterol synthesis by acting on the squalene epoxidase (Paltauf *et al.*, 1982). In contrast with the squalene epoxidase of *S. cerevisiae* and rat liver, that of *C. albicans* prefers NADH instead of NADPH as cofactor, furthermore the yeast enzymes seem to be independent of cytosolic fractions (Ryder 1990). An intensive derivatization programme based on naftifine led to the synthesis of the orally active antimycotic terbinafine (Ryder 1990). Both allylamines are reversible non-competitive inhibitors of the microsomal squalene epoxidase of *C. albicans* with Ki-values of 1.1 μM and 0.03 μM, respectively (Ryder,

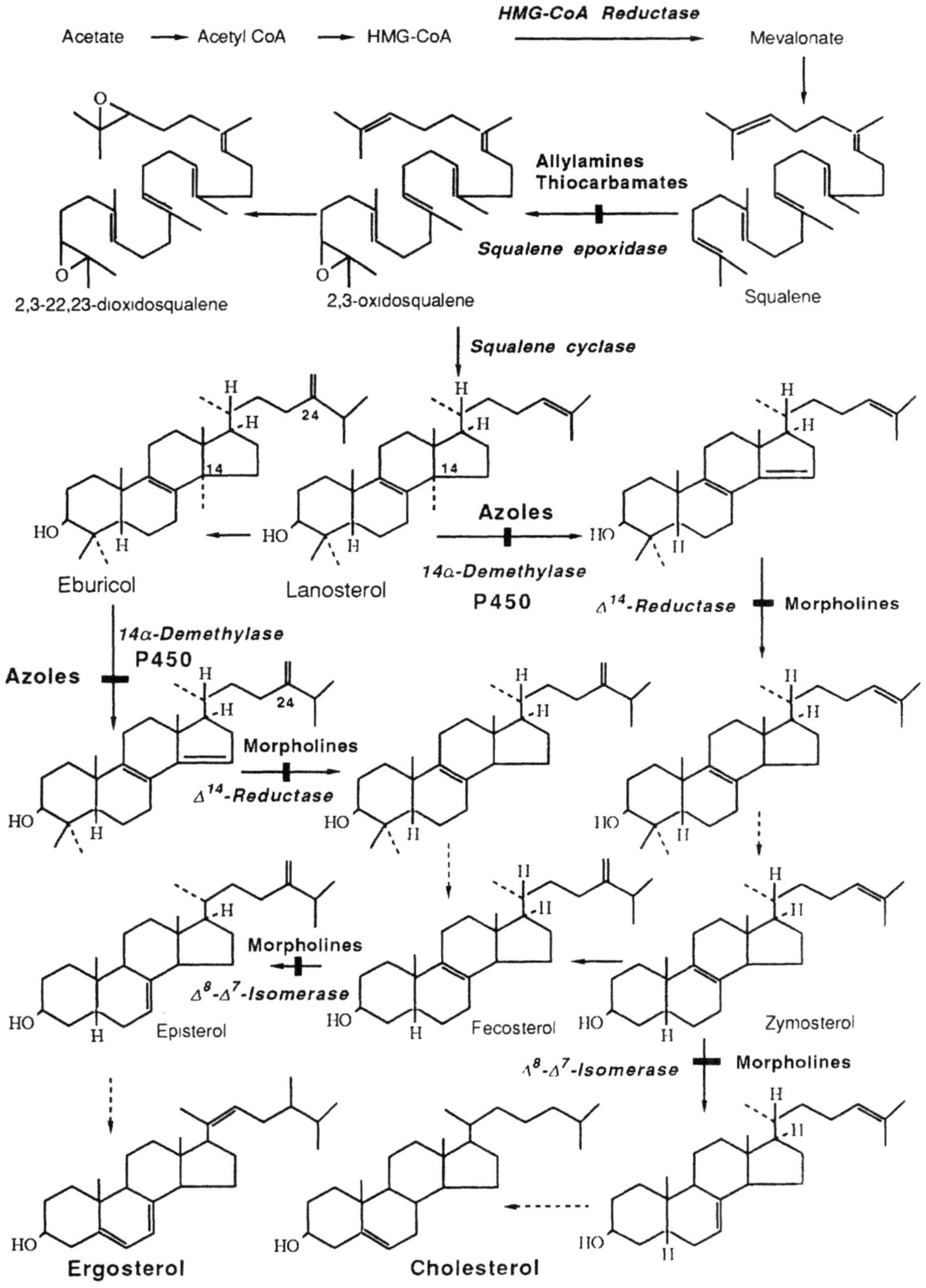

----------> multistep reaction

Figure 1. Ergosterol and cholesterol biosynthetic pathways showing the steps inhibited by antifungal agents.

1990). In contrast, terbinafine is a weak competitive inhibitor of the rat liver enzyme with a Ki of 77 μM, 50 % inhibition (IC_{50}-value) is reached at 93 μM) (Ryder, 1990, 1991). Ryder, 1987, 1988a, 1991) compared the inhibitory effects of terbinafine on rat and guinea pig liver microsomal squalene epoxidations, the latter was found to be about 23 times more sensitive. This indicates a significant degree of variation among mammalian epoxidases. Another allylamine derivative, NB-598 {(E) *N*-ethyl-*N*-(6,6-dimethyl-2-hepten-4-ynyl)-[(3,3'-bithiophen-5-yl) methoxy] benzenemethamine} was found to inhibit microsomal squalene epoxidase from Hep G2 cells in a competitive manner with a Ki of 0.68 nM (IC_{50}-value=0.75 nM) (Horie *et al.*, 1990). NB-598 also inhibits microsomal squalene epoxidase from dog liver (IC_{50}-value=2 nM) but is devoid of antifungal activity against *Trichophyton mentagrophytes* and *C.albicans* (Horie *et al.*, 1990). The investigators take this as a measure for selectivity. However, it is possible that NB-598 does not reach the target in these species.

The precise mechanism of selectivity of the allylamines is still not clear but may reside in intrinsic differences in the respective epoxidase enzymes. Recently, Ryder speculated that the high affinity of terbinafine for the fungal enzyme may result from binding of the molecule to two separate sites on the enzyme, namely the squalene-binding site (by its naphthalene ring?) and an adjacent lipophilic pocket (by its side-chain ?) (Ryder, 1991). However, as long as the amino acid sequences of the fungal and mammalian enzymes are not known it is hard to speculate on possible binding places.

The mechanism of antifungal action of the allylamines and the thiocarbamates (tolciclate and tolnaftate) has been related to their inhibitory effects on the squalene epoxidase (Barrett-Bee *et al.*, 1986, Ryder, 1988b). In dermatophytes there is a clear correlation between inhibition of the squalene epoxidase and inhibition of growth. It is well-known that filamentous fungi are particularly susceptible to ergosterol biosynthesis inhibition. For example, the filamentous form of *C. albicans* is more susceptible to allylamines than the yeast form. Mycelium formation was fully suppressed at a drug concentration causing only 80 % inhibition of ergosterol synthesis (Ryder 1987). However, although the enzymes of *C. albicans* (yeast form) and *C. parapsilosis* are almost equisensitive to terbinafine, the latter species is about 8 times more susceptible to growth inhibibition. This difference between sensitivity of ergosterol synthesis and growth is even more pronounced when *C. albicans* is compared with *C. glabrata*. Ergosterol synthesis in *C. glabrata* is about 5 times less sensitive whereas the minimum inhibitory concentration (MIC) is 33 times higher. Of course it is possible that these *Candida* species and dermatophytes differ in their sensitivity to the accumulating squalene. However, so far there is no evidence for fungitoxic effects of squalene. On the contrary, a number of studies suggest the opposite.

- Microgram amounts of squalene have been found in *Phytophthora cactorum* a fungus unable to epoxidize and cyclisize squalene to lanosterol (Gottlieb *et al.*, 1979, Nes *et al.*, 1982). It has been hypothesized that in *Phytophthora* squalene itself may be assuming the architectural role of sterols, allowing vegetative growth (Margalith 1986).
- Almost 24% of the total radioactivity measured in the lipid extracts from *T. mentagrophytes* incubated in the presence of [^{14}C]-acetate is found in the squalene fraction, this is only two times less than the radioactivity incorporated into ergosterol. The presence of squalene did not prevent growth of this dermatophyte (Vanden Bossche *et al.*, 1990a).

- The addition of increasing concentrations of squalene to *T. mentagrophytes* cultures resulted, after 48 h of growth, in an increased intracellular squalene and ergosterol content. At a squalene concentration of 12 mg per 100 ml medium the intracellular squalene and ergosterol contents were as high as 0.775 mg/100 ml cells and 0.113 mg/100 ml cells (unpublished results). The squalene/ergosterol ratio of 6.8 did not affect growth, on the contrary an increased fungal outgrowth is noted. A similar high squalene/ergosterol ratio was also found by Ryder in *T. mentagrophytes* incubated in the presence of terbinafine (Ryder, 1988b).
- Differential scanning calorimetry (DSC) of multilamellar vesicles of dipalmitoylphosphatidylcholine (DPPC) containing 10 to 35 mole% squalene did not show any effect on the transition temperature or the enthalpy of melting (Vanden Bossche and Marichal 1991). Thus squalene differs from its cyclic product lanosterol that even at 15 mole % decreased the enthalpy of melting by about 60 % (Vanden Bossche *et al.*, 1984a).
- Squalene has been reported to be a safe substance in feeding experiments and in some animals it is the predominant lipid in the liver (Horie *et al.*, 1990).

All these results suggest that the antifungal activity of the allylamines may originate from the decreased availability of ergosterol rather than from an accumulation of squalene. More refined studies are needed to clarify the fungitoxic consequences of squalene epoxidase inhibition.

A possible approach is the use of resistant mutants. Among 150 terbinafine-resistant laboratory isolates of *Ustilago maydis* (a fungal pathogen of maize) two were studied in detail (Orth and Sisler, 1990, Orth *et al.*, 1990). Terbinafine is a potent inhibitor of both growth (IC_{50}-value=19 ng/ml) and ergosterol synthesis (IC_{50}-value=5 ng/ml) in wild-type *U. maydis* sporidia (Orth *et al.*, 1990). Squalene epoxidase activity in cell-free extracts showed at 0.5 μg/ml 50 % inhibition of the formation of 2,3-oxidosqualene + 2,3-22,23-dioxidosqualene + lanosterol. The first mutant studied (AR 217) showed IC_{50}-values for growth and ergosterol biosynthesis at 0.93 μg/ml and 0.158 μg/ml, respectively; those found for the second mutant (AR 212) are 0.135 μg/ml and 28 ng/ml. The basis of resistance in AR 217 appears to be a target site change leading to reduced affinity for the fungicide (Orth *et al.*, 1990). AR 212 has a constitutively higher squalene epoxidase activity than the wild type which would explain the resistance of this isolate (Orth *et al.*, 1990). In both resistant mutants eburicol was found after terbinafine treatment. According to Orth *et al.*, the presence of this 14-methylsterol indicates a possible secondary site of action of terbinafine at the point of 14α-demethylation (Orth *et al.*, 1990). These studies do not eliminate the possibility that the accumulation of squalene is involved in the fungitoxicity of the allylamines. However, they focussed attention on an effect that may be secondary to the partial inhibition of the squalene epoxidase in for example *T. mentagrophytes* i.e. oxygen radical formation and lipid peroxidation. As already mentioned above, the epoxidation of squalene requires a reductase that, at least in rat liver, is identical with the NADPH-cytochrome P450 reductase. An inhibition of squalene epoxidase makes the reductase available for e.g. the transfer of an electron to ubiquinone to form semiquinone that reacts with oxygen to form superoxide (Sisler and Buchman-Orth, 1990). Superoxide can be converted to hydrogen peroxide and oxygen by the superoxide dismutase and the resulting hydrogen peroxide may be reduced by superoxide to hydroxyl radicals which can abstract H-atoms from polyunsaturated fatty acids (PUFAs) to form PUFA radicals. The latter can react with oxygen to form lipid peroxides leading to fungitoxicity (Sisler and Buchman-Orth, 1990). This, and the possible oxygen radical generation in mammalian cells should be investigated more thoroughly.

14α-demethylase inhibitors

Since the synthesis of miconazole, econazole and clotrimazole in the sixties a great number of imidazole (e.g. ketoconazole, bifonazole, tioconazole) and triazole (e.g. terconazole, itraconazole, fluconazole, saperconazole, ICI 195,739) antifungals became available. Some of them are used topically, other both topically and orally. All these azole antifungals belong to the class of 14α-demethylase inhibitors (Barrett-Bee *et al.*, 1988, Vanden Bossche, 1985, 1988, 1991, Vanden Bossche and Marichal, 1991, Vanden Bossche *et al.*, 1978, 1980, 1984a, 1987a,b, 1988a,b, 1990a,b). Their antifungal activity originates from binding to a cytochrome P450 ($P450_{14DM}$) (Yoshida, 1988) involved in the 14α-demethylation of lanosterol in *S. cerevisiae* (Vanden Bossche *et al* 1984b, Yoshida, 1988, Yoshida and Aoyama 1991) and *C. glabrata* (Vanden Bossche *et al.*, 1990b) or of eburicol (24-methylenedihydrolanosterol) in filamentous fungi (Vanden Bossche *et al.*, 1988b, 1990a), the yeast form of *Histoplasma capsulatum* (Vanden Bossche *et al.*, 1990b), *Cryptococcus neoformans* (unpublished results) and in a number of *C. albicans* isolates (Barrett-Bee *et al.*, 1988, Fryberg *et al.*, 1975, Vanden Bossche *et al.*, 1978) (Fig. 1). Affinity of some azole antifungals agents for $P450_{14DM}$ is very high. For example, binding studies with ketoconazole showed an apparant Kd of less than 10 nM (Yoshida and Aoyama, 1991). By using a reconstituted system consisting of $P450_{14DM}$ and NADPH-cytochrome P450 reductase (both isolated from *S. cerevisiae*) Yoshida and Aoyama found complete inhibition of the demethylase when ketoconazole was added at a concentration equal to the $P450_{14DM}$ content (Yoshida and Aoyama, 1986). This suggests that ketoconazole inhibits the 14α-demethylation by forming a stoichiometric complex with $P450_{14DM}$.

$P450_{14DM}$ (product of the *CYP 51* gene), purified to homogenity from rat liver (Trzaskos *et.al.* 1986) and *S. cerevisiae* (Yoshida, 1988, Yoshida and Aoyama, 1991) microsomes catalyzes three oxidative steps:

1. the hydroxylation of the C-32-methyl (14α-methyl) group of lanosterol.
2. 32-Hydroxylanosterol is converted to the 32-gem-diol derivative which may be in equilibrium with the C-32 aldehyde, 32-oxolanosta-$\Delta^{8,24}$-dien-3ß-ol (3ß-hydroxylanost-8-32-al)
3. The C-32 of the second intermediate is attacked by the 3rd $P450_{14DM}$-activated oxygen.

The C-C bond between C-14 and C-32 is cleaved and C-32 is eliminated together with hydrogen at C-15 as formic acid and a double bond is formed between C-14 and C-15. In *S. cerevisiae* and liver 4,4-dimethyl-$\Delta^{8,14,24}$-cholestatrienol is formed. In most fungi the end product of the three monooxygenations is 4,4-dimethyl-$\Delta^{8,14,24(28)}$-ergostatrienol.

Recently, Fischer *et al* isolated and characterized a third oxidative demethylation intermediate generated during the lanosterol demethylation reaction cycle in rat hepatic microsomes (Fischer *et al.*, 1991). 14α-Formyloxylanost-8-3β-ol is formed from 3β-hydroxylanost-8-32-al and is further metabolised with the formation of formic acid and 4,4-dimethyl-$\Delta^{8,14,24}$-cholestatrienol.

The amino acid sequences of the $P450_{14DM}$ from *C. albicans* (Lai and Kirsch 1989), *C. tropicalis* (Chen *et al.*, 1988) and *S. cerevisiae* (Kalb *et al.*, 1987) are known. Using a VGAP (alignment with a **v**ariable **gap** penalty) alignment programme (Moereels *et al.*, 1990) the *C. albicans* and *C. tropicalis* $P450_{14DM}$ share with $P450_{14DM}$ of *S. cerevisiae* 64.2 and 65.2% identical amino acids (Vanden Bossche *et al.*, 1989). $P450_{14DM}$ from *C. albicans* and *C. tropicalis* share 83% identical amino acids (Vanden Bossche *et al.*, 1989). The amino acid sequence of the $P450_{14DM}$ from liver is still not published. Since liver and *S. cerevisiae* $P450_{14DM}$ use lanosterol as substrate it would be of interest to see whether their sequences share more identical amino acids than when compared with $P450_{14DM}$s using eburicol as substrate.

So far the only cytochrome P450 species whose three-dimensional structure has been resolved is the soluble P450 ($P450_{cam}$) from *Pseudomonas putida* responsible for the 5-exo hydroxylation of the monoterpene *d*-camphor (Poulos *et al.*, 1987).

The studies of Poulos *et al.*, show that in $P450_{cam}$ cysteine-357 provides the axial sulfur ligand to Fe of the heme. $P450_{cam}$ divides into a helix-rich domain and a helix-poor domain containing a majority of antiparallel β-pairs (Poulos *et al.*, 1987). The majority of protein -substrate contact points in the helical domain occur within the distal helix, helix I, (residues 244 to 252) and in the helix-poor one with the inner strand of a β3 segment (residues 295 to 299). The central region of helix I intimately contacts the heme and camphor, it also contacts an oxygen molecule in the $P450_{cam}$-oxy complex. This helix I consists of 34 residues, all of which display normal helical hydrogen-bonding geometry with the exception of residues 248 (glycine) to 252 (threonine). Between these residues a local deformation of the helix results in a widening of the helix and places the side-chain hydroxyl group of threonine-252 in a position to donate a hydrogen bond to the carbonyl oxygen atom of glycine-248. This creates a binding pocket for oxygen between helix I and camphor. As shown by Poulos *et al.*, eukaryotic P450 species exhibit good sequence homology near the proposed oxygen-binding region of helix I (Poulos *et al.*, 1987). Examples of alignments of distal helices in the vicinity of the oxygen binding site are shown in Table 1. Poulos *et al.*, conclude that the location of

Table 1.
Alignment of amino acid sequences assignable to distal helices of $P450_{cam}$ and eukaryotic cytochromes P450

$P450_{cam}$	243	GLLLVGGLD	T	252	VVNFLSF	(Unger *et al.* 1986)
$P450_{14DM}$ S. c.	309	IGVLMGGQH	T	318	SAATSAW	(Kalb *et al.* 1987)
$P450_{14DM}$ SG1	309	IDVLMGGQH	T	318	SAATSAW	(Ishida *et al.* 1988)
$P450_{14DM}$ C. a.	302	IGILMGGQH	T	311	SASTSAW	(Lai and Kirsch 1989)
$P450_{14DM}$ C. t.	302	IGVLMGGQH	T	311	SASTSAW	(Chen *et al.* 1988)
$P450_{arom}$ h	301	LEMLIAAPD	T	310	MSVSLFF	(Harada 1988)
$P450_{17\alpha}$ h	297	GDIFGAGVE	T	306	TTSVVKW	(Chung *et al.* 1987)

S.c. = *S. cerevisiae*; C.a. = *C. albicans*; C.t .= *C. tropicalis*; h = human, arom = aromatase; 17α = 17α-hydroxylase the distal helix and the oxygen pocket are common structural features found in P450 species (Poulos *et al.*, 1987).

Yoshida *et al.*, purified a P450 ($P450_{SG1}$) from a nystatin resistant mutant, *S. cerevisiae* SG_1 (ATCC 46786, [*mata, his*-1, *erg*11]), which is defective in lanosterol 14α-demethylation (Yoshida *et al.*, 1985). The lack of 14α-demethylase activity was proven by analysing the sterols formed by SG_1 grown in PYG-medium (containing 10g polypeptone, 10g yeast extract and 40g glucose per litre). Only a small amount (1.4%) of a 14-desmethylsterol, $\Delta^{7,22}$ergostadienol, was detected, all the other sterols were 14-methylsterols i.e. lanosterol (71%), eburicol (4%), obtusifoliol (8%), 14-methylfecosterol (2%) and an unidentified 14-methylsterol (12.6%) (Vanden Bossche *et al.*, 1991).

A single nucleotide change resulting in substitution of the glycine-310 (G-310) residue of $P450_{14DM}$ by an aspartic acid (D) residue was found to have occured in $P450_{SG1}$ (Table 1; Ishida *et al.*, 1988). In this protein the 6th ligand to the heme iron is a histidine residue (Ishida *et al.*, 1988) instead of the hydroxyl group of water or of a serine, tyrosine or a threonine residue, which are the most likely candidates for the 6th ligand in normal P450s. Alignment (Table 1) of the primary structure of $P450_{SG1}$ with the $P450_{cam}$ and other P450 sequences suggests that glycine-310 may locate at the distal helix (Ishida *et al.*, 1988). One histidine residue (H-317) is found near the mutation point (G-310). Therefore, Yoshida and Aoyama assume that this histidine comes to interact with the heme iron and that this interaction may be caused by the conformational change induced by the glycine to aspartic acid replacement (Yoshida and Ayoama, 1991).

From these studies it can be deduced that the region containing the conserved threonine is involved in ligand binding. This is further supported by the results of molecular modelling of androstenedione with the heme and the corresponding distal helix of $P450_{arom}$ and site-directed mutagenesis in this region (Graham-Lorence *et al.*, 1991). Changing glutamic acid-302 (E) to alanine (A), valine (V) or glutamine (Q) resulted in proteins devoid of aromatase activity. Changing E-302 to aspartic acid (D) or mutating proline-308 (P) to valine reduced activity to about one-third of the wild-type. Similar studies on $P450_{14DM}$ would provide important information on the active site and azole-binding place(s).

The $P450_{SG1}$ not only lost its catalytic activity but also did not interact with the pyridyl antifungal, buthiobate (Aoyama *et al.*, 1983). Hence, the field strength of the native sixth ligand (histidine) is stronger than that of pyridine. This *S. cerevisiae* mutant is also less sensitive to ketoconazole and itraconazole. Fifty % inhibition of growth (24h cultures) was reached with 2 μM of ketoconazole, whereas, with the parent strain (D587), this inhibition was already achieved at 7 nM. Itraconazole did not affect growth of SG_1 at concentrations up to 5 μM. Fifty % inhibition of the growth of strain D587 was achieved at 1.9 nM (Vanden Bossche *et al.*, 1990c).

In contrast with the pyridine derivative, ketoconazole and itraconazole can replace histidine from their binding place i.e. the 6th coordination position of the heme iron. It has been suggested that nitrogen heterocycles with large hydrophobic *N*-1 substituents not only bind to the heme iron but also interact with the apoprotein (Yoshida, 1988). Spectrophotometric studies suggest that the affinity of both azole antifungals for the apoprotein of this P450 is much lower than that for the apoprotein of the parent and other azole

Table 2.
Sterols present in lipid extracts from fungal cells after incubation of intact cells with itraconazole*

Species	C.a.	C.g.	Cr.n.	H.c.	P.o.	T.m.	A.f.
Medium	CYG	CYG	PYG 1	GY	Dix	PYG 4	Sab
Incubation time	24h	24h	16h	48h	48h	48h	16h
Itraconazole	30 nM	10 μM	30 nM	10 nM	0.3 μM	1 μM	1 μM
Sterols present†							
I. *Desmethyl - Sterols*							
ergosterol	0	14	18	7	0	14	2
brassicasterol	0	0	0	1	0	0	0
fungisterol+22-dihydro-ergosterol	0	0	2	0	0	0	0
II. *14-Methylsterols*							
3,6-diol	71	57	4	4	0	26	21
tetran	1	1	0	0	0	4	0
14-MF	7	0	0	0	0	2	10
4,14-dimethyl-zymosterol	0	6	0	0	0	0	0
obtusifoliol	8	0	6	2	7	0	8
eburicol	4	0	31	45	93	37	47
lanosterol	8	21	0	2	0	8	12
III. *3-Ketosteroids*							
14-MFone	0	0	1	0	0	0	0
obtusifolione	0	0	31	39	0	0	0
IV. *Precursor*							
squalene	0	1	7	0	0	9	0
IC_{50}-values (nM)	5	22	11	3	173	120	90

*Itraconazole concentrations were those giving more than 80% inhibition of ergosterol synthesis. Results are expressed as % of radioactivity incorporated from [14C]-acetate into squalene + sterols + 3-ketosteroids. C.a.= *C. albicans* (yeast form), C.g.= *C. glabrata*, Cr.n.= *Cryptococcus neoformans*, H.c.= *Histoplasma capsulatum*, P.o.= *Pityrosporum ovale*, T.m.= *T. mentagrophytes*, A.f.=*Aspergillus fumigatus*. Except for T. m. (30°C), all cells were incubated at 37°C. †3,6-diol= 14α-methyl-ergosta-8,24(28)-dien-3β, 6α-diol, tetraen= 14α-methyl-ergosta-5,7,22,24(28)-tetraen-3β-ol, 14-MF= 14α-methylfecosterol, 14-MFone= 14α-methylfecosterone. Sterols, 3-ketosteroids and squalene were separated by HPLC. IC_{50}-value= concentration needed to reach 50 % inhibition of ergosterol synthesis.
CYG= casein hydroxylate: yeast extract: glucose (5:5:5 g/l), PYG= polypeptone: yeast extract: glucose (PYG 1= 10:10:10 g/l; PYG 4= 10:10:40 g/l), GY= glucose: yeast extract: cysteine (20:10:0.3 g/l), Dix= malt extract: ox gall: Tween 40: glycerol (40:20 g/l: 10:25 ml/l, Sab= neopeptone: dextrose (10:20 g/l)

sensitive fungi (Vanden Bossche *et al.*, 1990c). This is in accord with the idea that the conformation of the apoprotein of $P450_{SG1}$ is significantly altered in the substrate-binding place and/or heme environments (Aoyama *et al.*, (1987), sites where the azole antifungals probably interact with.
The results discussed here suggest that the antifungal activity of the azole derivatives depends on the sensitivity of the 14α demethylase. This was further proven by studying a *C. glabrata* isolate that became resistant to fluconazole

after 9 days of treatment (Warnock *et al.*, 1988). As compared with the parent, this mutant has a higher P450 content resulting in an increased ergosterol synthesis and about 5 times higher ergosterol content (manuscript in preparation). This increased enzyme activity may explain, at least partly, the greater tolerance to fluconazole and other azole antifungals.

The first consequence of inhibition of the 14α-demethylase is a decreased availability of ergosterol. For example, incubation of *C. albicans* in the presence of 30nM itraconazole resulted in a complete block of ergosterol synthesis (Table 2). At 100 nM itraconazole the ergosterol level per cell (total amount measured at 280nm) is below the detection limit (< 5 fg/cell). Since ergosterol plays an essential regulatory function in processes of cell proliferation (Dahl and Dahl 1988) the itraconazole induced ergosterol depletion alone will already result in an inhibition of cell division. Furthermore, studies of Long *et al.*, (1988) prove that ergosterol-depleted membranes are more permeable to glucose and other small polar molecules and ions.

As mentioned above a *S. cerevisiae* mutant defective in lanosterol 14α-demethylation accumulates 14-methylsterols as found in fungi incubated in the presence of azole antifungals.

Table 2 gives an overview of sterols, 3-ketosteroids and sterol precursors found in fungal cells after incubation in the presence of itraconazole. For example, when *C. albicans* is incubated with nanomolar concentrations of itraconazole, radioactivity derived from [^{14}C]-acetate is found in 14α-methylfecosterol, obtusifoliol, eburicol, lanosterol, 14α-methyl-ergosta-5,7,22,24(28)-tetraen-3β-ol and especially in 14α-methyl-ergosta-8,24(28)-dien-3β,6α-diol (3,6-diol). The 6-hydroxylation involved in the synthesis of the latter sterol may be part of the $\Delta^{5,6}$ double bond insertion, one of the last reactions in ergosterol synthesis. The accumulation of the 3,6-diol in azole antifungal treated *C. albicans* may originate from the decreased ability of the $\Delta^{5,6}$ -desaturase to use 14-methylsterols as substrates. In contrast with *C. albicans*, *C. glabrata* incubated in the presence of itraconazole, does not accumulate eburicol. This suggests that, as in *S. cerevisiae*, lanosterol may be the substrate for the 14α-demethylase. However, the accumulation of 3,6-diol indicates that the *S*-adenosyl-L-methionine-Δ^{24}-sterol methyltransferase is able to use 14-methylsterols as substrate. The presence of 3,6-diol and small amounts of its precursor, 14α-methylfecosterol indicates that *C. albicans* is able to remove C-4 methyl groups before the removel of the 14α-methyl group. *T. mentagrophytes* and *A. fumigatus* are also accumulating 3,6-diol whereas *P. ovale* is not, and *C. neoformans* and *H. capsulatum* are accumulating less than 5 % (Table 2). In itraconazole-treated *P. ovale* the main 14-methylsterol found is eburicol. The latter sterol together with obtusifolione are the major products found in itraconazole-treated *H. capsulatum* and *Cr. neoformans*. The results summarized in this table further prove that (1) itraconazole is a potent inhibitor of ergosterol synthesis in yeast, dimorphic and filamentous fungi (IC50-values from 5nM to 173 nM) and (2) is a potent inhibitor of the sterol 14α-demethylase. These results also indicate that in *H. capsulatum* and *C. neoformans* the last step in the demethylation at C-4 is blocked. Both methyl groups at C-4 (C-30 and C-31) are removed as carbon dioxide (Mercer, 1984). The first group to be removed is the 4α-methyl. This reaction requires NADPH, molecular oxygen, a cyanide-sensitive factor and cytochrome b_5. The first intermediate formed is a C-4 alcohol. This 4α-hydroxymethyl derivative is oxidated to a C-4 aldehyde and further to 4α-carboxy-4ß-methyl-3-

ketosterol, which is then decarboxylated by a NAD-dependent sterol 4-decarboxylase to yield the 4α-methyl-3-ketosteroid, obtusifolione. The final step in the 4-demethylation reaction is the reduction of the 3-keto to a 3β-hydroxyl group by the NADPH-dependent 3-ketosteroid reductase. The second demethylation at C-4 is mechanically identical and catalysed by the same microsomal enzymes (Mercer, 1984). The intermediate formed is 14-methylfecosterone. The accumulation of 3-ketosteroids in itraconazole-treated *H. capsulatum* and *C. neoformans* points to an inhibition of the 3-ketosteroid reductase. Thus, it is possible that for both fungi the 3-ketosteroid reductase is next to the 14α-demethylase a target for itraconazole and, as shown previously, also for ketoconazole (Vanden Bossche *et al.*, 1990b). The interaction of ketoconazole and itraconazole with two targets, the 14α-demethylase and the 3-ketosteroid reductase, in the ergosterol biosynthetic pathway might explain the high sensitivity of *C. neoformans* and *H. capsulatum* to these azole antifungals.

Prerequisites for both the condensing (increase in the chain order of the phospholipids in the liquid crystalline state of the membrane) and liquefying effect (decrease in the chain order in the gel state) of sterols are a planar ring system, a long flexible chain at C-17 and a 3β-hydroxyl group (Vanden Bossche, 1990, Marichal *et al* 1990). Since hydrogen bonding occurs between the carbonyl oxygen of the phospholipid-acyl chain and the 3β-hydroxyl group of sterol this hydroxyl group is important for the interaction with phospholipids (Cooper and Strauss, 1984). 3-Ketosteroids strongly destabilize the lipid bilayer structure, inhibit the growth of sterol-requiring mycoplasmas and when incorporated in erythrocytes they greatly increase the permeability and fragility of the membranes (Gallay and De Kruijff, 1982).

Thus the interaction (direct or indirect) of ketoconazole and itraconazole with the 3-ketosteroid reductase, in the ergosterol biosynthetic pathway might already explain the high sensitivity of *C. neoformans* and *H. capsulatum* to these azole antifungals.

Not only 3-ketosteroids destabilize membranes also the 14-methylated sterols, such as lanosterol and eburicol, are not suitable for membrane stability and function (Nes *et al.*, 1978; Buttke and Bloch, 1981; Marichal *et al.*, 1990).

Studies of Kelly *et al* (1991) point to the inability of 14α-methyl-ergosta-8,24(28)-dien-3β,6α-diol to support growth of *S. cerevisiae* even in the presence of ergosterol. As shown in Table 2, this 3,6-diol is the main accumulating sterol in *C. albicans* and *C. glabrata*. The accumulation of this 14α-methylsterol together with the observed depletion of ergosterol (observed at 0.1 μM) may be at the origin of itraconazole's candicidal effect observed after 24 h of contact with 10 μg/ml (Van Cutsem, 1989). Fungicidal activity is obtained when *T. mentagrophytes* is incubated with 1 μg itraconazole/ml (Van Cutsem 1989). In this dermatophyte eburicol, lanosterol, 3,6-diol and 14α-methylfecosterol are accumulating. The fungicidal activity observed is to a certain extent in conflict with the believe that 14α-methylfecosterol (Table 2) is functional in fungal membranes even in the absence of ergosterol (Orth and Sisler, 1990). For example, a mutant of *U. maydis* that accumulates 14α-methylfecosterol has been found to be viable. However, the doubling time of the mutant is about 6 1/2 h compared to 2 1/4 h for the wild-type (Sisler *et al.*, 1983). The residual slow growth rate of *U. maydis* sporidia observed in the presence of 14α-demethylase inhibitors has also been related to the accumulation of 14 α-methylfecosterol (Orth and Sisler, 1990). However, this hypothesis is based on studies of Buttke and Bloch who used 14 α-methyl-Δ^7-cholestenol and not the 24-alkylsterol, 14α-methylfecosterol (Buttke and Bloch, 1981). Furthermore, studies of Taylor *et al.*, and Rodriguez *et al.*, indicate that although 14 α-methylfecosterol can satisfy "bulk" membrane requirements it does not fulfill the highly specific role of ergosterol ("critical domain" function) in *S. cerevisiae cells* (Taylor *et al.*, 1983; Rodriguez *et al.*, 1985). Finally, 14α-methylfecosterol is not found in itraconazole-treated *C. glabrata* which is less sensitive to itraconazole than *C. albicans* in which this sterol is accumulating (Table 2). Further studies are needed to evaluate the differences between the accumulating 14α-methylsterols in their ability to maintain growth of fungal cells treated with ergosterol biosynthesis inhibitors. It should be noted that itraconazole's cidal effect does not originate from its direct effect on membranes. Indeed, in contrast with for example miconazole (Vanden Bossche *et al* 1982), itraconazole does not induce significant changes in the lipid membrane organization parameters as measured by DSC, infrared spectroscopy and leakage of entrapped [^{14}C]-glucose (Vanden Bossche, 1985, Brasseur *et al* 1991).

Changes in sterol structure might lead to an alteration in fatty acid composition. For example, an increase in palmitate was found in *C.albicans*, grown in the presence of miconazole, clotrimazole, econazole, or ketoconazole (Vanden Bossche 1985, Georgopapadakou *et al.*, 1987). The increased synthesis of saturated fatty acids suggests an effect on the Δ^9 desaturase, a microsomal enzyme which requires phospholipids for full activity. The requirement for phospholipids indicate that this enzyme, as other membrane-bound enzymes, is only active at a defined fluidity of the environment. It is thus possible that the azole-induced ergosterol depletion and accompanying accumulation of 14α-methylsterols alter the fluidity in such a way that the desaturase is inhibited. From antagonistic effects of unsaturated fatty acids it can be deduced that the decreased availability of unsaturated fatty acids may contribute to the antifungal activity of azole antifungals (Yamaguchi, 1977, Georgopapadakou *et al.*, 1987).

From the azole antifungal-induced ergosterol depletion, accumulation of 14α-methylsterols and consequent changes in the fatty acids, important alterations in the properties of fungal membranes can be expected. Studies of Sancholle *et al.*, (1984) on the effects of the azole antifungal propiconazole on *Taphrina deformans* indicate that the observed inhibition of the 14α-demethylase results in a steady loss of radioactive substances from cells that had been incubated with ^{32}P. Miconazole, inhibits, at concentrations lower than those affecting growth, the uptake of purines by *C. albicans* (Vanden Bossche, 1974). Measuring the effects of sterol structures on viscosity changes in the lipid layers and on enzyme activity in yeast mitochondria Parks *et al.*, showed that structural changes in the sterols impart substantial effects not only on the physical but also on enzymatic properties of membranes (Parks *et al.*, 1984). An example is the chitin synthetase (for reviews see Vanden Bossche 1985, 1990). This enzyme is inhibited at high ergosterol levels (Chiew *et al.*, 1982) and mutants of *C. albicans*, with a low ergosterol content, showed increased activity of this synthetase (Pesti *et al.*, 1981). Thus, ergosterol biosynthesis inhibitors should increase chitin synthesis. Indeed, azole antifungals increase this synthesis in for example *C. albicans* (Vanden Bossche 1985) and *A. fumigatus* (Vanden Bossche *et al.*, 1988a,b) resulting in an irregular distribution of patches of chitin. Other membrane-bound enzymes which activities are altered by 14 α-demethylase inhibitors are e.g. cytochrome oxidase and peroxidase and ATPases of both mitochondria and plasma membrane (for reviews see Vanden Bossche 1985, 1990, Kerridge *et al.*, 1988, Kelly *et al.*, 1990). A consequence of ergosterol biosynthesis inhibition may also be the observed mitochondrial petite induction in *S. cerevisiae* grown in the presence of ketoconazole at concentrations below the minimal inhibitory concentrations (Kelly *et al.*, 1990, 1991). It is not surprising that fungal mitochondria are affected. Indeed, in contrast to animal cells high amounts of sterols (ergosterol) are found in fungal mitochondria (Vanden Bossche 1985).

In summary, the interaction of azoles with the fungal $P450_{14DM}$ is at the origin of a cascade of perturbations, all together leading to the antifungal activity.

Δ^{14}-reductase and Δ^{8-7}-isomerase inhibitors

N-Substituted morpholine fungicides (e.g. dodemorph, tridemorph, fenpropimorph) have been marketed for the control of plant pathogenic fungi. The fenpropimorph derivative amorolfine also is highly active against phytopathogenic fungi (Mercer, 1991) but has been developed as a nail lacquer for the treatment of onychomycosis (Polak and Zaug, 1990). Excellent reviews on the mode of action of morpholine antifungals are available (Kerkenaar, 1987, Mercer, 1988, 1991, Polak, 1988, 1990). Therefore, in this paper just a short summary will be presented. Similar to the azole antifungals (Vanden Bossche, 1985), morpholine antifungals do not affect fungi unable to synthesize ergosterol (Mercer, 1991). This pinpoints inhibition of ergosterol synthesis as their primary site of action.

In *C. albicans*, amorolfine (31.5 μM, 24h) causes accumulation of 5 α-ergosta-8,14-dienol (ignosterol) (Polak, 1988) suggesting that this fenpropimorph derivative interacts with ergosterol synthesis at the Δ^{14}-reductase catalysed step (Fig. 1). As pointed out by Mercer, although no accumulation of Δ^{8}-sterols was detected, amorolfine might share with fenpropimorph its inhibitory effects on the $\Delta^{8} \rightarrow \Delta^{7}$-isomerase (Mercer, 1988). Indeed, such an inhibition could only be detected if sufficient sterol has leaked through the Δ^{14}-reductase catalysed step. Incubation of *T. mentagrophytes* for 24 h in the presence of ≥ 2.8 μM amorolfine resulted in the accumulation of ignosterol, 5 α-ergosta-8,14-24(28)-trienol and squalene (Polak 1988). These results suggest that amorolfine inhibits next to the $\Delta^{8} \rightarrow \Delta^{7}$-isomerase and Δ^{14}-reductase also the squalene epoxidase. The latter enzyme was also inhibited in the plant pathogen *Nectria haematococcus* treated with fenpropimorph, an effect not seen with fenpropidin or tridemorph (Ziogas *et al.*, 1991). The reason why these fenpropimorph-related fungicides do not cause accumulation of squalene is not clear. The fact that these antifungals also induce accumulation of 5 α-ergosta-8,14-24(28)-trienol and/or ignosterol excludes the possibility that the observed inhibition of the squalene epoxidase results from accumulation of the Δ^{8}- and $\Delta^{8\text{-}14}$-sterol intermediates (Ziogas *et al.*, 1991). Although amorolfine seems to inhibit three enzymes in the ergosterol biosynthetic pathway it is not able to block completely ergosterol synthesis. For example, after 24 h of incubation of *T. mentagrophytes* with about 85 μM amorolfine the sterol +squalene extract still contains 7.9 % ergosterol (Polak 1988).

As discussed above it is not yet clear whether or not squalene is toxic for fungal cells. Studies of Steel *et al.*, suggest that inhibition of cell growth by fenpropimorph and related fungicides is the consequence of modifications of membrane properties by the accumulating Δ^8- and $\Delta^{8\text{-}14}$-sterol intermediates (Steel *et al.*, 1989). However, a more recent study cast doubt on the concept that the fungitoxic effect originates from the accumulation of these sterol intermediates. Marcireau *et al.*, (1990) used a *S. cerevisiae* wild-type (FL100) and a mutant (FK*aux 30*) that is able to synthesize ergosterol *de novo* but, unlike the wild-type is permeable to exogenous sterols. In the wild-type no growth inhibition occurred even when the ergosta-8-enol (Δ^8- sterol) level was maximal and growth inhibition was almost complete before ignosterol ($\Delta^{8\text{-}14}$-sterol) began to appear. Growth inhibition took place when the ergosterol content was <15% of the total sterol content. As expected, ergosterol supplementation (4μg/ml) relieved growth inhibition of the wild-type only slightly (MIC increased from 0.3 μM to 0.9μM). The FK*aux 30* mutant has, in the presence or absence of inhibitor, a sterol composition similar to that of the wild-type. Supplementation of the medium with 4 μM of ergosterol increased the MIC value from 0.3 μM up to >60 μM. In the presence of 0.6 μM fenpropimorph growth is blocked but the cells remained viable and started to proliferate and bud after the addition of 0.1μg/ml of ergosterol. When grown in the presence of ergosterol the FK*aux 30* mutant accumulated amounts of both ergosta-8-enol and ignosterol similar to those accumulating in nonsupplemented, fenpropimorph-treated cultures. These findings suggest, that at least in *S. cerevisiae*, the fungistatic action of fenpropimorph originates from ergosterol deprivation rather than from the accumulation of Δ^8- and $\Delta^{8\text{-}14}$-sterol intermediates.

Fungi, grown in inhibitory concentrations of fenpropimorph and of the 14-demethylase inhibitor imazalil, showed the same morphological alterations (Kerkenaar and Barug, 1984). Both imazalil and fenpropimorph also caused an irregular deposition of β-1,3 and β-1,4 polysaccharides, probably chitin, in *Ustilago maydis* and *Penicillium italicum* (Kerkenaar and Barug, 1984, Kerkenaar *et al.*,, 1984). This suggests that the decreased availability of ergosterol is already enough to induce an uncoordinated synthesis of chitin.

Conclusions

It is not surprising that most of the available antifungal agents belong to the class of the ergosterol biosynthesis inhibitors (EBIs). Indeed, this pathway offers many target enzymes for the development of selective broad spectrum antifungals. The present available EBIs can be divided in two classes: (1) those of which the fungitoxicity may originate from ergosterol deprivation rather than from the accumulation of sterol intermediates (e.g. amorolfine and terbinafine); (2) those inhibiting ergosterol synthesis with the concommitent accumulation of precursors that either can not fulfill the functions of ergosterol or have proven to have substantial effects not only on the physical but also on enzymatic properties of membranes (e.g. itraconazole). Both ergosterol deprivation and accumulation of the 14-methylated sterols may be at the origin of e.g. itraconazole's broad spectrum activity against yeasts, dimorphic and filamentous fungi. Some of the azole antifungals inhibit (directly or indirectly) the 3-ketosteroid reductase in *H. capsulatum*

and *C. neoformans* resulting in the accumulation of 3-ketosteroids that strongly destabilize the lipid bilayer structure and greatly increase the permeability and fragility of membranes. The interaction with this second target may explain the high efficacy of itraconazole against both fungi.
The development of the EBIs stimulated studies on the enzymes of the ergosterol biosynthesis pathway. However, many questions have also been generated. For example it is still not well understood why the fungal 14α-demethylase, squalene epoxidase or Δ^{14}-reductase are mostly much more sensitive than their mammalian counterparts. Sequencing, molecular modeling and site-directed mutagenesis studies may help to unravel the peculiarities of the host and fungal enzymes. The expression of N-terminal truncated fungal and liver $P450_{14DM}$ in yeast or *E. coli* may enable the production of a soluble $P450_{14DM}$ of which crystals can be prepared to perform crystallographic and molecular modeling studies. This approach may be applied to other membrane-bound enzymes of the sterol biosynthesis pathway. These studies may provide insight into the molecular structures needed to interact with the substrate binding place(s) or to modify the polypeptide conformation.
This paper focussed on antifungals interfering with a limited number of enzymes from the ergosterol biosynthetic pathway. The targets listed in the introduction and the other enzymes involved in sterol synthesis also need to be studied since they may provide us with potent and selective antifungal agents.

References

Aoyama Y, YoshidaY, Hata S, Nishino T, Katsuki H, Maitra US, Mohan VP, Sprinson DB (1983) Altered cytochrome P-450 in a yeast mutant blocked in demethylating C-32 of lanosterol. J. Biol. Chem. 258: 9040-9042.

Aoyama Y, Yoshida Y, Nishino T, Katsuki H, Maitra US, Mohan VP, Sprinson DB (1987) Isolation and characterization of an altered cytochrome P-450 from a yeast mutant defective in 14α-demethylation. J. Biol. Chem. 262: 14260-14264.

Barrett-Bee KJ, Lane AC, Turner RW (1986) The mode of antifungal action of tolnaftate. J. Med. Vet. Mycol. 24: 155-160.

Barrett-Bee KJ, Lees J, Pinder P, Campbell J, Newboult L (1988) Biochemical studies with a novel antifungal agent, ICI 195,739. Ann. N. Y. Acad. Sci. 544: 231-244.

Brasseur R, Goormaghtigh E, Ruysschaert J-M, Duquenoy PH, Marichal P, Vanden Bossche H (1991) Lipid-itraconazole interaction in lipid model membranes. J. Pharm. Pharmacol. 43: 167-171.

Buttke TM, Bloch K (1981) Utilization and metabolism of methyl-sterol derivatives in the yeast mutant strain GL7. Biochemistry 20: 3267-3272.

Chen C, Kalb VF, Turi TG, Loper JC (1988) Primary structure of the cytochrome P-450 lanosterol 14α-demethylase gene from *Candida tropicalis*. DNA 7: 617-626.

Chiew YY, Sullivan PA, Shepherd MG (1984) The effects of ergosterol and alcohols on germ-tube formation and chitin synthase in *Candida albicans*. Can. J. Biochem. 60: 15-20.

Chung B-C, Picado-Leonard J, Haniu M, Bienkowski M, Hall PF, Shively J E, Miller W L (1987). Cytochrome P450C17 (steroid 17α-hydroxylase/17,20 lyase): cloning of human adrenal and testis cDNAs indicates the same gene is expressed in both tissues. Proc. Nat. Acad. Sci. USA, 84: 407-411.

Cooper RA, Strauss JF (1984) Regulation of cell membrane cholesterol. In: Shinitzky M (ed.) Physiology of membrane fluidity, vol.1, CRC Press, Inc., Boca Raton, Florida, pp. 73-97.

Dahl C, Dahl J (1988) Cholesterol and cell function. In: Yeagle PL (ed.), Biology of cholesterol, CRC Press, Boca Raton, Florida, pp. 147-171.
Ferguson JB, Bloch K (1977) Purification and properties of a soluble protein activation of rat liver squalene epoxidase. J. Biol. Chem. 252: 5381-5385.
Fischer RT, Trzaskos JM. Magolda RL, Ko SS, Brosz CS, Larsen B (1991) Lanosterol 14α-demethylase. Isolation and characterization of the third metabolically generated oxidative demethylation intermediate. J. Biol. Chem. 266: 6124-6132.
Fryberg M, Oehlschlager AC, Unrau AM (1976) Sterol biosynthesis in antibiotic sensitive and resistant *Candida*. Arch. Biochem. Biophys. 173: 171-177.
Galley J, De Kruijff (1982) Correlation between molecular shape and hexagonal H_{II} phase promoting ability of sterols. FEBS Lett. 143: 133-143.
Georgopapadakou NH, Dix BA, Smith SA, Freudenberger J, Funke PT (1987) Effect of antifungal agents on lipid biosynthesis and membrane integrity in *Candida albicans*. Antimicrob. Ag. Chem. 31: 46-51.
Gottlieb D, Knaus RJ, Wood SG (1979) Differences in the sterol synthesizing pathways of sterol-producing and non-sterol producing fungi. Phytopathol. 68: 1168-1169.
Graham-Lorence S, Khalil MW, Lorence MC, Mendelson CR, Simpson ER (1991) Structure-function relationships of human aromatase cytochrome P-450 using molecular modeling and site-directed mutagenesis. J. Biol. Chem. 266: 11939-11946.
Harada N (1988). Cloning of a complete cDNA encoding human aromatase : immunochemical identification and sequence analysis. Biochem. Biophys. Res. Comm. 156:725-732.
Horie M, Tsuchiya Y, Hayashi M, Iida Y, Iwasawa Y, Nagata Y, Sawasaki Y, Fukuzumi H, Kitani K, Kamei T (1990) NB-598: a potent competitive inhibitor of squalene epoxidase. J. Biol. Chem. 265: 18075-18078.
Ishida NY, Aoyama Y, Hatanaka R, Oyama Y, Imajo S, Ishiguro M, Oshima T, Nakazato H, Noguchi T, Maitra US, Mohan VP, Sprinson DB, Yoshida Y (1988) A single amino acid substitution converts cytochrome $P450_{14DM}$ to an inactive form, cytochrome $P450_{SG1}$: complete primary structures deduced from cloned DNAs. Biochem. Biophys. Res. Commun. 155: 317-323.
Kalb VF, Woods CW, Turi TG, Dey CR, Sutter TR, Loper LC(1987) Primary structure of the P450 lanosterol demethylase gene from *Saccharomyces cerevisiae*. DNA 6: 529-537.
Kelly SL, Kenna S, Bligh HFJ, Watson PF, Stansfield I, Ellis SW, Kelly DE (1990) Lanosterol to ergosterol- enzymology, inhibition and genetics. In: Kuhn PJ, Trinci AP, Jung MJ, Goosey MW, Copping LG (eds) Biochemistry of cell walls and membranes in fungi, Springer-Verlag, Berlin, pp.223-243.
Kelly SL, Rowe J, Watson PF (1991) Molecular genetic studies on the mode of action of azole antifungal agents. Biochem. Soc. Transact. 19: 796-798.
Kerkenaar, A (1987) The mode of action of dimethylmorpholines. In: Fromtling JR (ed) Recent Trends in the Discovery, Development and Evaluation of Antifungal Agents, JR Prous Science Publishers, S.A., Barcelona, pp. 523-542.
Kerkenaar A and Barug D (1984) Fluorescence microscope studies of *Ustilago maydis* and *Penicillium italicum* after treatment with imazalil or fenpropimorph. Pestic Sci. 15: 199-205.
Kerkenaar A, van Rossum JM, Versluis GG and Marsman JW (1984) Effect of fenpropimorph and imazalil on sterol biosynthesis in *Penicillium italicum*. Pestic Sci. 15: 177-187.
Kerridge D, Fasoli M, Wayman FJ (1988) Drug resistance in *Candida albicans* and *Candida glabrata*. Ann. N. Y. Acad. Sci. 544: 245-259.
Kerridge D, Vanden Bossche H (1990) Drug discovery: a biochemist's approach. In: Ryley JF (ed) Chemotherapy of fungal infections, Springer Verlag, berlin, pp 31-76.
Langcake P, Kuhn PJ, Wade M (1983) The mode of action of systemic fungicides. In: Hutson DH, Roberts TR (eds) Progress on pesticide biochemistry and toxicology, vol 3, John Wiley & Sons Ltd, Chicester, England, pp. 1-109.
Long MT, Steel CC, Mercer EI (1988) Location of squalene accumulation and

physiological effects of ergosterol depletion in naftifine-grown yeast. Biochem. Soc. Transact. 16: 1044-1045.
Margalith PZ (1976) Steroid microbiology. Charles C Thomas, Springfield, p. 99.
Marcireau C, Guilloton M, Karst F (1990) In vivo effects of fenpropimorph on the yeast *Saccharomyces cerevisiae* and determination of the molecular basis of the antifungal property. Antimicrob. Ag. Chemother. 34: 989-993.
Marichal P, Vanden Bossche H, Moereels H, Brasseur R (1990) Mode of insertion of azole antifungals and sterols in membranes. In: Brasseur R (ed) Molecular description of biological membranes by computer aided conformational analysis, vol 2, CRC Press, Boca Raton, pp. 27-42.
Mercer EI (1984) The biosynthesis of ergosterol. Pestic Sci. 15: 133-155.
Mercer EI (1988) The mode of action of morpholines. In: Berg D, Plempel M (eds).Sterol biosynthesis inhibitors. Pharmaceutical and Agrochemical aspects, Ellis Horwood Ltd., Chichester, England, pp. 120-150.
Mercer EI (1991) Morpholine antifungals and their mode of action. Biochem. Soc. Transact. 19: 788-793.
Moereels H, De Bie L, Tollenaere JP (1990) CGEMA and VGAP: a colour graphics editor for multiple alignment using a variable gap penalty. Application to the muscarinic acethylcholine receptor. J. Computer-aided Mol. Des. 4: 131-145.
Nes WR, Sekula BC, Nes WD, Adler JH (1978) The functional importance of structural features of ergosterol in yeast. J. Biol. Chem. 253: 6218-6225.
Nes WD, Saunders GA, Heftman E (1982) Role of steroids and triterpenoids in the growth and reproduction of *Phytophthora cactorum*. Lipids 17: 178-183.
Ono T, Nakazono K, Kosaka H (1982) Purification and partial characterization of squalene epoxidase from rat liver microsomes. Biochim. Biophys. Acta 709: 84-90.
Ono T, Ozasa S, Hasagawa F, Imai Y (1977) Involvement of NADPH-cytochrome c reductase in the rat liver squalene epoxidase system. Biochim. Biophys. Acta 486: 401-407.
Orth AB, Henry MJ, Sisler HD (1990) Mechanism of resistance to terbinafine in two isolates of *Ustilago maydis*. Pestic Biochem. Physiol. 37: 182-191.
Orth AB, Sisler HD (1990) Mode of action of terbinafine in *Ustilago maydis* and characterization of resistant mutants. Pestic Biochem. Physiol. 37: 53-63.
Paltauf F, Daum G, Zuder G, Högenauer G, Schulz G, Seidl G (1982) Squalene and ergosterol biosynthesis in fungi treated with naftifine, a new antimycotic agent. Biochim. Biophys. Acta 712: 268-273.
Parks LW, Bottema DK, Rodriguez RJ (1984) Physical and enzymatic function of ergosterol in fungal membranes. In: Nes WD, Fuller G, Tsai K-S (eds) Isopentenoids in plants. Biochemistry and function. Marcel Dekker, Inc, pp 433-452.
Pesti M, Campbell JM, Peberdy JF (1981) Alteration of ergosterol content and chitin synthase activity in *Candida albicans*. Current Microbiol. 5: 187-190.
Polak A (1988) Mode of action of morpholine derivatives. Ann. N. Y. Acad. Sci. 544: 221-228.
Polak A (1990) Mode of action studies. In: Ryley JF (ed) Chemotherapy of fungal infections, Springer Verlag, Berlin, pp 152-182.
Polak A, Zaug M (1990) Amorolfine. In: Ryley JF (ed) Chemotherapy of fungal infections, Springer Verlag, Berlin, pp. 505-521.
Poulos TL, Finzel BC, Howard AJ (1987) High-resolution crystal structure of cytochrome $P450_{cam}$. J. Mol. Biol. 195: 687-700.
Rodriguez RJ, Low C, Bottema CDK, Parks LW (1985) Multiple functions for sterols in *Saccharomyces cerevisiae*. Biochim. Biophys. Acta 837: 336-343.
Ryder NS (1987) Squalene epoxidase as the target of antifungal allylamines. Pestic Sci. 21: 281-288.
Ryder NS (1988a) Mechanism of action and biochemical selectivity of allylamine antimycotic agents. Ann. N. Y. Acad. Sci. 544: 208-220.
Ryder NS (1988b) Mode of action of allylamines. In: Berg D, Plempel M (eds) Sterol biosynthesis inhibitors, Ellis Horwood Ltd, Chichester, England, pp. 151-167
Ryder NS (1991) Squalene epoxidase as a target for the allylamines. Biochem. Soc. Transact 19: 769-774.

Sancholle M, Weete JD, Montant C (1984) Effects of triazoles on fungi: I. growth and cellular permeability. Pestic Biochem. Physiol. 21: 31-44.
Sisler HD, Walsh RC, Ziogas BN (1983) Ergosterol biosynthesis: a target of fungitoxic action. In: Matsunaka S, Hutson DH, Murphy SD (eds) Pesticide chemistry: Human welfare and the environment. Vol 3, Mode of action, metabolism and toxicology. Pergamon Press, New York, pp 129-134.
Sisler HD, Buchman-Orth AM (1990) Oxygen radical generation as a basis of fungitoxicity. In: Tagugungs Bericht No. 291, Systemic fungicides and antifungal compounds. Akademie der Landwirtschaftswissenschaften, Berlin, pp. 17-23.
Steel CC, Baloch RI, Mercer EI, Baldwin BC (1989) The intracellular location and physiological effects of abnormal sterols in fungi grown in the presence of morpholine and functionally related fungicides. Pestic Biochem. Physiol. 33: 101-111.
Taylor FR, Rodriguez RJ, Parks LW (1983) Requirement for a second sterol biosynthesis mutation for the viability of a sterol C-14 demethylation defect in *Saccharomyces cerevisiae*. J. Bacteriol. 155: 64-68.
Trzaskos J, Kawata S, Gaylor JL (1986) Microsomal enzymes of cholesterol biosynthesis. Purification of lanosterol 14α-methyl demethylase cytochrome P-450 from hepatic microsomes. J. Biol. Chem. 261: 14651-14657.
Unger BP, Gunsalus IC, Sligar SG (1986). Nucleotide sequence of the *Pseudomonas putida* cytochrome P-450_{cam} gene and its expression in *Escherichia coli*. J. Biol. Chem. 261: 1158-1163.
Van Cutsem J (1989) The in vitro antifungal spectrum of itraconazole. Mycoses 32 (suppl 1): 7-13.
Vanden Bossche H (1974) Biochemical effects of miconazole on fungi-I. Effects on the uptake and/or utilization of purines, pyrimidines, nucleosides, amino acids and glucose by *Candida albicans*. Biochem. Pharmacol. 23: 887-899.
Vanden Bossche H (1985) Biochemical targets for antifungal azole derivatives: hypothesis on the mode of action. In: McGinnis MR (ed) Current topics in medical mycology, vol 1, Springer Verlag, New York, pp. 313-351.
Vanden Bossche H (1988) Mode of action of pyridine, pyrimidine and azole antifungals. In: Berg D, Plempel M (eds) Sterol biosynthesis inhibitors, Ellis Horwood Ltd, Chichester, England, pp. 79-119.
Vanden Bossche H (1990) Importance and role of sterols in fungal membranes. In: Kuhn PJ, Trinci AP, Jung MJ, Goosey MW, Copping LG (eds) Biochemistry of cell walls and membranes in fungi, Springer-Verlag, Berlin, pp.135-157.
Vanden Bossche H (1991) Anti-*Candida drugs*- Mechanisms of action. In; Tümbay E, seelliger HPR, Ang Ö (eds) *Candida* and candidamycosis, Plenum Press, New York, pp. 82-95.
Vanden Bossche H, Willemsens G, Van Cutsem J (1978) Biochemical effects of miconazole on fungi. II. Inhibition of ergosterol biosynthesis in *Candida albicans*. Chem. Biol. Interact. 21: 59-78.
Vanden Bossche H, Willemsens G, Cools W, Cornelissen F, Lauwers WF, Van Cutsem J (1980) In vitro and in vivo effects of the antimycotic drug ketoconazole on sterol synthesis. Antimicrob. Ag. Chemother. 17: 922-928.
Vanden Bossche H, Ruysschaert J-M, Defrise-Quertain F, Willemsens G, Cornelissen F, Marichal P, Cools W, Van Cutsem J (1982) The interaction of miconazole and ketoconazole with lipids. Biochem. Pharmacol. 31: 2609-2617.
Vanden Bossche H, Lauwers W, Willemsens G, Marichal P, Cornelissen F, Cools W (1984a) Molecular basis for the antimycotic and antibacterial activity of N-substituted imidazoles and triazoles: the inhibition of isoprenoid biosynthesis. Pestic Sci. 15: 188-198.
Vanden Bossche H, Willemsens G, Marichal P, Cools W, Lauwers W (1984b) The molecular basis for the antifungal activities of N-substituted azole derivatives, focus on R 51211. In: Trinci APJ, Ryley JF (eds) Mode of action of antifungal agents, Cambridge University Press, Cambridge, pp. 321-341.
Vanden Bossche H, Willemsens G, Marichal P (1987a) Anti-*Candida* drugs- The biochemical basis for their activity. CRC Crit. Rev. Microbiol. 15: 57-72.
Vanden Bossche H, Marichal P, Gorrens J, Bellens D, Verhoeven H, Coene M-C,

Lauwers W, Janssen PAJ (1987b) Interaction of azole derivatives with cytochrome P-450 isozymes in yeast, fungi, plants and mammalian cells. Pestic Sci. 21: 289-306.
Vanden Bossche H, Marichal P, Gorrens J, Geerts H, Janssen PAJ (1988a) Mode of action studies- basis for the search of new antifungals. Ann. N. Y. Acad. Sci. 544: 191-207.
Vanden Bossche H, Marichal P, Geerts H, Janssen PAJ (1988b) The molecular basis for itraconazole's activity against *Aspergillus fumigatus*. In: Vanden Bossche H, Mackenzie DWR, Cauwenbergh G (eds) *Aspergillus* and aspergillosis, Plenum Press, New York, pp. 171-197.
Vanden Bossche H, Marichal P, Gorrens J, Coene M-C, Willemsens G, Bellens D, Roels I, Moereels H, Janssen PAJ (1989) Biochemical approaches to selective antifungal activity. Focus on azole antifungals. Mycoses 32 (suppl 1): 35-52.
Vanden Bossche H, Marichal P, Willemsens G, Bellens D, Gorrens J, Roels I, Coene M-C, Le Jeune L, Janssen PAJ (1990a) Saperconazole: a selective inhibitor of the cytochrome P-450-dependent ergosterol synthesis in *Candida albicans*, *Aspergillus fumigatus* and *Trichophyton mentagrophytes*. Mycoses 33: 335-352.
Vanden Bossche H, Marichal P, Gorrens J, Bellens D, Coene M-C, Lauwers W, Le Jeune L, Moereels H, Janssen PAJ (1990b) Mode of action of antifungals of use in immuno-compromised patients. Focus on *Candida glabrata* and *Histoplasma capsulatum*. In: Vanden Bossche H, Mackenzie DWR, Cauwenbergh G, Van Cutsem J, Drouhet E, Dupont B (eds) Mycoses in AIDS patients, Plenum Press, New York, pp. 223-243.
Vanden Bossche H, Marichal P, Gorrens J, Bellens D, Moereels H, Janssen PAJ (1990c) Mutation in cytochrome P-450-dependent 14α-demethylase results in decreased affinity for azole antifungals. Biochem. Soc. Transact. 18: 56-59.
Vanden Bossche H, Marichal P (1991) Mode of action of anti-Candida drugs: focus on terconazole and other ergosterol biosynthesis inhibitors. Am. J. Obstet. Gynecol. 165: 1193-1199.
Vanden Bossche H, Moereels H, Janssen PAJ (1991) Role of cytochrome P450 in the anabolism and catabolism of endobiotics. In: Arinç E, Schenkman JB, Hodgson E (eds) Molecular aspects of monooxygenases and bioactivation of toxic compounds, Plenum Press, New York, pp. 305-330.
Warnock DW, Burke J, Cope NJ, Johnson EM, von Fraunhofer NA, Williams EW (1988) Fluconazole resistance in *Candida glabrata*. The Lancet, December 3, 1310.
Yamaguchi, H (1977) Antagonistic action of lipid components of membranes from *C. albicans* and variuos other lipids on two imidazole antimycotics. Antimicrob. Ag. Chem. 12: 16-25.
Yoshida Y (1988) Cytochrome P450 of fungi: primary target for azole antifungal agents. In: McGinnis MR (ed) Current topics in medical mycology, vol 2, Springer Verlag, New York, pp. 388-418.
Yoshida Y, Aoyama Y, Nishino T, Katsuki H, Maitra US, V.P. Mohan VP, Sprinson DB (1985) Spectral properties of a novel cytochrome P-450 of a *Saccharomyces cerevisiae* mutant SG1. A cytochrome P-450 species having a nitrogenous ligand trans to thiolate. Biochem. Biophys. Res. Comm. 127: 623-628.
Yoshida Y, Aoyama Y (1986) Interaction of azole fungicides with yeast cytochrome P-450 which catalyzes lanosterol 14α-demethylation, In: Iwata K, Vanden Bossche H (eds) In vitro and in vivo evaluation of antifungal agents, Elsevier Science Publishers, Amsterdam, pp. 123-134.
Yoshida Y, Aoyama Y (1991) Cytochromes P450 in ergosterol biosynthesis.In: Ruckpaul K, Rein H (eds) Frontiers in biotransformation, vol 4, Microbial and plant cytochromes P450: biochemical characteristics genetic engineering and practical applications, Akademie Verlag, Berlin, pp. 127-148.
Ziogas BN, Oesterhelt G, Masner P, Steel C, Furter R (1991) Fenpropimorph: a three site inhibitor of ergosterol biosynthesis in *Nectria haematococcus* var. *cucurbitae*. Pestic Biochem. Physiol. 39: 74-83.

Molecular studies on azole sensitivity in fungi

S.L. Kelly and D.E. Kelly
Krebs Institute for Biomolecular Research
Department of Molecular Biology and Biotechnology
Sheffield University
Sheffield
UK

Introduction

Azole inhibitors of fungal sterol 14α-demethylase have provided a rich source of drugs and pesticides and new compounds are continuing to be evaluated for efficacy. The inhibition of the cytochrome P450 enzyme mediating this reaction was indicated by the effect on ergosterol biosynthesis, with the accumulation of C14-methylated sterols (figure 1; for review; Vanden Bossche, 1985). Despite their importance many of the molecular details of azole tolerance and resistance are only now beginning to be addressed and revealed. Compounds found to be active as inhibitors of sterol 14α-demethylase include pyrimidines, piparazines, pyridines, imidazoles and triazoles (Kato, 1986). The imidazoles and particularly the triazoles have been successfully developed as orally active drugs (figure 1). Molecular genetic techniques and molecular modelling may assist the design of further drugs. Increased importance for the development of new anti-fungals has resulted from the susceptibility to fungal infections of increasing numbers of immuno-compromised patients. Fungal resistance has also been reported and a molecular understanding is desirable here also.

The yeast *Saccharomyces cerevisiae* has been the main organism used in our studies as a model and as a host for the heterologous expression of P450. Modern molecular genetics provides an opportunity to extend the analysis of the effect of azole antifungal drugs in concert with the more traditional methods applied previously.

Mode of action studies with azole-resistant sterol mutants

S. cerevisiae mutants resistant to azole antifungal agents have been useful in probing the mode of action. Mutants of the sterol biosynthetic pathway were isolated on the basis of ergosterol auxotrophy, or resistance to polyene antibiotics which bind to ergosterol in the membrane (Henry, 1982). A few sterol 14α-demethylase defective strains were isolated in mutants which were already defective in sterol $\Delta^{5,6}$ desaturation. These are JR4 (Taylor *et al.*, 1983), SG1 (Trocha *et al.*, 1977) and NYS P-100 (Pierce *et al.*, 1978). Such mutants were found to be resistant to azole antifungal agents (Kenna *et al.*, 1989) without a commensurate change in cellular azole content resulting from a change in membrane composition.

NATO ASI Series, Vol. H 69
Molecular Biology and its Application to Medical Mycology
Edited by B. Maresca, G. S. Kobayashi and H. Yamaguchi

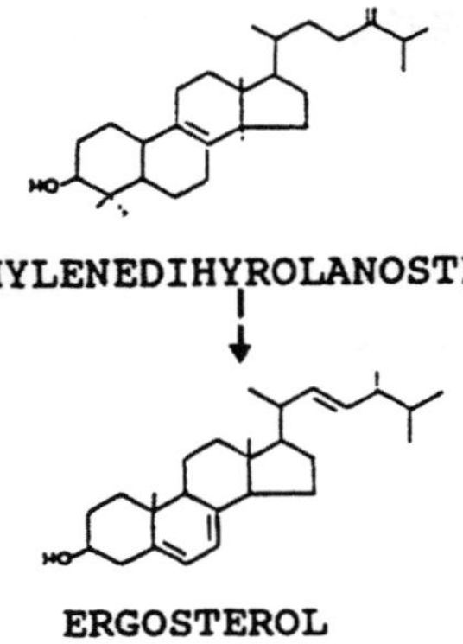

24-METHYLENEDIHYROLANOSTEROL

ERGOSTEROL

Figure 1. Chemical structure for the substrate, 24-methylene dihydrolanosterol, for many sterol 14α-demethylases and ergosterol, the product of the pathway.

Taylor *et al.*, (1983) were unable to segregate in JR4 the sterol 14α-demethylation block from a sterol $\Delta^{5,6}$ desaturase mutant (*erg3*) background, although *erg3* mutants were obtained following sporulation on their own. They suggested that the second defect was required for viability due to the lethal nature of 14α-methylated sterols when $\Delta^{5,6}$ desaturated. Other data indicated this may not be so for all sterol 14α-demethylase mutants (King *et al.*, 1985), but a lethal phenotype for stringent mutants is to be expected for this P450 as an antifungal target.
Studies on azole-resistant mutants of *S. cerevisiae* isolated directly have elucidated the basis of the association of the two defects and pointed to the mechanism of growth arrest following azole treatment. Over twenty independent isolates resistant to ketoconazole and fluconazole have been analysed and found to be defective in sterol $\Delta^{5,6}$ desaturation (Watson *et al.*, 1988; Watson *et al.*, 1989, Kelly *et al.*, 1990). The effect of chemical inhibition of sterol 14α-demethylation on growth appeared, as with a genetic block, to be suppressed in cells defective in sterol $\Delta^{5,6}$ desaturation. Study of the sterols

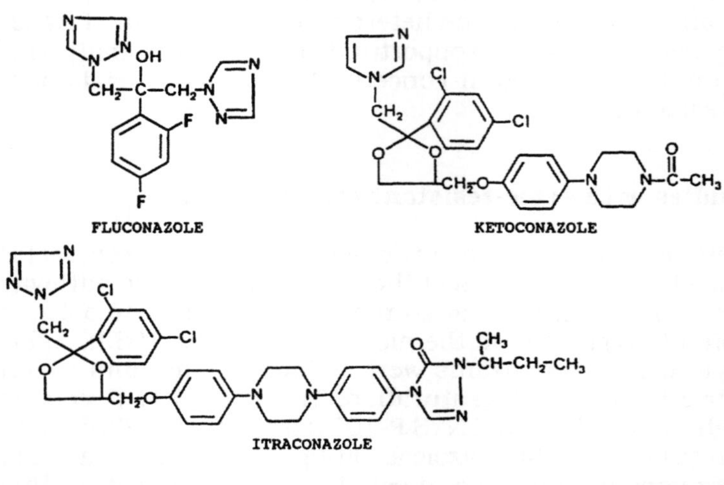

FLUCONAZOLE

KETOCONAZOLE

ITRACONAZOLE

Figure 2. The chemical structures of some azole antifungal drugs.

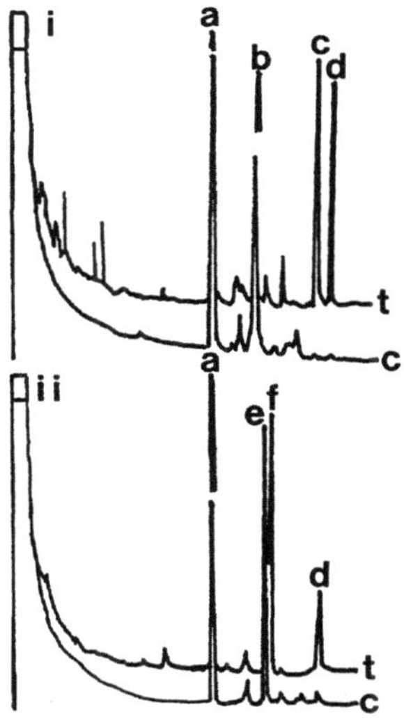

Figure 3. Sterols analysed by capillary GC from (i) sterol wild-type *S. cerevisiae* strain XY-729-5a and (ii) stringent *erg3* mutant derivative of XY-729-5a, A2. Shown are sterols from untreated cultures (c) and cultures treated with 10^{-4} M fluconazole (t). The sterols are: a: internal standard (cholesterol), b: ergosterol, c:lanosterol, d: 14-methyl-3,6-diol, e: ergosta-7,22-dien-3B-ol, f: 14-methylfecosterol.

present after treatment in the resistant and wild-type cells revealed the critical changes (Watson *et al.*, 1989; Kelly *et al.*, 1990). The wild-type parental strain accumulated lanosterol and 14α-methylergosta-8,24(28)-dien-3B, 6α-diol (14-methyl-3,6-diol) in contrast to a stringent *erg3* mutant where lanosterol and 14α-methylergosta-8,24(28)-dien-3B-ol (14-methylfecosterol) accumulated (figure 3). These mutants contained ergosta-7,22-dien-3Δ-ol when untreated.

The results indicated that formation of 14-methyl-3,6-diol in the wild-type required sterol $\Delta^{5,6}$ desaturase activity and that mutants blocked in its formation were able to grow on 14-methylfecosterol. Thus not all 14-methylated sterols are incompatible with growth. These findings can also be applied to interpret results for the P450 mutants described earlier where the *erg3* mutants interfere with formation of 14-methyl-3,6-diol and cause

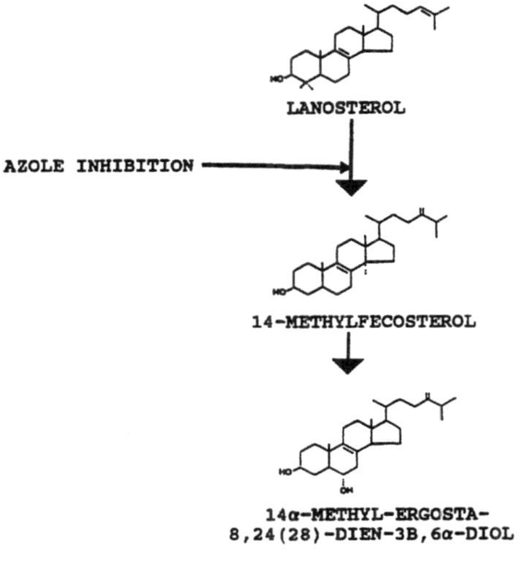

Figure 4. Sterols accumulating in ketoconazole treated *S. cerevisiae*.

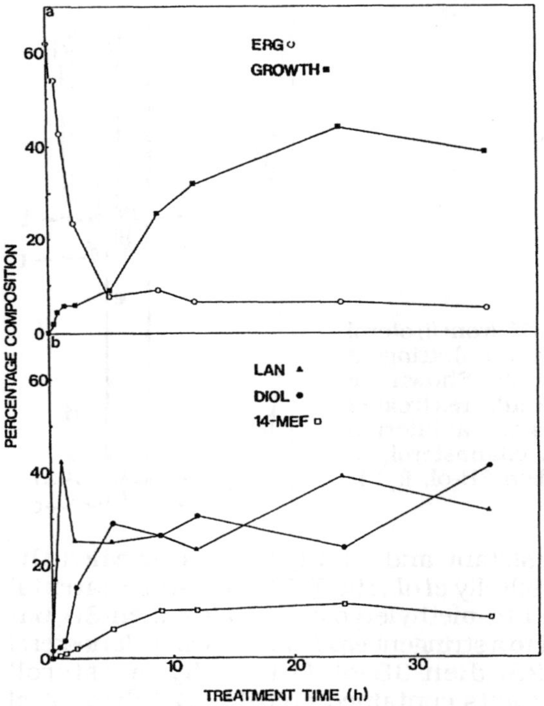

Figure 5. Timecourse of percentage changes during treatment of *S. cerevisiae* with the minimum inhibitory concentration of ketoconazole. (a) Dry weight increase (▪) and ergosterol content relative to total sterols (○); (b) lanosterol (▲), 14-methylfecosterol (□) and 14α-methyl-3,6-diol (●) increases in relation to total sterols.

retention of sufficient 14-methylfecosterol for growth. The sterol structures are shown in figure 4 indicating the point of inhibition of sterol metabolism in *S. cerevisiae* i.e. lanosterol 14α-demethylation. Subsequently lanosterol may be subject to C4 demethylation and C24 methylation to give 14-methylfecosterol from which it is proposed 14-methyl-3,6-diol is derived. The formation of the 6-hydroxylated product is probably due to interference in the $\Delta^{5,6}$ desaturation reaction.

Studies of the sterol pattern during cell growth arrest have also reflected the role of 14-methyl-3,6-diol as it reaches a maximum at the point of cell growth arrest (Kelly *et al.*, 1991). Such studies also detect the retention of significant quantities of ergosterol indicating that absolute starvation for this cell component is not the cause of growth arrest. Figure 5 shows the timecourse for sterol perturbation following treatment of *S.cerevisiae* with the MIC of ketoconazole (Quail *et al.*, 1992). Ergosterol became depleted during treatment with initially lanosterol accumulating. Subsequently, 14-methyl-3,6-diol was produced and reached a maximum at the point of growth arrest.

Gene disruption studies provide useful indicators that the absence of a gene product is lethal and worthy of consideration as a target for antifungal drugs. In the case of the lanosterol 14α-demethylase of *S. cerevisiae* gene disruption produced ergosterol auxotrophs which, due to sterol uptake only occuring anaerobically, were obligate anaerobes (Kalb *et al.*, 1987). Such strains were reported to give rise to mutants capable of aerobic growth.

Figure 6 shows a petri dish with the meiotic products of a lanosterol 14α-demethylase disruption experiment cultured on medium aerobically. Two of the progeny cultures containing the P450 disruption do not exhibit aerobic growth as previously reported and mutant colonies arise from them. Analysis

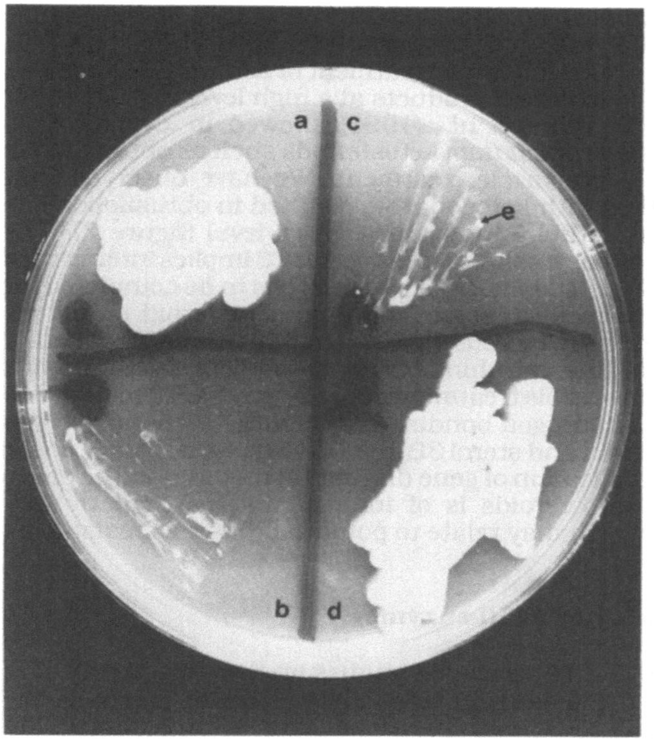

Figure 6. Photograph showing cultures derived from the sporulation of a diploid yeast with one allele of the lanosterol 14α-demethylase disrupted. 2:2 segregation is shown for cultures capable of aerobic growth (a,d) and incapable (c,d) with the disrupted strains c and d giving suppressor mutants (e).

of the sterols of the respective strains confirmed the previous results of classical genetic analysis. The non-disrupted strains (a,d) contained ergosterol and the disrupted strains (b,c) when left to incubate aerobically for a day accumulated predominantly 14-methyl-3,6-diol. A strain derived from the disruptants that was capable of aerobic growth (e) accumulated predominantly 14-methylfecosterol consistant with a secondary *erg3* block (Kelly *et al.*, in preparation).

The relevance of these findings to other pathogenic fungi is discussed below. The use of *S. cerevisiae* as an easily manipulated model has enabled the identification of key events resulting in cell growth arrest. However, studies with other fungi indicate the important conversion of 14-methylfecosterol to 14-methyl-3,6-diol identified using *S. cerevisiae* may not apply to all situations. The presence of 14-methyl-3,6-diol, following azole treatment in fungi, was first reported by Ebert *et al.*, (1983) for *Ustilago maydis*. Various other fungi have been identified as accumulating the sterol to a high level under azole treatment including *Candida albicans*, *C. glabrata* and *Aspergillus fumigatus* (Vanden Bossche *et al.*, 1990). Also the analysis of a *C. albicans* strain defective in sterol 14α-demethylation has revealed the same requirement for sterol $\Delta^{5,6}$ desaturation in the formation of 14-methyl-3,6-diol (Shimokawa *et al.*, 1989). The genetic system of *S. cerevisiae* allowed the identification of leaky *erg3* mutants causing resistance to azoles (Watson *et al.*, 1989) and such mutations may be present as suppressors of other P450 mutants of *C. albicans*. Gene disruption of the *C. albicans* sterol 14α-demethylase gene remains an important objective to identify the viability and potential suppressor mechanisms for

such a genotype. These studies would also have importance for resistance mechanisms to azole antifungals in *C. albicans*.
Investigation of the sterol profile following treatment of some other fungi has led to the identification of ketosteroid products at a high level instead of 14-methyl-3,6-diol. Vanden Bossche *et al.*, (1990) showed in *Cryptococcus neoformans* and *Histoplasma capsulatum* ketosteroids such as obtusifolione and 14-methylfecosterone after azole treatment. We have observed that treatment of *Botrytis cinerea* with ketoconazole resulted in obtusifolione as a major product with 14-methyl-3,6-diol at a lower level (figure 7). The formation of the ketosteroids possessing keto groups at C3 implies interference in the C4 demethylation step (Mercer, 1984) which needs to be completed for 14-methyl-3,6-diol to be formed. In some fungi the ketosteroid reductase portion of C4 demethyation may be sensitive to retention of the C14 methyl group unlike others where this is unhindered and 14-methyl-3,6-diol can accumulate. Both products of azole treatment could be envisaged to interfere with sterol function where hydrogen bonding between the carbonyl oxygen of the phospholipid-acyl chain and sterol 3B-hydroxyl group occurs (Cooper and Strauss, 1984). The application of gene disruption methods in the cases of pathogens producing ketosteroids is of interest and particularly any suppressor mechanisms which may relate to potential routes to resistance.

Studies on the multiplicity of P450 enzymes

Yoshida and Aoyama (1974) first developed studies on the enzymology of *S. cerevisiae* P450 and found the electron transport system to resemble the

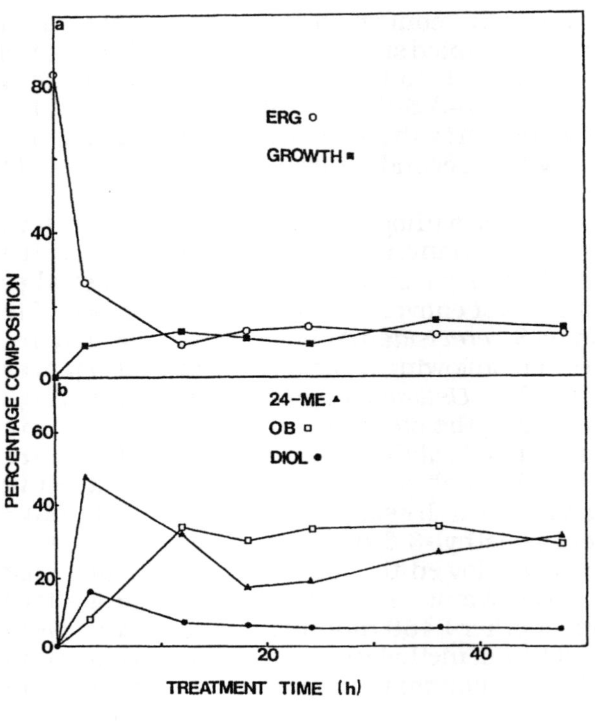

Figure 7. Timecourse for the percentage changes in sterol composition and dry weight during MIC treatment of *B. cinerea* with ketoconazole. (a) dry weight increase (■) and ergosterol content (○) relative to other sterols; (b) 24-methylenedihydrolanosterol (▲), obtusifolione(□) and 14-methyl-3,6-diol (●) increases. Figure 7c shows the molecular structure of obtusifolione.

hepatic microsomal system. Only one P450 was identified, that responsible for lanosterol 14α-demethylation in vegetatively growing cells (Yoshida and Aoyama, 1984). This first P450 has been cloned and sequenced (Kalb *et al.*,1987) and assigned the gene identification CYP51A1, but the P450 gene superfamily is more extensive than this in fungi (Nebert et al, 1991).

Other P450's has been identified including alkane hydroxylases (CYP52), benzoate-para-hydroxylase (CYP53), cycloheximide-inducible P450 in *Neurospora crassa* (CYP54), a soluble P450 of *Fusarium oxysporum* (CYP55), sporulation-specific P450 of *S. cerevisiae* (CYP56) and pisatin demethylase (CYP57; for review see Nebert *et al.*,1991). Further P450's are likely including the fungal steroid hydroxylases (Brevskar *et al.*,1987) and possibly the other forms of P450 identified in P450 purification studies with *C. albicans* (Hitchcock *et al.*,1989). With the ability to extract active sterol 14α-demethylase from filamentous fungi this approach may also be extended (Ballard *et al.*,1990). Overall identification of the multiplicity of P450's present will be important for interpreting their contribution to azole tolerance as all P450's bind azole as a sixth ligand of their haem, but vary in the affinity of binding due to the differing apoprotein interaction with the N-1 substituent groups (Vanden Bossche, 1985). It is also possible that some of the P450s could be targets for new drugs.

A P450 benzo(a)pyrene hydroxylase has previously been purified from vegetative cells of *S. cerevisiae* (King *et al.*, 1984) with a molecular mass of 55500Da. This contrasts to the lanosterol 14α-demethylase molecular mass of 58000Da (Yoshida and Aoyama, 1984ab) which did not exhibit such an activity. However, both were reported as the single P450 form present under semi-anaerobic growth. Upon purification of the benzo(a)pyrene hydroxylase of strain NCYC754 we observed two proteins on electrophoresis using a 50cm

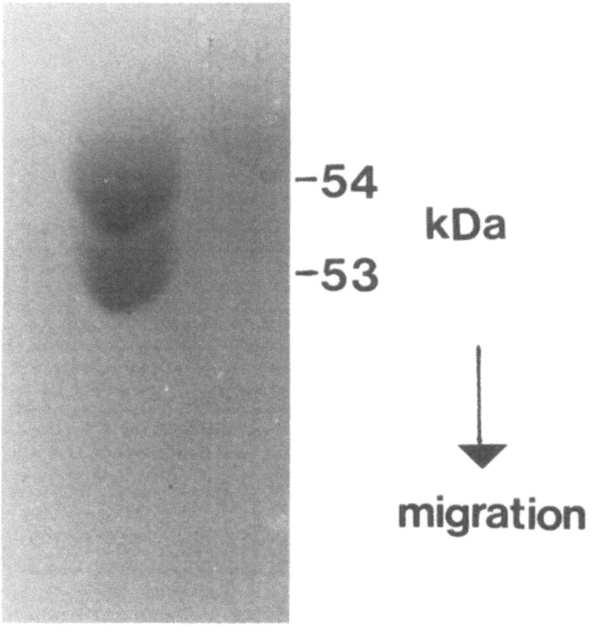

Figure 8. Separation of the two purified *S. cerevisiae* P450 proteins from NCYC 754 by SDS-PAGE.

apparatus for separation (Fig. 8).
The two proteins appeared very similar in V8 peptide mapping experiments and were not therefore likely to represent different families (Stansfield, 1990). Western blots with side-by-side comparisons indicated that this sample and the purified lanosterol 14α-demethylase of Yoshida and Aoyama, (1984) were proteins which have the same molecular weight (Fig. 9) and the immunological similarity has been confirmed using double-immunodiffusion experiments (Yoshida, personal communication). It appeared likely that both benzo(a)pyrene hydroxylase and lanosterol 14α-demethylase were the same P450, perhaps with allelic differences.
Recent evidence suggests further work is needed to resolve these differences. It might be expected from the results above that microsomes from *S. cerevisiae* lanosterol 14α-demethylase gene disruptants would not exhibit typical P450 Type I substate binding spectra for benzo(a)pyrene as observed in wild-type microsomal fractions (Kelly *et al.*, 1985). Figure 10 shows that Type I substrate binding spectra were obtained from microsomes derived from colony e (figure 6) upon addition of benzo(a)pyrene.
Further, a classical reduced carbon monoxide difference spectrum was observed indicating the presence of additional P450 species to the lanosterol 14α-demethylase. This may represent the benzo(a)pyrene hydroxylase or perhaps the sterol Δ^{22} desaturase which may also be a P450 (Hata *et al.*, 1987) and whose regulation may have been affected because of the lanosterol 14α-demethylase gene disruption.

Expression studies on *C. albicans* sterol 14α-demethylase

The activity of azole antifungals results from binding to the haem of P450 as

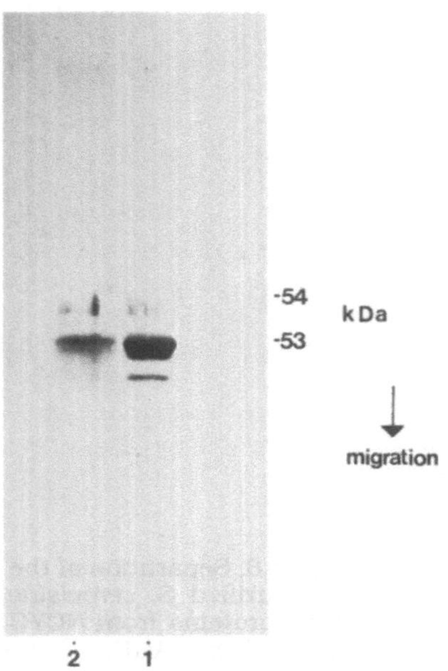

Figure 9. Western blot of P450 purified from NCYC 754 (2) and lanosterol 14α-demethylase (1) probed with anti-P450(NCYC 754) polyclonal antibodies.

a sixth ligand via the N-3 of imidazole and N-4 of triazole groups. Relative affinities for P450s are determined by the interaction of apoprotein and N-1 substituent group of the azole antifungals as explained previously. To date *in vitro* structure / activity analysis of azoles has employed both microsomal fractions and purified enzyme (Vanden Bossche, 1985; Yoshida and Aoyama, 1990). For analysis of the interaction of azole antifungals and the target enzyme of pathogens at the molecular level it is clear that heterologous gene expression provides an invaluable approach.

Previous studies have reported the expression and function of the *C. albicans* sterol 14α-demethylase in *S. cerevisiae* from it's own promoter (Kirsch *et al.*,1988). Only a relatively low level of expression could be obtained using this system, for instance 18 pmol.mg^{-1} microsomal protein in strain AH22 using pEMBLYe30 (a *leu2-3,2-112 his4-519*). This strain of *S. cerevisiae* forms the basis of a useful system for heterologous expression studies for enzyme production as no endogenous *S. cerevisiae* P450 is detectable using spectrophotometric techniques. The lanosterol 14α-demethylase gene disrupted strains produced contain the P450 shown in Figure 10 which may confuse investigation of P450/azole interaction. Attempts to elevate production of the *C. albicans* protein have achieved a yield of 101.7±17.5 pmol.mg^{-1} microsomal protein in strain AH22 (Fig. 11a). For this a *Bam* H1 restriction site was introduced into the *C. albicans* gene close to the initiation codon allowing the ligation of the *C. albicans* CYP51A1 gene behind the strong phosphoglycerate kinase promoter obtained from pMA91 to give pW91P (Fig. 11b; Mellor *et al.*,1983). The PGK promoter and the P450 coding sequence and terminator were then cloned into pEMBLYe30.

The P450 observed in the microsomal fraction of AH22::pW91P is predominantly the heterologous *C. albicans* protein and has enabled studies on the effect of expression on resistance and on the interaction of the *C. albicans* target protein with azoles without the requirement for purification. Figure 12

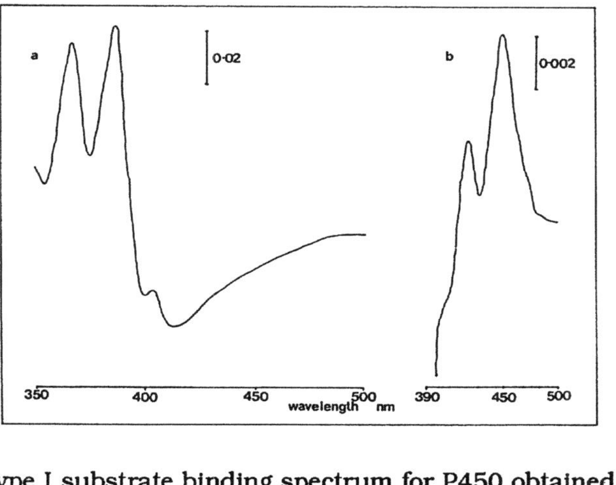

Figure 10. (a) Type I substrate binding spectrum for P450 obtained on addition of benzo(a) pyrene to microsomes of a strain containing a disrupted lanosterol 14α-demethylase gene; (b) Reduced CO difference spectrum for the same microsomal preparation and still exhibiting a classical P450 response.

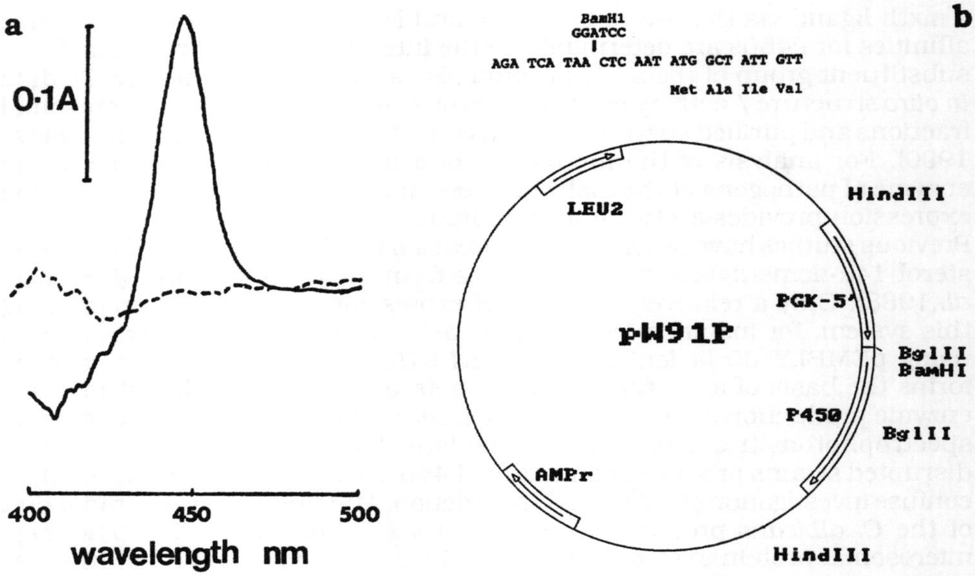

Figure 11. (a) Reduced CO difference spectrum showing the typical P450 spectral maximum obtained for microsomal fractions of AH22::pW91P transformants (–) and for AH22::pMA91 transformants where no P450 is detectable (---); (b) pW91P and the *Bam* H1 site introduced by *in vitro* mutagenesis close to the ATG of the gene.

shows the effect of expression on resistance comparing the AH22::pW91P, AH22::pWIG and AH22::pEMBLYe30 transformants. The construct pWIG contains the *C. albicans* CYP51A1,including it's promoter, cloned into the yeast shuttle vector pEMBLYe30 and produces 18 pmol P450.mg^{-1} microsomal protein in AH22. Heterologous expression produces approximately five fold resistance to both ketoconazole and fluconazole, but not to an equivalent extent to the increase in enzyme specific content. Also surprisingly there was no distinction between the azole sensitivity of transformants containing the different expression constructs.

The resistance of sterol 14α-demethylase defective strains containing pEMBLYe30 and pWIG has also been examined to investigate whether

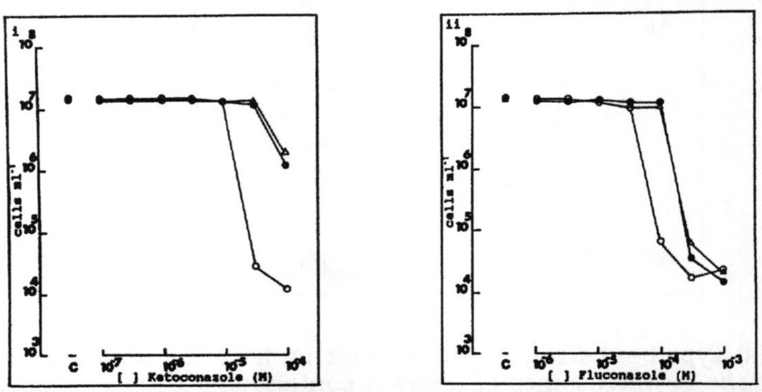

Figure 12. Cells/ml of *S. cerevisiae* AH22::pEMBLYe30 ( ), AH22::pWIG ( ) and AH22::pW91P ( ) grown for 72h in YM with (i) ketoconazole and (ii) fluconazole

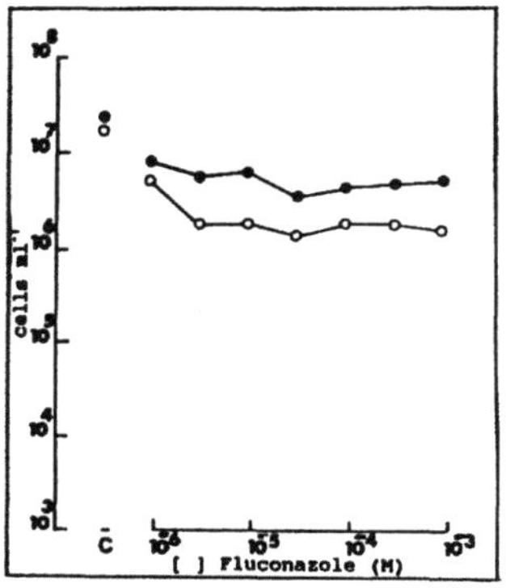

Figure 13. Cells/ml of *S. cerevisiae* DH4::pEMBLYe30 ( )and DH4::pWIG ( )grown for 72h with fluconazole.

complementation of the sterol 14α-demethylase defect changes the resistance. This can in any case be conferred by the *erg3* mutation in these strains and such experiments can indicate any contribution of the sterol 14α-demethylation defect to resistance. Figure 13 shows that the sensitivity of transformants of the strain DH4 (derived from JR4, Taylor *et al.*, 1983) remained unchanged to fluconazole (Kelly *et al.*,in preparation).

The selectivity of azole inhibition can also be examined through the sensitivity of transformants. Figure 14 shows the sensitivity of AH22::pMA91 and AH22::pCK1. The latter transformants express human P4501A1 at 33.3±10.8 pmol.mg^{-1} microsomal protein (Ching *et al.*, 1991) and yet do not detectably change resistance in transformants due to low affinity for these azoles.

The selectivity observed for the interaction of azole N-1 substituent group and apoprotein of the P450 can also be investigated through the use of these transformants. The investigation of enantiomers of antifungal azoles has previously indicated differing activities (Yoshida and Aoyama, 1990). Recently we identified an extreme example for two enantiomers obtained from Schering-Plough (Fig. 15i). Examination of the sensitivity of AH22::pEMBLYe30 and AH22::pW91P indicated that only one of the enantiomers had activity and the other was completely inactive (Fig. 15ii).

The presence of a convenient source of the *C.albicans* protein from AH22::pW91P allowed the basis of this difference to be examined. Azole

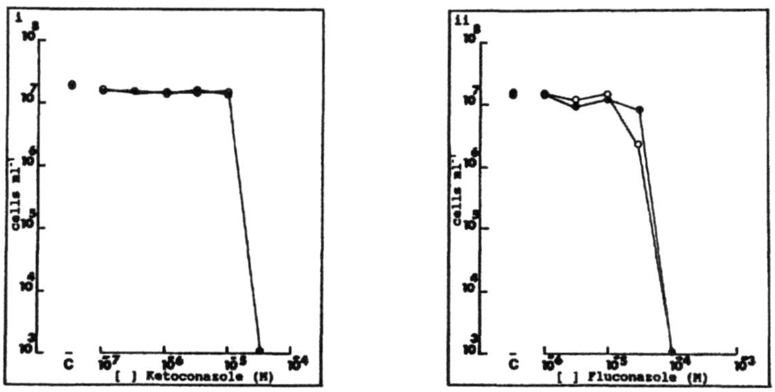

Figure 14. Cells/ml of *S. cerevisiae* AH22::pCK1 () and AH22::pMA91 () grown for 72h with (i)ketoconazole and (ii) fluconazole.

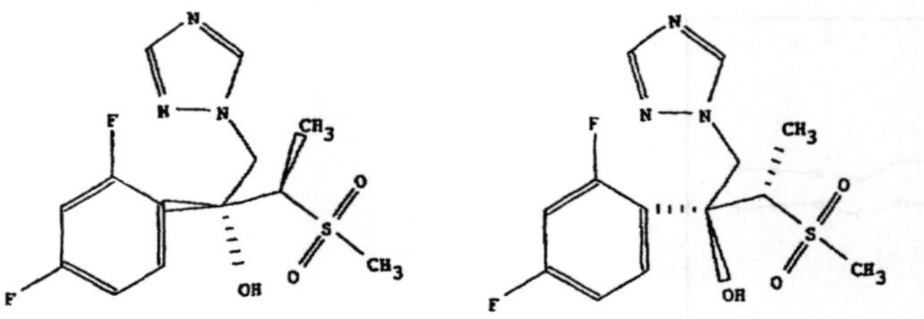

1

SCH-42427
(-)-(2R,3R)-2-(2,4-difluoro phenyl)-3-methylsulphonyl 1-(1,2,4-triazol-1-yl)-butan-2-ol

2

SCH-42426
(+)-(2S,3S)-2-(2,4-difluoro phenyl)-3-methyl sulphonyl 1-(1,2,4-triazol-1-yl)-butan-2-ol

ii

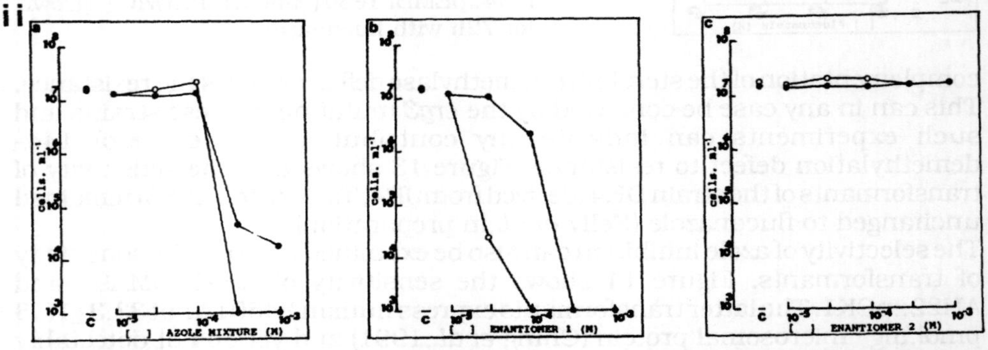

Figure 15. (i) Structures of the enantiomers- SCH42427 and SCH42426; (ii) Cells/ml of *S. cerevisiae* AH22::pEMBLYe30 ( ) and AH22::pW91P ( ) grown for 72h with (a) SCH39304 (mixture of the enantiomers), (b) SCH42427 and (c) SCH42426.

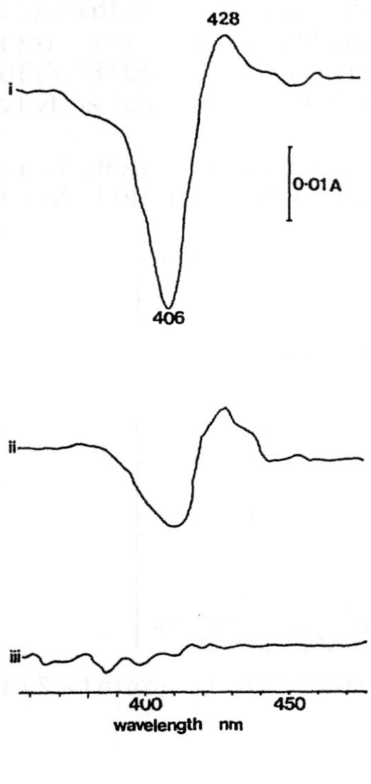

Figure 16. The spectra observed on addition of equimolar concentrations of (i) SCH42427, (ii) SCH39304 and (iii) SCH42426 to suspension of microsomes prepared from AH22::pW91P transformants containing 0.5μM P450.

binding to P450 results in a Type II spectrum which increases and becomes saturated at equimolar concentrations(for instance, Yoshida and Aoyama, 1990). The spectrum has a maximum at approximately 430nm and a minimum at approximately 410nm and refects the binding of azole as a sixth ligand to haem of a low spin state P450. Figure 16 shows the spectra obtained using 0.5μM P450 on addition of an equimolar concentration of SCH39304 (the mixture of both enantiomers), SCH42427 and SCH42426. A Type II spectrum was observed with a maximum at 428nm and minimum at 426nm for SCH39304 and SCH42427, but not SCH42426. The mixture also required approximately double the addition of azole required for SCH42427 to achieve saturation. This indicated that SCH42426 was inactive due to a faliure to bind to the sterol 14α-demethylase rather than representing a ligand with poor affinity. Such differences will prove invaluable in establishing a molecular model for the structure of this P450.

Conclusions and prospects

The application of molecular biology in studies on azole sensitivity will increase in the coming years in many of the areas described. For resistance studies the emphasis will be on the molecular basis of resistant isolates of pathogens from clinical or laboratory investigations. Also the multiplicity of P450s in pathogens will be addressed.
The expression systems in *S. cerevisiae* will, however, continue to be exploited as a molecular recognition approach to understanding azole/P450 interaction is undertaken. Crystallisation of eukaryotic membrane-bound P450 is still some way off and a molecular modelling approach offers the best prospect of a deeper understanding of this interaction. The ability to select for changes in azole sensitivity in transformants, coupled to the rich collection of elegant chemical probes which the azole antifungals represent, give some extra advantage to studies on this P450 over others. A model of the *S. cerevisiae* P450 has already been developed (Morris and Richards,1991) and this will be refined in conjunction with *in vitro* mutagenesis studies.

References

Ballard SA, Ellis SW, Kelly SL and Troke PF (1990) Ergosterol biosynthesis by a cell-free preparation of *Aspergillus fumigatus* and its inhibition by azole antifungal agents. J. Med. Vet. Mycol. 28:335-344.

Brevskar KK, Cresnar B and Hudnik-Plevnik T (1987) Resolution and reconstitution of cytochrome P450 containing steroid hydroxylation system of *Rhizopus nigricans*. J. Steroid Biochem. 26:499-501.

Cooper RA and Strauss JF (1984) Regulation of cell membrane cholesterol, In: Shinitzky M.(Ed) Physiology of Membrane Fluidity. Vol 1 CRC Press, Boca Raton, Florida.

Ebert E, Gaudin J, Muecke W, Ramsteiner K and Vogel C (1983)Inhibition of ergosterol biosynthesis by etaconazole in *Ustilago maydis*. Z. Naturforsch. 38c:28-34.

Henry SA (1982) Membrane lipids of yeast: Biochemical and genetic studies. In: Strathern J., Jones E., Broach J. (Eds) The molecular biology of the yeast *Saccharomyces cerevisiae*, Vol2, Cold Spring Harbor Press pp 101-158.

Hitchcock CA, Dickenson K, Brown SB, Evans EGV and Adams DJ (1989a) Purification and properties of cytochrome P450 dependant 14α-demethylase from *Candida albicans*. Biochem. J. 263:573-579.

Hata S, Nishino T, Katsuki A, Aoyama Y and Yoshida Y (1987) Characterisation of Δ^{22} desaturation in ergosterol biosynthesis of yeast. Agric. Biol. Chem. 51:1349-1354.

Hitchcock CA, Brown SB, Evans EGV and Adams DJ (1989b) Cytochrome P450 dependant 14α-demethylation of lanosterol in *Candida albicans*. Biochem. J. 260:549-556.

Kalb VF, Woods CW, Turi TG, Dey CR, Sutter TR and Loper JC (1987) Primary structure of P450 lanosterol demethylase gene from *Saccharomyces cerevisiae*. DNA 6:529-537.

Kato T (1986) Sterol biosynthesis in fungi, a target for broad spectrum fungicides. In: Haug G., Hoffman H. (Eds) Chemistry of Plant Protection, Vol 1. Springer-Verlag pp1-24.

Kelly SL, Watson PF, Kenna S, Ellis SW, Rose ME and Kelly DE (1990) Molecular genetic studies on yeast cytochrome P450 in relation to azole inhibition.In: Lyr H. (Ed) Systemic Fungicides and Antifungal Compounds.Tagungsbericht, Akademie der Landwirtschaftswis-senschaften, DDR, Berlin. 291:71-76.

Kenna S, Bligh HFJ, Watson PF and Kelly SL (1989) Genetic and physiological analysis of azole sensitivity in *Saccharomyces cerevisiae*. J. Med. Vet. Mycol. 27:397-406.

King DJ, Azari MR and Wiseman A (1984) Studies on the properties of highly purified cytochrome P448 and its dependant activity in benzo(a)pyrene hydroxylase from *Saccharomyces cerevisiae*. Xenobiotica 14:187-206.

Kirsch DR, Lai MH and O'Sullivan J (1988) Isolation of the cytochrome P450 LIA1(lanosterol 14α-demethylase) from *Candida albicans*. Gene 68:229-237.

Mellor J, Dobson MJ, Roberts NA, Tuite MF, Emtage JS, White S, Lowe PA, Patel T, Kingsman AJ and Kingsman SM (1983) Efficient synthesis of enzymatically active calf chymosin in *Saccharomyces cerevisiae*. Gene 24:1-14.

Mercer EI (1984) The biosynthesis of ergosterol. Pestic. Sci. 15:133-155.

Morris GM and Richards WG (1991) Molecular modelling of the sterol C-14 demethylase of *Saccharomyces cerevisiae*. Biochem. Soc. Trans. 769:793-795.

Nebert DW, Nelson DR, Coon MJ, Estabrook RW, Feyereisen R, Fuji-Kuriyama Y, Gonzalez FJ, Guenguerich FP, Gunsalus IC, Johnson EF, Loper JC, Sato R, Waterman MR and Waxman DJ (1991) The P450 superfamily: Update on new sequences, gene mapping, and recommended nomenclature. DNA and Cell Biology 10:1-14.

Pierce AM, Mueller RB, Unrau AM and Oehlschlager AC (1978) Metabolism of Δ^{24} sterols by yeast mutants blocked in the removal of the C-14 methyl group. Can. J. Biochem. 56:794-800.

Quail MA, Arnoldi A, Moore DJ, Goosey MW and Kelly SL (1992) Ketoconazole mediated growth inhibition in *Botrytis cinerea* and *Saccharomyces cerevisiae*. Phytochemistry (in press).

Shimokawa O, Kato Y, Kawano K and Nakayama H (1989) Accumulation of 14α-methylergosta-8,(24)-dien-3B,6α-diol in 14α-demethylation mutants of *Candida albicans*: genetic evidence for the involvement of 5-desaturase. Biochem. Biophys. Acta 1003:15-19.

Stansfield IS (1990) Purification and regulation of P450 in yeast. PhD thesis. University of Sheffield.

Taylor FR, Rodriguez RJ and Parks LW (1983a) Requirement for a second sterol biosynthetic mutation for viability of a sterol C-14 demethylation defect in *Saccharomyces cerevisiae*. J. Bacteriol. 155:64-68.

Trocha PJ, Rodriguez RJ and Sprinson DB (1977) Yeast mutants blocked in removing the methyl group of lanosterol at C-14. Seperation of sterol by high pressure liquid chromatography. Biochemistry 16:4721-4726.

Vanden Bossche H (1985) Biochemical targets for antifungal azole derivatives: hypothesis on the mode of action. In: M.R. McGinnis (Ed) Curr. Topics Med. Mycol., Vol1. Springer-Verlag, pp313-350.

Vanden Bossche H, Marichal P, Gorrens J, Bellens D, Coene M, Lauwers W, Jeune LL, Moereels H and Janssen PAJ (1990) Mode of action of antifungals of use in immunocompromised patients. Focus on *Candida glabrata* and *Histoplasma capsulatum.* In: Vanden Bossche H., Mackenzie D.W.R., Cauenberg G., Van Cutsum J., Drouhet E. and Dupont B. (Eds.) Mycoses in AIDS Patients. Plenum pp223-243.

Watson PF, Rose ME and Kelly SL (1988) Isolation and characterisation of ketoconazole resistant mutants of *Saccharomyces cerevisiae.* J. Med. Vet. Mycol. 26:153-162.

Watson PF, Rose ME, Ellis SW, England HE and Kelly SL (1989) Defective sterol C5-6 desaturation and azole resistance: A new hypothesis on the mode of action of azole antifungals. Biochem. Biophys. Res. Commun. 164:1170-1175.

Yoshida Y and Aoyama Y (1984) Yeast cytochrome P450 catalysing lanosterol 14α-demethylation. I. Purification and spectral properties. J. Biol. Chem. 259:1655-1660.

Yoshida Y and Aoyama Y (1990) Stereoselective interaction of an antifungal agent with its target, lanosterol 14α-demethylase. Chirality 2: 10-15.

Yoshida Y, Kumaoka H and Sato R (1974) Studies on the microsomal electron transport system of anaerobically grown yeast. J. Biochem. 78: 1201-1210.

Antifungal chemotherapy and drug strategies: Projected clinical needs

Gerald Medoff
Department of Medicine and Molecular Microbiology
Infections Diseases Division Box 8051
Washington University School of Medicine
660 South Euclid
St. Louis, Missouri 63110
USA

Introduction

In recent years, the incidence of systemic fungal infections has increased tremendously, particularly in severely ill and immunocompromised patients. Adequate treatment of these infections when they occur in the normal host is available. Amphotericin B (AmB) is very effective, but its toxicity and inconveniences associated with its intravenous administration are important negative factors associated with its use. Present attempts to decrease toxicity of AmB by combining it with other drugs and/or by developing unique delivery systems are interesting but the results are only preliminary. Azoles are also effective and ketoconazole is particularly attractive because it can be administered orally. The triazoles, itraconazole and fluconazole, represent promising agents in this group because they have a broader range of action, are well absorbed through the gastrointestinal tract, have a prolonged half life, have fever side effects and also penetrate well into the central nervous system and urine in an active form. Despite these advances, significant problems exist. None of the antifungal agents appear to be curative in patients with AIDS. Consequently, life-long suppressive therapy is necessary to prevent recurrence of disease. Drug resistance is an increasing problem with the potential for cross-resistance among the presently available agents which all affect the fungal cell envelopes. Clearly, there is a need to develop antifungal agents which have completely different sites of action to avoid cross resistance. Different classes of drugs affecting different cellular metabolic processes might interact with polyenes and azoles synergistically to enhance therapeutic effects. There is also a need to understand the basis for susceptibility to systemic fungal infections and to devise strategies based on these insights to prevent them.

My plan in this review is to briefly review clinically relevant data on the presently available antifungal agents and to comment on their strengths and limitations. I will then attempt to evaluate the role of each of these agents in the treatment of systemic fungal infections.

NATO ASI Series, Vol. H 69
Molecular Biology and its Application to Medical Mycology
Edited by B. Maresca, G. S. Kobayashi and H. Yamaguchi

Presently available antifungal agents

Amphotericin B. AmB is the only polyene antifungal that can be used systemically in patients probably because its relative binding specificity for ergosterol, the sterol in fungal cytoplasmic membranes is greater than cholesterol, the sterol in animal cell membranes. This binding to membrane sterols is probably the basis for its antifungal effects (Brajtburg, et al., 1990). AmB is fungicidal and this property in addition to its broad spectrum of action against many different fungi have made AmB the most effective antifungal agent that has been available. There are, however, fungal infections which fail to respond to AmB. Invasive aspergillosis in markedly immunocompromised patients and coccidioidal meningitis are examples of these limitations. In addition, there have been troublesome reports of infections with fungi inately resistant to AmB such as *Candida lusitaniae* and *Pseudoallescheria boydii.*
The major disadvantage of the conventional clinical formulation of AmB (Fungizone) is the high frequency of adverse reactions. All of the patients treated with Fungizone experience acute reactions such as chills, fever, and even hypotension. Decreased renal function is invariable in essentially every patient receiving Fungizone, but fortunately it is reversible when the drug is discontinued. Although it is possible to diminish the adverse reactions or to treat through them, their frequency and severity has stimulated considerable effort to develop different formulations with reduced toxicity. Initially, these efforts were focussed on attempts to alter the AmB molecule, but any such changes were associated with decreases in antifungal effects. The methyl ester of AmB was the most intensively studied of these analogues because it appeared to decrease toxicity to animal cells more than to fungi (Hoeprich *et al.*, 1987, Howarth *et al.*, 1975). Animal studies confirmed decreased nephrotoxicity, and based on these studies, several patients with central nervous system coccidioidomycosis were treated with the methyl ester of AmB. Unfortunately, an unanticipated toxicity of this analogue of AmB to central nervous system ended further clinical studies. The mechanism of this toxicity is unknown.
More recent attempts to decrease the toxicity of AmB have focussed on altering the carrier or delivery vehicle of the drug (Brajtburg, et al., 1990). These new formulation have utilized liposomes, lipo-proteins, phospholipids or detergents and all serve to alter the physical properties of AmB. The most promising of these new formulations appear to significantly decrease the toxicity of AmB without altering its antifungal properties. Most of these preparations are still being tested in animals, but several have been given to humans in early studies of toxicity and compassionate use for intractable fungal infections (Lopez-Berestein *et al.*, 1987). Although these initial results are promising, it is too early to judge whether or not these new formulations will be clinically useful.

5-fluorocytosine

Flucytosine is an oral antifungal agent and acts by inhibiting RNA and DNA synthesis (Medoff *et al.*, 1983). In addition to its convenience as an oral agent, it also penetrates well into central nervous system and is excreted in an active form in the urine. However, its spectrum of action is limited to yeast, and unfortunately resistance is common either as a primary event or developing during treatment. Toxicity can also be a problem because blood levels above 100 μg/ml can suppress bone marrow function. For this reason blood levels must be followed closely and doses have to be adjusted for compromised renal function. All of these factors significantly limit its usefulness and normally flucytosine is only used in combination with AmB to treat cryptococcal meningitis and invasive candida infections. The value of combination therapy to treat cryptococcal meningitis in non-AIDS patients has been established by two large clinical studies (Bennett *et al.*, 1979; Dismukes *et al.*, 1987).

Azoles

This family contains two main groups - the imidazoles and the triazoles (Fromtling, 1988). Only miconazole and ketoconazole (imidazoles) and fluconazole and itraconazole (triazoles) are currently available clinically for the treatment of systemic fungal infections. The azoles work by inhibiting the cytochrome P-450 dependent enzyme causing the 14-α demethylation of lanosterol to ergosterol in the fungal cell membrane. The affinity for the fungal enzyme is greater than for analogous mammalian enzymes involved in adrenal androgen pathways, but the specificity is not complete. This may result in interference with synthesis of these hormones in the host (Carver, 1991). Other toxicities include rare cases of hepatitis, but in general these are remarkably non-toxic agents. The most important problems associated with this family of drugs is that: they are fungistatic, they have a relatively narrow spectrum and resistance in fungi has developed after long-term use. The fungistatic effects may severely limit chances of cure of infection in immunocompromised patients. The relatively narrow spectrum compromises uses of these agents in broad spectrum empiric therapy and also in prophylaxis. Resistance in fungi is still an emerging problem, but the fact that it already has occurred is disturbing. The development of resistance appears to be dependent on at least two mechanisms. One involves decreased uptake into the fungi and the other bypasses the effectiveness of the blockade of ergosterol synthesis by substituting another sterol in the fungal cell membrane. The latter may also result in resistance to AmB, a particularly disturbing prospect (Watson *et al.*, 1989). The mechanism of this cross resistance may be based on a decreased affinity of AmB to the sterol substituted for ergosterol in the azole resistant fungus.

Miconazole can only be given to patients intravenously and therefore is as inconvenient as AmB (Bennett and Remington, 1981). With the exception of the rare infection due to *Pseudallescheria boydii*, miconazole is not as effective as AmB and therefore is rarely used.

Ketoconazole is an oral agent which requires gastric acidity for absorption. Variable blood levels due to poor absorption, inactivation by the liver with no active drug in the urine, poor penetration into central nervous system and a relatively narrow spectrum of action against fungi severely limit its usefulness (Fainstein *et al.*, 1987).

Fluconazole can be given both orally and intravenously. It is well absorbed after oral administration and is excreted unchanged in the urine. It penetrates into central nervous system in therapeutic concentrations and has been used successfully to treat cryptococcal meningitis in patients with AIDS. It is not possible to cure AIDS patients of this infection and fluconazole must be given for life to prevent relapse (Hay, 1991).

The other new triazole, itraconazole, is given orally and is less well absorbed than fluconazole. It is effective against *Candida sp.*, *Cryptococcus neoformans* and the endemic mycoses (histoplasmosis, blastomycosis, coccidioidomycosis) and is probably also an effective agent in the treatment of systemic aspergillosis (Graybill and Craven, 1988). It has not yet been released for general use by the FDA in the United States so that clinical experience with this agent is still rather limited.

Combination therapy

The combination of AmB and flucytosine has been shown to be more effective than either drug alone in the treatment of cryptococcal meningitis in non-AIDS patients (Bennett *et al.*, 1979). *In vitro* and animal data indicate that this combination may also be beneficial in the treatment of disseminated candidal infection but there is no adequate clinical data to confirm this. This is also true for AmB plus flucytosine and AmB plus rifampin in candidal and cryptococcal infections and also for other systemic fungal infections such as aspergillosis and mucormycosis (Medoff *et al.*, 1983).

The *in vitro* and animal data using AmB and several different azoles are mixed with some results showing synergism and others antagonism. In theory, the results showing antagonism should be more believable since inhibition of ergosterol synthesis by azoles should decrease AmB binding to fungal cell membranes and consequently decrease its effectiveness. Recently there is animal data suggesting fluconazole and flucytosine may have synergistic effects but these experiments should be confirmed. If these results are substantiated, clinical trials should proceed rapidly.

Projected clinical needs

AmB has been and continues to be a remarkably effective antifungal agent. Its broad spectrum of action and fungicidal effects make it the "gold standard" by which other agents have to be compared. The inconvenience of intravenous administration and its toxicity are important negative factors in its use. The development of new preparations of AmB such as the lipid formulations offers some promise of decreased toxicity.

The azoles, particularly the triazoles, are important additions to the spectrum of antifungal therapy because they can be given orally and because of the low level of toxicity. Their efficacy and their role are still being studied, but the fact that they are only fungistatic is discouraging. Perhaps newer azoles or use of these agents in combination with AmB or flucytosine will improve their antifungal effects.

The major challenges in antifungal therapy in the future will be in dealing with the increase in resistance of fungi to the presently available antifungal agents. Understanding the basis of this resistance and also development of antifungal agents with other unique mechanisms of action will be important in dealing with this problem. The recent work on cell wall inhibitors, myristylation analogs and inhibitors and protease inhibitors is interesting and it is vital to continue these lines of investigation.

A more important challenge relates to the nature of the patient who is the unfortunate victim of most cases of intractable invasive fungal infections. These patients are severely immunocompromised because of the nature of their underlying disease (e.g. AIDS, malignancy) and/or the therapy used to treat these diseases (e.g., organ transplants, immunosuppressive therapy, antitumor agents). It may be that no antifungal agent will be effective in these circumstances. Here a more imaginative approach is in order which concentrates on improving therapy for these underlying diseases so that less immunosuppression results. The recent use of hematopoietic growth factors and bone marrow stem cell infusions to decrease the time of bone marrow suppression in patients treated for malignancies is an example of this strategy. Ultimately, it is likely that combining efforts to improve antifungal agents and decrease immunosuppression will have the biggest impact on these devastating infections.

References

Bennett JE, Dismukes WE, Duma RJ, Medoff G, *et al.* (1979) A comparison of amphotericin B alone and combined with flucytosine in the treatment of cryptococcal meningitis. N. Engl. J. Med. 301:126-f131.

Bennett JE, Remington JS (1981) Miconazole in cryptococcosis and systemic candidiasis: a word of caution. Ann. Int. Med. 94:708-709.

Brajtburg J, Powderly WG, Kobayashi GS, Medoff G (1990) Amphotericin B: Current understanding of mechanisms of action. (Review) Antimicrob. Ag. Chemother. 34:183-188.

Brajtburg J, Powderly WG, Kobayashi GS, Medoff G (1990) Amphotericin B: Delivery systems. (Review) Antimicrob Ag Chemother 34:381-384.

Carver PL (1991) Clinically relevant drug interactions and adverse effects of antimicrobial agents. Curr. Opin. Inf. Dis. 4:749-756.

Dismukes WE, Cloud G, Gallis HA, Kerkering TM, Medoff G et al (1987) Treatment of cryptococcal meningitis with combination amphotericin B and flucytosine: Results of a prospective, randomized clinical trial. N. Engl. J. Med. 317:334.

Fainstein V, Bodey GP, Elting L, Makymink A, Keating M, McCredie KB (1987) Amphotericin B or ketoconazole therapy of fungal infections in neutropenic cancer patients. Antimicrob. Ag. Chemother. 31:11-15.

Fromthing RA (1988) Overview of medically important antifungal azole derivatives. Clin. Micro. Rev. 1:187-217.

Graybill JR, Craven PC (1983) Antifungal agents used in systemic mycoses: activity and therapeutic use. Drugs 25:41-62.

Hay RJ (1991) Overview of the treatment of disseminated fungal infections. J. Antimicrob. Chemother. 28, Suppl B, 17-25.

Hoeprich PD, Flynn NM, Kawachi MM, Lee KK, Lawrence RM, Heath LK, Schaffner (1987) Treatment of fungal infections with semisynthetic derivatives of amphotericin B. Ann. N.Y. Acad. Sci. 544:517-546.

Howarth WR, Tewari RP, Solotorovsky M (1975) Comparative *in vitro* antifungal activity of amphotericin B and amphotericin B methyl ester. Antimicrob. Ag. Chemother. 7:58-63.

Lopez-Berestein G, Bodey GP, Frankel LS, Mehta K (1987) Treatment of hepatosplenic fungal infections with liposomal amphotericin B. J. Clin. Oncol. 5:310-317.

Medoff G, Brajtburg J, Kobayashi GS, Bolard J (1983) Antifungal agents useful in

therapy of systemic fungal infections. Ann. Rev. Pharmacol. 23:303-330.
Weston PF, Rose ME, Ellis SW, England H, Kelly SL (1989) Defective sterol C5-C6 desaturation and azole resistance: a new hypothesis for the mode of action of azole antifungals. Biochem. Biophys. Res. Comm. 164:1170-1175.

Funding medical mycology: Strategies for attracting the private sector

John N. Galgiani
Department of Medicine (111)
Veterans Affairs Medical Center
University of Arizona College of Medicine
6th and Ajo Streets
Tucson, Arizona 85723
USA

Introduction

In past years, fundamental scientific research has enjoyed support from a variety of governmental and philanthropic sources. In the United States, for example, these have included investigator-initiated and centrally-initiated research programs through the National Institutes of Health, research agencies of various medical specialties such as the American Lung Association, investigators' own institutions, and research funds distributed by individual states including New York, Texas, and Arizona. Although such support continues, it has not expanded to accommodate the number of scientists currently interested in pursuing meritorious research programs to their fullest. It is therefore useful to consider alternative means of financial support. The purpose of this essay is to reconsider the option of industrial or private-sector support of medical mycology.

If private-sector investment in fundamental research is underexploited, it may be the result of a lack of understanding between independent scientists and those within industry. In general, independent investigators generally perceive industrial interests as directed to explicit goals, almost always related to a marketable product, and with the primary emphasis on producing a product. In contrast, those planning industrial programs may perceive academic research as exploratory, highly theoretical, and very distant from practical applications. Although there are many examples in both the private sector and in academic studies to reinforce these perceptions, there are also areas of overlapping interests that are often overlooked. If better recognized, these common interests might result in mutually advantageous collaborations.

Applied nature of medical mycology

Although general biologic patterns may be learned from studying pathogenic fungi, most do not make ideal model biologic systems for study. For example, many do not mate. For others that do, mating is difficult to control. In addition, work with pathogens is commonly dangerous. Elaborate containment procedures often are needed for even the simplest experiments. Moreover, since very few fungi are pathogens, it is axiomatic to view them as exceptions,

NATO ASI Series, Vol. H 69
Molecular Biology and its Application to Medical Mycology
Edited by B. Maresca, G. S. Kobayashi and H. Yamaguchi

in many ways not at all representative of the entire fungal kingdom. Thus, lessons learned in studying pathogens may provide counterpoint rather than the most general biologic patterns. These are but a few of the characteristics of pathogenic fungi that make them problematic for study. Nonetheless, some investigators have chosen to study them anyway because they provide opportunities to study organisms that impact on human health. The questions pursued by a medical mycologist may or may not clearly be seeking a specific intervention. However, it is conceivable that new understanding about any aspect of medically relevant fungi might suggest medically useful interventions. If viewed in this light, *most* research in medical mycology is potentially valuable to industrial concerns and financial support of such studies as long term investments could be profitable.

If one accepts the premise that there are areas of mutual interest between fundamental medical mycologists and the private sector, then consideration can shift to optimal exploitation of the overlap. Certainly it is possible for large corporations to channel block grants to support centers of excellence within the academic community. However, more relevant to the majority of individual investigators who are outside of such centers are strategies that involve their own specific lines of research, and the remainder of this discussion will expand upon this theme.

Indirect funding

One method of recruiting private sector support for medical mycological research has been to develop indirect funding arrangements. It is common for industrial sponsors to identify, in their own terms, specific research needs and then to offer contracts to investigators outside of their own company to meet those needs. As part of such arrangements, investigators earn a certain amount of "profit" for their efforts. These funds then become discretionary for support of other work not specifically relevant to the interests of the corporate sponsor.

Indirect support of this type has several advantages. It is perhaps the most common means of private sector support of medical mycology and therefore has many examples to use for developing future relationships of the same sort. Preclinical antifungal drug evaluations, clinical trials of antifungal drug efficacy, and laboratory support of clinical trials are specific examples. Another advantage is that funding of this nature is governed by market forces. Budgets are set realistically based on real costs of personnel time and supplies. Often such budgets are more realistic than those funded by governmental or philanthropic agencies. Finally, support for such work is typically flexible. How fastor for what category of expense funds are expended is usually not of particular interest to the sponsor.

On the other hand, there are also potential disadvantages in these sorts of arrangements. Since support is for one project while actually conducting another, success and productivity of the indirectly supported project has no bearing on the level of continued support. This uncoupling may have a negative effect on the project in the long run. A second potential risk involves the time commitment to the industrially-sponsored project. Indirect funding works to investigator's best advantage when the corporate contract is well within the investigator's broadest interest. To the extent that corporate contracts actually detract from investigator's personal interests, the indirect funding strategy may represent a significant hinderance. Finally, research supported in this way is not subjected to peer review and as a result, receives neither peer criticism nor peer encouragement. As a result, an important mechanism of constructive criticism is bipassed.

Direct funding

An alternative method of private sector participation in medical mycology research is to receive direct support for fundamental studies the results of which might suggest new ideas for marketable products. Research of this sort is conducted within some companies in an on going fashion. However, much more research is conducted outside of corporations in academic institutions and these results could be of value to industry. The exact outcome from such pursuits are not as predictable as directed goal-oriented programs. On the other hand it is almost a certainty that fundamental research will lead to profitable insights of one form or another. Therefore, this strategy is most analogous to "venture capital" relationships which have both high risk and potentially high gain.

In considering direct support of fundamental studies by private industry, the advantages seem clear. The investigator receives funding for exactly what he or she wishes to pursue, and the sponsor has early access to information that could lead to innovative and lucrative products. However, there are possible disadvantages as well. These might include restrictions by a sponsor on the reporting of findings and conflicts concerning legal ownership of any discoveries. Additionally, there are not very many examples of direct funding of independent investigators and therefore there are few standard practices to guide either scientists or private sector interests.

Summary

Private sector support of medical mycology is a logical and potentially valuable collaboration for both independent scientists and corporations. Both medical mycologists and those in the private sector have substantial interest in the "clinical relevance" of research with pathogenic fungi; this common ground could provide the basis for developing research support. Hopefully, both medical mycologists and those in the private sector will increasingly realize these areas of mutual interest and as result funds from the private sector will increase.

Fungal Morphogenesis

Chromosomal organization of *Histoplasma capsulatum*

P.E. Steele and R. L. Himes
Department of Pathology and Laboratory Medicine
University of Cincinnati
231 Bethesda Avenue
Cincinnati, Ohio 45267
USA

Large-DNA electrophoresis of strains of *Histoplasma capsulatum* has revealed up to seven chromosomes, with marked variation in the electrophoretic mobility of the chromosomes among different strains (Steele *et al.*, 1989, 1991). The size of the chromosomal bands is in the megabase range, extending from the size of the largest *Saccharomyces cerevisiae* chromosomes to beyond that of the largest *Schizosaccharomyces pombe* chromosomes. Resolution of these bands required low field strength electrophoresis (0.5 - 1.0 volt/cm), with long switching times (up to 1 hour) and long gel runs (up to 1 week). In the Downs strain of *H. capsulatum*, six bands were resolved with contour-clamped homogeneous electric field (CHEF) gels, whereas five bands were resolved with field-inversion gel electrophoresis (FIGE). The use of chromosomal band-specific probes (most of which were random sequences) permitted the determination that at least seven chromosome-sized DNAs were present in the Downs strain. The resolution of these megabase bands by FIGE was illustrative of a double-valued relation between mobility and size described by Carle *et al*, (1986) in the original description of FIGE, in which larger DNAs migrate with increasingly greater mobility and thus co-migrate with much smaller DNAs. In the Downs, the four largest chromosomal DNAs migrated in this fashion. Although sizing of DNAs is problematic with FIGE, the degree of resolution of very large molecules can be facilitated with the use of FIGE by exploiting this property.

The Downs strain of *H. capsulatum* was studied in greater depth than other strains simply because it exhibited the largest number of resolvable bands. The applicability of the results from the Downs strain to other *H. capsulatum* strains was uncertain, since restriction fragment length polymorphism (RFLP) studies had revealed that the Downs strain appeared to be unique, unlike other commonly used North and Central American isolates (Vincent *et al*,, 1986). However, electrophoretic karyotyping of clinical strains from Barnes Hospital in St. Louis revealed that a clinical isolate from a patient with AIDS and disseminated histoplasmosis appeared similar to the Downs strain. An RFLP study of such isolates was carried out by Spitzer *et al*, (1990) which revealed Downs-like isolates among AIDS patients with histoplasmosis. It is nevertheless clear that the Downs strain electrophoretic karyotype is different from other commonly used laboratory strains, which are themselves different from each other (Steele, 1991). The G217B strain contains two chromosomal DNA bands in the 3.5 to 5.0 megabase (mb) range and then at least one very large band which migrates more slowly than the largest S.

NATO ASI Series, Vol. H 69
Molecular Biology and its Application to Medical Mycology
Edited by B. Maresca, G. S. Kobayashi and H. Yamaguchi

pombe chromosome. The G186B strain contains a very small chromosomal band at 0.5 mb, two or three bands at 4.0-4.6 mb, and at least two very large DNAs. The G184B strain, which is similar to G186B by RFLP analysis (Vincent *et al.*, 1986) appears very different by electrophoretic karyotype analysis, including the fact that it does not contain the 0.5 mb chromosome. In order to probe the basis of the electrophoretic karyotype dissimilarity between strains of *H. capsulatum*, we carried out a study of the small chromosome of G186B. Use of the small chromosome as a probe of a genomic library of G186B DNA revealed mostly repetitive sequences (see below). However, one clone obtained hybridized only to the small chromosome and one other band, the very slowly migrating chromosomal band of G186B. Use of contiguous restriction fragments encompassing the entire insert of this clone as probes against Southern blots of chromosomal DNA revealed that the entire insert hybridized in this manner (as opposed to a repetitive element contained within the recombinant insert).

A culture of G186B which had been multiply passaged was found to have lost the 0.5 mb band. Restriction of this G186B subculture DNA and newly obtained G186B DNA from the American Type Culture Collection (which contains the 0.5 mb band) revealed two restriction length polymorphisms involving *Sal* I.

This data suggests that a duplication of sequences on the large chromosome of G186B resulted in the formation of the smaller chromosome. Thus, gross genomic rearrangement or alteration appears to characterize these *H. capsulatum* strains. Such results may have implications for the use of parasexual genetic techniques such as protoplast fusion to study genetic complementation in this organism. Further study of the structure of the small chromosome of G186B may provide clues to genomic elements which could serve in a vector for introducing DNA into this organism.

Some of the clones obtained by using the entire small chromosome as a probe of the G186B genomic library contained inserts derived from other chromosomes but which contained repetitive DNAs present on the small chromosome. One such element was studied in detail and found to be at least 5 kb long, with termini which hybridized to one another. This feature is reminiscent of the structure of transposable elements and thus this clone may represent such an element.

The copy number of this element appears to differ among strains, with fewer copies in the Downs strain.

References

Carle GF, Frank M, Olson MV (1986) Electrophoretic separations of large DNA molecules by periodic inversion of the electric field. Science 232:65-68.

Spitzer ED, Keath EF, Travis SJ, Painter AA, Kobayashi GS, Medoff G (1990) Temperature-sensitive variants of *Histoplasma capsulatum* isolated from patients with acquired immunodeficiency syndrome. J. Infect. Dis. 162:258-261.

Steele PE, Carle GF, Kobayashi GS, Medoff G (1989) Electrophoretic karyotype of *Histoplasma capsulatum* chromosomal DNA. Mol. Cell. Biol. 9:983-987.

Steele PE (1991) Current concepts of fungal virulence. Adv. Pathol. Lab. Med. 4:107-119.

Vincent RD, Goewert R, Goldman WE, Kobayashi GS, Lambowitz AM, Medoff G (1986) Classification of *Histoplasma capsulatum* isolates by restriction fragment polymorphisms. J. Bacteriol. 165:813-818.

Chitin, chitin synthase and chitin synthase conserved region homologues in *Wangiella dermatitidis*

Paul J. Szaniszlo and Michelle Momany
Department of Microbiology
The University of Texas at Austin
Austin, TX 78712
USA

Introduction

The phenomenon of polarized growth leading to yeast and hyphal phenotypes in fungi is dependent upon the coordinated expression of genes and the complex interactions of their products (Drubin, 1991). While certain aspects of polarized growth are beginning to be understood in *Saccharomyces cerevisiae*, study of this fungus alone may not lead to a complete appreciation of the developmental decisions made by most medically important fungi, because this nonpathogen is essentially monomorphic. In contrast, *Wangiella dermatitidis* is an exquisitely suitable model for studies of polarized growth and cellular differentiation among pathogenic fungi, because it is polymorphic and can grow as a budding yeast, a hyphomycetous conidiogenesis hyphal organism or as a multicellular mass. In fact, most of what is known about the growth of fungi in a multicellular phenotype comes from studies of this melanized agent of phaeohyphomycosis of humans (reviewed in Szaniszlo *et al.*, 1983a; 1983b; Geis and Jacobs, 1985; Dixon *et al.*, 1991). It is now clear that multicellular-form induction in *W. dermatitidis* is a cell-cycle-linked event, resulting directly from the inhibition of polarized bud expansion or hyphal tip elongation, without the inhibition of growth, DNA synthesis, mitosis and cytokinesis (Roberts and Szaniszlo, 1978; 1980; Roberts *et al.*, 1979; 1980; Jacobs and Szaniszlo, 1982). Our previous findings that chitin content increases and deposition is delocalized during yeast-to-multicellular form transition (Szaniszlo *et al.*, 1983a; Cooper *et al.*, 1984; Jacobs *et al.*, 1985; Harris and Szaniszlo, 1986), coupled with the discovery of multiple chitin synthase enzymes in *S. cerevisiae* (Bulawa *et al.*, 1986; Orlean, 1987; Cabib *et al.*, 1988; Silverman, *et al.*, 1988), prompted our search for multiple chitin synthases in *W. dermatitidis*. Our preliminary results suggest that multiple chitin synthases are involved in the complex phenotypic transitions associated with this fungus.

Vegetative polymorphism and the cell cycle in *Wangiella*

An initial critical discovery related to vegetative polymorphism in *W. dermatitidis* was our finding that uniformly-aged yeast cells undergo nearly

NATO ASI Series, Vol. H 69
Molecular Biology and its Application to Medical Mycology
Edited by B. Maresca, G. S. Kobayashi and H Yamaguchi

quantitative transitions to hyphal forms only after acquisition of spore-like cytoplasm and thick walls (Oujezdsky *et al.*, 1973). However, yeasts that have not acquired these spore-like characteristics, i.e., those aged for shorter periods on agar medium surfaces, or to early stationary phase in liquid medium, simply resume growth by yeast budding when inoculated into fresh medium.

A second critical discovery in this field was our finding that incubation of *W. dermatitidis* yeasts under acidic conditions induces their quantitative conversion to a multicellular phenotype (Szaniszlo *et al.*, 1976). This conversion to the multicellular morphology at 25°C is within two to five days after inoculation of logarithmically growing yeasts into media acidified to pH 2.5. Subsequent return of the multicellular bodies, or their enlarged thick-walled precursors, to media with a pH near neutrality allows hyphal outgrowth. These hyphae then produce lateral hyphal buds that regenerate the yeast phase. Thus, the simple manipulation of pH can be used to induce both multicellular and hyphal morphologies from yeast cells of *W. dermatitidis*. The hyphal morphologies are ultrastructurally identical to those induced from aged yeasts. However, the multicellular morphology seems phenotypically arrested between the yeast and hyphal phenotypes and consists of many cells having one or more nuclei that are separated by septa (Oujezdsky *et al.*, 1973; Szaniszlo *et al.*, 1976). Some septa are nonperforate and similar to those formed between yeast mother cells and buds, whereas others are similar to the septa seen in hyphae and have a central pore and associated Woronin bodies.

Because the conversion of yeasts to the multicellular phenotype appeared to result from the cessation of bud emergence without the inhibition of growth, nuclear division or cytokinesis, we anticipated that the acidic conditions leading to the multicellular morphology are inhibiting an essential cell cycle event leading to normal yeast growth. An analogous transition from blastic to isotropic development had been reported for the terminal phenotype of the *cdc*24 mutant of *S. cerevisiae* (Hartwell *et al.*, 1973). Later, similar terminal phenotypes were reported for *cdc*42 and *cdc*43 (reviewed in Pringle and Hartwell, 1981). Thus, we initiated a search for similar *cdc* strains in *W. dermatitidis*. This search yielded a number of *cdc* mutants designated multicellular (Mc) strains (Roberts and Szaniszlo, 1978). At pH 6.5 these strains grow as yeasts at the permissive temperature (25°C), but quantitatively convert to the multicellular morphology at the restrictive temperature (37°C). The parental wild type (wt), on the other hand, grows at pH 7.6 as yeasts at both permissive and restrictive temperatures. However, unlike the *cdc24, 42,* and *43* mutants of *S. cerevisiae*, the Mc strains of *W. dermatitidis* are long lived and give rise to visible flat colonies at 37°C consisting of multicellular forms that develop because of the ability of *W. dermatitidis* to form transverse and vertical septa during isotropic growth.

Studies of yeast-to-multicellular form transition in the Mc mutants incubated at 37°C revealed that different strains show more lethality or commit to conversion at different times in the yeast cell cycle (Roberts and Szaniszlo, 1978). So that the execution points of the ts lesions in the mutants could be established exactly, we characterized the cell cycle events at the permissive temperature (Roberts and Szaniszlo, 1980). Microfluorometric analysis revealed that the yeast cell cycle phases of this polymorphic fungus are somewhat longer than those reported for *S. cerevisiae* (Slater et al., 1977).

Studies of synchronously and asynchronously dividing cells revealed that bud emergence in *W. dermatitidis* occurs in G2 under conditions of slow growth (Roberts *et al.*, 1979; 1980; Roberts and Szaniszlo, 1980), but under conditions of more rapid growth is most often associated with late G1 and S phase cells (unpublished results). These investigations collectively established that bud emergence in *W. dermatitidis* is not dependent on DNA synthesis, and cells inhibited in DNA synthesis arrest as budded forms with a nucleus in mitosis at the isthmus between the mother cell and daughter bud (Roberts and Szaniszlo, 1980).

Our observations of budding growth in *W. dermatitidis* led us to propose that the yeast cell cycle of this polymorphic pathogen is basically similar to that of *S. cerevisiae*. Consequently we initially presented a two-pathway model for the cell cycle consisting of a DNA-nuclear division pathway and a bud formation, nuclear migration pathway, which diverge in G1 or S and converge at cytokinesis (Roberts and Szaniszlo, 1980). Further studies of the effects of the microtubule inhibitors nocodazole and MBC provided data suggesting that the cycle also contains a third pathway responsible for nuclear migration and chromosome segregation (Jacobs and Szaniszlo, 1982). The execution points for the Mc mutations were temporally mapped in this newer work to two different positions before bud emergence in the bud emergence-bud growth pathway. Subsequent complementation studies using protoplast fusion methodology have recently confirmed that the mutations in some Mc strains are in different *CDC* genes (Cooper and Szaniszlo, unpublished data).

Chitin changes associated with multicellular form development

Conversion of yeasts to the multicellular phenotype by the *cdc* mutant Mc3 follows the same two-stage process noted for multicellular form development induced by acidity (Cooper *et al.*, 1984). Stage I is arbitrarily marked by the formation of swollen, unbudded cells having multiple nuclei and thickened cell walls, whereas Stage II forms have one or more transverse septa. These morphological changes are accompanied by major changes in the cell wall polymers chitin, ß-glucan, and DHN melanin, with the relative amount of chitin and melanin in the cell wall dramatically increasing, while that of ß-glucan decreases (Szaniszlo *et al.*, 1983a; Geis and Jacobs, 1985). When logarithmic phase wild-type or Mc3 yeasts were assayed, they were found to contain about 0.4 - 0.8 µg of chitin per mg dry mass during 12 h incubation at 25°C (Figure 1). However, when yeasts of either strain were incubated at 37°C, the chitin content of the cells increased, but in different manners. In the case of the wild-type, the increase began almost immediately in an approximately linear fashion. In contrast, in the mutant, when related to either dry mass or cell number, chitin content did not change much during the first 3 h, but thereafter began to increase exponentially. The chitin content in the mutant equaled that of the wild type at about 9 h, and thereafter exceeded it. Isolated cell walls of multicellular forms that developed after 36 to 48 h at the restrictive temperature similarly have at least 5 times more N-acetylglucosamine than have cell walls of yeasts of either the wild type or Mc3 grown for the same time at 25°C, and about 4 times more than have yeast cell walls of the wild type grown at 37°C (Szaniszlo *et al.*, 1983a). Chitin deposition also becomes dramatically delocalized during yeast-to-

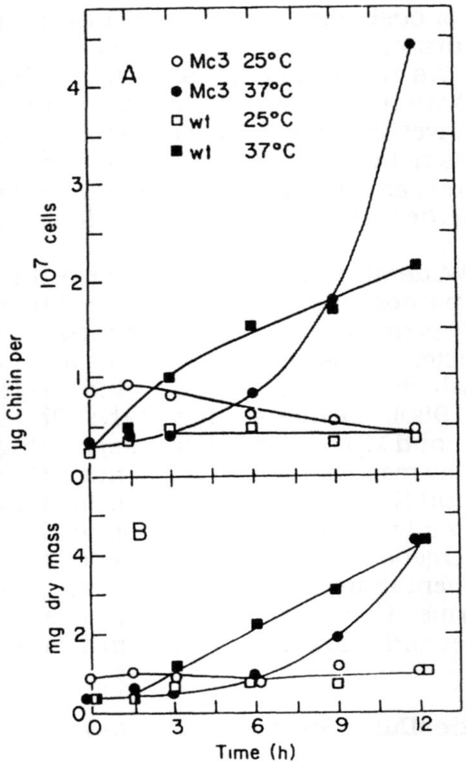

Figure 1. Changes in absolute chitin content of cells of wild-type and Mc3 mutant strains during the first 12 h incubation of cells at 25°C or 37°C in relation to cell number (A) or dry mass (B). A modification of the method of Ride and Drysdale (1972) was used to estimate chitin content (Jacobs and Szaniszlo, unpublished data).

multicellular form transition, shifting from the bud/birth scar regions to an inner wall layer during Stage I development (Figure 2) and to the inner wall layers and transverse septa during Stage II development (Figure 3) (Cooper *et al.*, 1984; Harris and Szaniszlo, 1986; Jacobs *et al.*, 1985). Inhibition of chitin synthesis in the mutant at 37°C by incubation with polyoxin (Figure 4A and B), an inhibitor of chitin synthases, or resumption of budding by Stage I cells after shift from the restrictive temperature to 25°C (Figure 4C), decreases the staining with calcofluor of the developing multicellular forms and the new buds, respectively (Cooper *et al.*, 1984; Jacobs *et al.*, 1985). Continued incubation of the multicellular forms in the presence of polyoxin causes increased cellular death due to lysis (Cooper *et al.*, 1984). By comparison, when wt yeasts are incubated in the presence of polyoxin, chains of yeasts form. These yeast chains have aberrant septa similar to those sometimes observed in *chs* mutants of *S. cerevisiae* (Silverman *et al.*, 1988), but little lysis occurs.

Interpretation of our cell wall chitin data can be made based on current models of fungal cell-wall growth, which suggest that cell-wall synthases and hydrolases, or their specific activators and inhibitors, are at least in part delivered to desired sites of cell wall synthesis by an intracellular transit system (Cabib *et al.*, 1988; 1990). In a dimorphic or polymorphic fungus, such as *W. dermatitidis*, a series of morphogenetic events results in the establishment of polarized growth leading to hyphal extension and branching, or yeast bud formation (Szaniszlo *et al.*, 1983b; Szaniszlo, 1985).

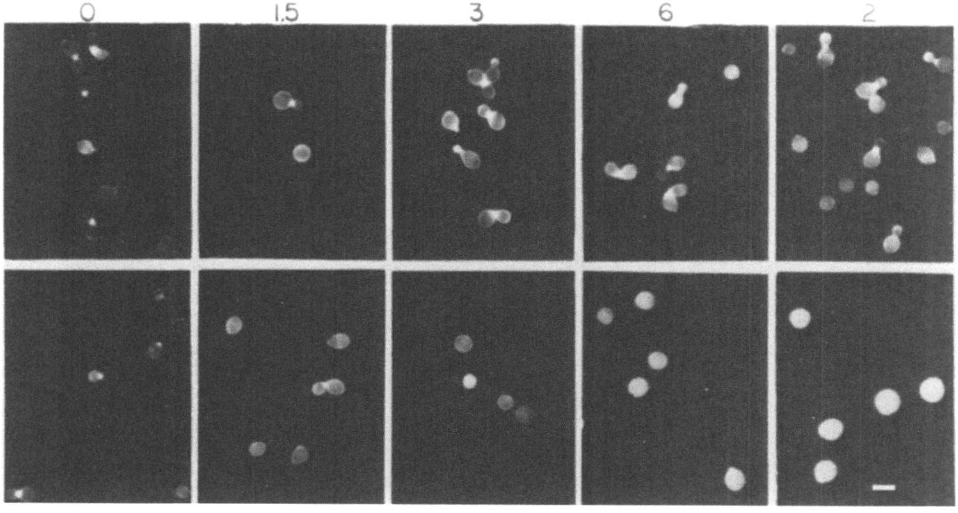

Figure 2. Calcofluor staining patterns of wild type (top) and Mc3 mutant (bottom) during the first 12 h following shift of yeasts to 37°C. Samples were stained at 0, 1.5, 3, 6, and 12 h after temperature upshift as indicated. Scale bar: 10 μm. (Jacobs and Szaniszlo, unpublished micrographs).

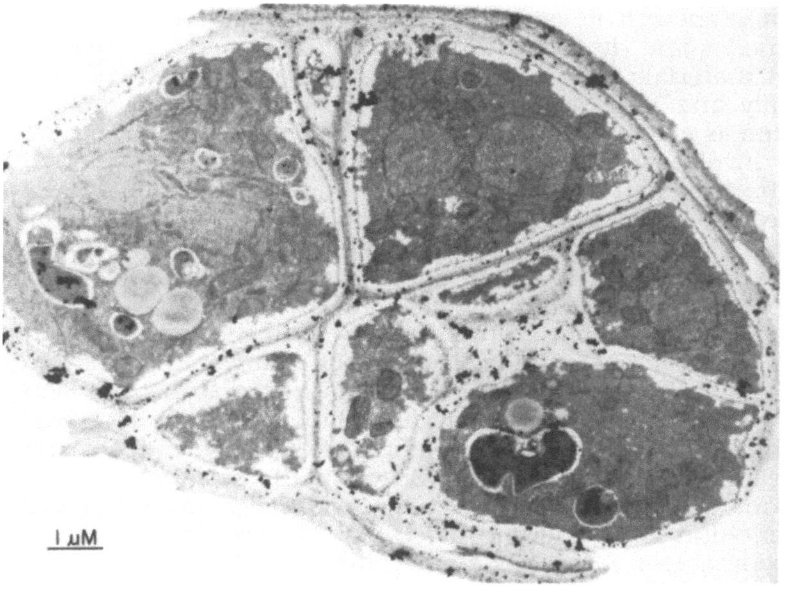

Figure 3. Multicellular form thin-section stained with colloidal gold-chitinase (Note darkened particles of gold are predominantly associated with wall and septal regions). This multicellular form of Mc3 developed after a number of days at 37°C (Harris and Szaniszlo, unpublished micrograph).

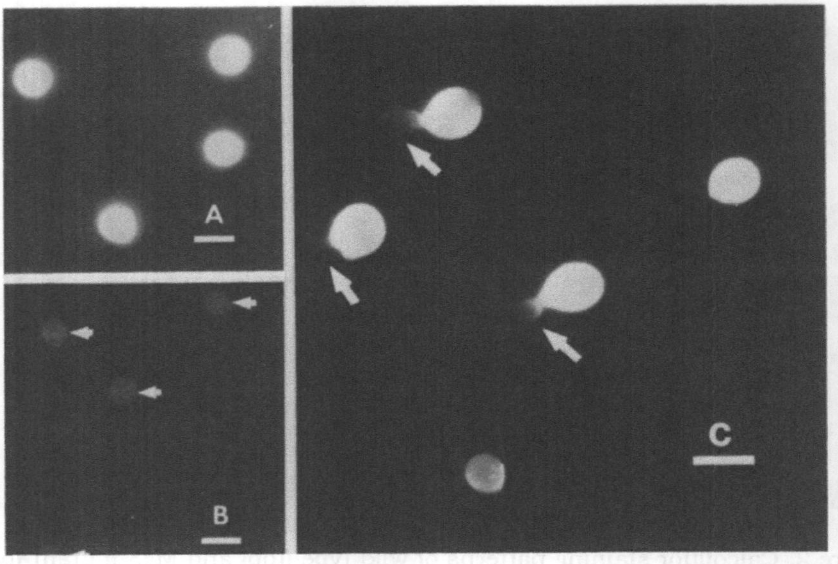

Figure 4. Calcofluor staining of isotropic Mc forms incubated in the absence (A) or presence (B) of polyoxin or shifted to 25°C after 12 h (C) to allow resumption of budding (from Cooper *et al.*, 1984; Jacobs *et al.*, 1985). Note decreased staining (arrows) in the presence of polyoxin (B) or in the new buds (C). Scale bars: 10 μm.

Concommitant with the establishment of polarity is the appropriate positioning of organelles and the localized incorporation outside the plasmalemma of cell-wall materials. However, during isotropic growth, the cell expands more uniformly and little polarized synthesis, except in the areas of septum formation, is necessary. As in other fungi, these morphogenetic events seem likely to involve the action of cytoskeletal elements, such as the actin, myosin and tubulin systems (reviewed in Harold, 1990; Drubin, 1991). Although there are disagreements on some specifics, such as the intracellular locations of inactive chitin synthase and its activators (Cabib *et al.*, 1990), this general model is accepted for budding growth in *S. cerevisiae* and most likely is appropriate for a vegetatively polymorphic fungus like *W. dermatitidis* (Szaniszlo, *et al.*, 1983b; Szaniszlo, 1985). In the latter fungus, dysfunction of a hypothetical "polarization director" could prevent the precise delivery of required components of the wall synthesis system to discrete loci on the cell envelope, and thus immediately prohibit bud emergence or apical extension. Random delivery of these same components, because of a reorientation of cytoskeletal structure or a loss of focus by the polarization director would then result in the delocalized synthesis of normally localized polysaccharides such as chitin (Geis and Jacobs, 1985). The recent discovery of multiple chitin synthases in *S. cerevisiae*, and *C. albicans* (Cabib *et al.*, 1990; Au-Young and Robbins, 1990), and our preliminary identification of at least three chitin synthase homologues in *W. dermatitidis* (Momany and Szaniszlo, unpublished), together with our prior finding that microtubules probably are not the polarization director for yeast bud growth in either *W. dermatitidis* or *S. cerevisiae* (Jacobs and Szaniszlo, 1982; Jacobs *et al.*, 1988), suggest that studies of chitin synthase and actin cytoskeletal interactions will be particularly useful for establishing correlates for morphogenesis among all fungal pathogens.

Chitin synthase activities

The relevance of chitin to form transition in *W. dermatitidis* has begun to be further investigated by assaying for chitin synthase (Chs) activity associated with membranes. The chitin synthase assay measures the incorporation of radiolabeled monomer (N-acetyl-D-glucosamine) into acid insoluble polymer (chitin). Some forms of the enzyme are zymogenic, requiring trypsin treatment before activity is seen *in vitro* (reviewed in Cabib, 1987; Cabib *et al.*, 1988; 1990). In *S. cerevisiae* three *CHS* genes have been cloned (Bulawa *et al.*, 1986; Silverman *et al.*, 1988; Valdivieso *et al.*, 1991; Bulawa and Wasco, 1991). Strains constructed by disruption of these genes have allowed individual Chs activities to be characterized. The three Chs activities of *S. cerevisiae* differ in regard to pH and cation optima and to trypsin sensitivity (Orlean, 1987; Bulawa and Osmond, 1990). Similarly, two Chs activities which differ in pH optima and response to proteolysis have been reported in *C. albicans* (Au-Young and Robbins, 1990).

Chitin synthase activities have been assayed in membranes from yeasts and developing multicellular forms of *W. dermatitidis* strain Mc3W-14 (Table 1; Momany and Szaniszlo, unpublished data). The total Chs activity from *W. dermatitidis* yeast form membranes fell between the activity levels reported for *S. cerevisiae* (Orlean, 1987; Sburlati and Cabib, 1986) and the yeast form of *C. albicans* (Au-Young and Robbins, 1990). Both trypsin- and nontrypsin-activated activities from each pheonotype of *W. dermatitidis* could be measured, with the trypsin-activated activities always being much higher than the nontrypsin-activated activities. As expected, the activities associated with the multicellular phenotype were also consistently higher than those associated

Table 1.
Chs activities[a] of Mc3W-14 yeast (Y) and multi-cellular (Mc) forms.[b]

Form and Temperature	Trypsin	pH and cation				
		7.0/Mg	7.5/Mg	8.0/Mg	7.5/Mn	7.5/Co
Y (25°C)	-	24	13	18	10	11
	+	169	151	139	117	36
Mc (39°C)	-	21	28	16	33	12
	+	298	268	235	183	71

[a]nmoles GlcNAc incorporated h^{-1} mg $protein^{-1}$
[b]After inoculation with stationary-phase yeast cells, cultures were incubated 48 h at 25°C or 39°C. Cell disruption, membrane isolation, and assay were as described by Kang *et al.*, 1984. Membranes were activated by incubation with 50 µg/ml trypsin.

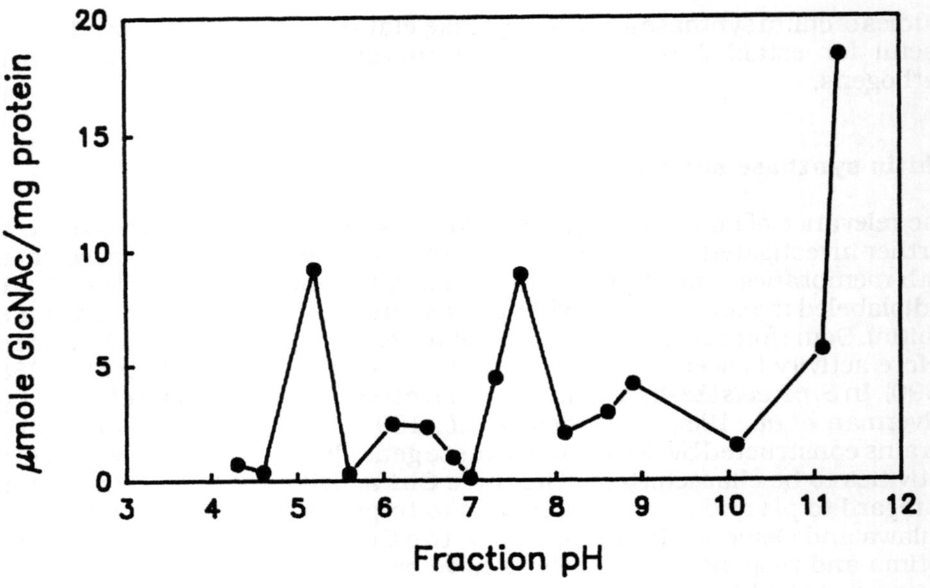

Figure 5. Isoelectric focusing of digitonin-solubilized, trypsin-activated Chs activity from Mc 3W-14 grown at 39°C for 48 h. Membrane isolation, solubilization and assay were as described by Kang *et al.*, 1984. Bulk isoelectric focusing was carried out using the Rotofor system (BioRad Laboratories, Melville, N.Y.).

with the yeast phenotype, although this could simply reflect the increased activity associated with incubation of cells at 39°C. In addition, subtle differences were observed among the pH optima and the cation effects on the activated and nonactivated enzymes from the two morphologies, inviting speculation that multiple chitin synthases in different amounts are contributing to the total activity associated with the different forms.
Additional suggestive evidence for multiple Chs enzymes in *W. dermatitidis* comes from our attempts to purify Chs activity using bulk isoelectric focusing. These preliminary studies showed that the Chs activity from digitonin-solubilized multicellular-form membranes can be eluted in a number of separated peaks: two very well defined peaks at pH 5 and pH 7.5 as well as two smaller, broader peaks at pH 6 and pH 9 (Fig. 5). While it is possible that the apparent pIs associated with these multiple activity peaks resulted from proteolysis during isolation, or some other artifact of membrane preparation, these results, along with the membrane assay data seem again to point toward multiple Chs enzymes in *W. dermatitidis*. Unfortunately, our present inability to construct *chs* deletion strains, and thus isolate individual Chs enzyme activities, plagues any definitive interpretation of our data. Therefore, to circumvent this problem, we decided to try to clone *CHS* genes from *W. dermatitidis*. Using this approach, we hoped to find out how many different chitin synthases *W. dermatitidis* might possess and examine those activities individually.

```
WDCHS1    EIEFTRTMHGIMRNIAHFCSRTRSRTW----GKDGWQKIVVCVIADGRQKVHPRTLNALAAMGVYQDGIAKNVVNQKEV
WDCHS2    EFLFARTMIGVFKNIEFMCNRSSSKTW----GKEAWKKIVVCIVSDGRAKINPRTRAVLAGLGVYQDGIAKQQVNGKDV
WDCHS3    KVLTARTLHGVMQNIRDIVNLKKSEFWNKGG--PAWQKIVVCLVFDGIDPCDKNTLDVLATIGVYQDGVMKKDVDGKET
SCCHS1    HILLGRTLKGIMDNVKYMVKKKNSSTW----GPDAWKKIVVCIISDGRSKINERSLALLSSLGCYQDGFAKDEINEKKV
SCCHS2    KYSLARTIHSIMKNVAHLCKREKSHVW----GPNGWKKVSVILISDGRAKVNQGSLDYLAALGVYQEDMAKASVNGDPV
CACHS1    EVAFARTMHGVMKNIAHLCSRHKSKIW----GKDSWKKVQVIIVADGRNKVQQSVLELLTATGCYQENLARPYVNNSKV
               **      *         *  *        * *  *    **           *   * **

WDCHS1    TAHVYEYTTQVSLDETLKFKGAEKGIV------PCQMIFCLKEKNKKKLNSHRWFFNAFGRALIPNVCILLDVGTKPDS
WDCHS2    TAHIYEYTTQVGLE----LKGTQVSLK-PRSATPVQLLFCLKEKNQKKINSHRWFFQAFGRVLDPNICVLIDAGTKPGK
WDCHS3    VAHIFEYTTQLSVTANQQLIRPNDND--ATSLPPAQMIFCLKQKNSKKINSHRWLFNAFGRILNPEVCILLDAGTKPGS
SCCHS1    AMHVYEHTTMINITN---ISESEVSLECNQGTVPIQLLFCLKEQNQKKINSHRWAFEGFAELLRPNIVTLLDAGTMPGK
SCCHS2    KAHIFELTTQVSINADLDY--VSKDIV------PVQLVFCLKEENKKKINSHRWLFNAFCPVLQPTVVTLVDVGTRLNN
CACHS1    NAHLFEYTTQISIDENLKFKGDEKNLA------PVQVLFCLKELNQKKINSHRWLFNAFCPVLDPNVIVLLDVGTKPDN
            *  * **                        * *  ****  * ** ***** *  *   * *    * * **

WDCHS1    KALYHLWKAFDQNSNVAGAAGEIKADKGKGWL---GLLNPLVAS
WDCHS2    DSIYQLWKAFDLEPMCGGACGEIKVMLDHG-K---KLLNPLVAT
WDCHS3    KSLMALWQAFYNDKDLGGACGEIHAMLGPGGVFGR
SCCHS1    DSIYQLWREF-RNPNVGGACGEIRTDLGKRFV---KLLNPLVAS
SCCHS2    TAIYRLWKVFDMDSNVAGAAGQIKTMKGKWGL---KLFNPLVAS
CACHS1    HAIYNLWKAFDRDSNVAGAAGEIKAMKGKGWI---NLTNPLVAS
               **  *       ** * *             * *****
```

Figure 6. Derived amino acid sequences of *CHS* conserved regions from *W. dermatitidis* (*WDCHS1*, *WDCHS2*, *WDCHS3*), *S. cerevisiae* (*SCCHS1*, *SCCHS2*), and *C. albicans* (*CACHS1*). Astericks indicate residues conserved among all six sequences (adapted from Bowen *et al.*, 1992).

Evidence for chitin synthase homologues

The easiest route to the identification of *CHS* genes from *W. dermatitidis* should have been by cross-hybridization to *CHS* genes cloned from other organisms. However, low stringency Southern blotting experiments probing genomic DNA of *W. dermatitidis* with *S. cerevisiae CHS 1* or *CHS 2* or *C. albicans CHS 1* showed no cross-hybridization (Momany and Szaniszlo, unpublished results). Curiously, none of the *CHS* genes cloned to date cross-hybridize with each other (Bulawa *et al.*, 1986; Silverman *et al.*, 1988; Au-Young and Robbins, 1990; Valdivieso *et al.*, 1991). Despite this apparent lack of DNA level conservation, the derived amino acid sequences of the *CHS* genes do share a region of very high homology (Silverman, 1989; Au-Young and Robbins, 1990). Based on this region, Chen-Wu and Robbins (Bowen *et al.*, 1992) designed degenerate PCR primers. These primers were used to amplify 600 bp *CHS* homologues from *W. dermatitidis*. The PCR products were cloned into M13 and screened using a single dideoxynucleotide sequencing reaction. M13 clones representing three unique patterns were fully sequenced. A comparison of the deduced amino acid sequences of these three *W. dermatitidis* clones with the same region of *CHS* genes from *S. cerevisiae* and *C. albicans* revealed considerable homology (Figure 6). Each of the three clones hybridized to unique PstI and XbaI fragments of *W. dermatitidis* DNA, but did not hybridize to *S. cerevisiae* DNA. In order to detect any additional *CHS* homologues in *W. dermatitidis* we developed a method we call "Denegerately Primed Multi-probe Southerns" (DPM Southerns). Label was introduced during a PCR reaction using the degenerate *CHS* primers with *W. dermatitidis* genomic DNA as template. The resulting labeled 600 bp product, which should contain all sequences amplified from the PCR primers, was excised from a gel and used as probe in Southern blotting experiments (Figure 7). By this method, the

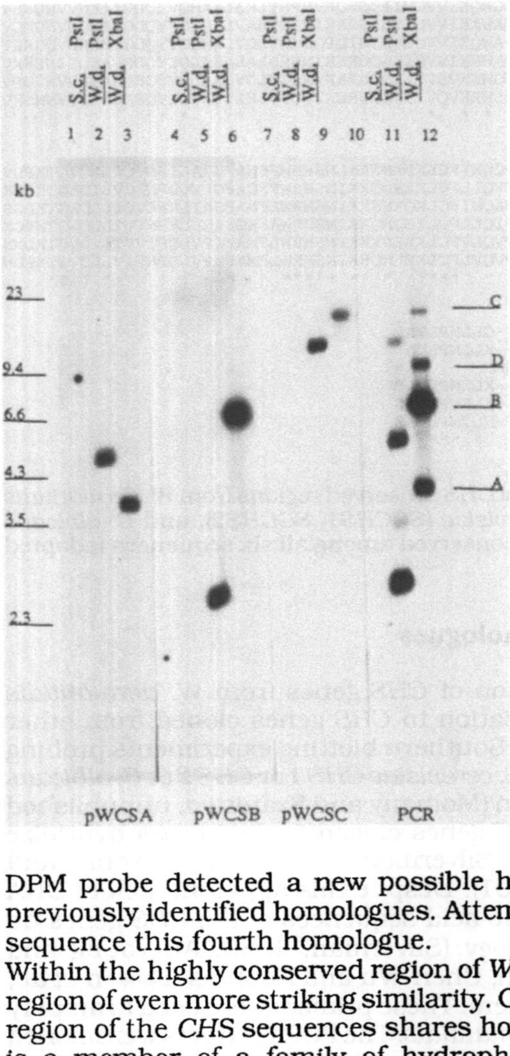

Figure 7. "Traditional" (with M13 probes) vs. DPM (with degenerately primed PCR probes) Southerns. An auto-lumigraph of a hybridization experiment is shown. Lanes 1-9 were probed with M13 clones (pWCSA, pWCSB and pWCSC) labeled by random priming. Lanes 10-12 were probed with degenerately primed probes labeled during PCR. *WCSA*, *B* and *C* are equivalent to *WDCHS1*, *2* and *3*, respectively. (Momany and Szaniszlo, unpublished data).

DPM probe detected a new possible homologue, in addition to the three previously identified homologues. Attempts are now under way to clone and sequence this fourth homologue.

Within the highly conserved region of *WDCHS1*, *WDCHS2* and *WDCHS3* is a region of even more striking similarity. Computer searches revealed that this region of the *CHS* sequences shares homology with *Ara G* (Figure 8). *Ara G* is a member of a family of hydrophilic membrane components of the periplasmic permeases (Mimura *et al.*, 1991). The members of this group have been shown to bind ATP and GTP, presumably as part of the active transport system. The region that the *CHS* homologues share with *Ara G* is highly conserved among the permeases and is thought to define part of the ATP binding site (Mimura *et al.*, 1991). Like the permeases, the *WDCHS* sequences also have homology to the Walker motifs of adenylate kinase, (Fry *et al.*, 1986), which define ATP binding sites (Walker *et al.*, 1982). In adenylate kinase the ATP is thought to nestle between an alpha helix and beta sheet at the center of a classic Rossman nucleotide binding fold (Rossman *et al.*, 1975). Secondary structural predictions for this portion of the *WDCHS* conserved sequences are consistent with such a fold (Momany, Momany and Szaniszlo, unpublished data).

```
                    ______α______                    ______β______
WDCHS1    PCQMIFCLKEKNKKKLNSHRWFFNAFGRALIPNVCILLDVG
WDCHS2    PVQLLFCLKEKNQKKINSHRWFFQAFGRVLDPNICVLIDAG
WDCHS3    PAQMIFCLKQKNSKKINSHRWLFNAFGRILNPEVCILLDAG
ARAG      AEQLIMNLSGGNQQKAILGRWLSEEMKVILLDEPTRGIDVG
ADK                                       PTLLLYVDAG
```

Figure 8. Homology among *WDCHS1*, *WDCHS2*, *WDCHS3*, *AraG* (Mimura *et al.*, 1991), and the second "Walker motif" of adenylate kinase (ADK) (Fry *et al.*, 1986; Walker *et al.*, 1982) α and β indicate predicted α helix and β sheet regions, respectively (Momany and Szaniszlo, unpublished data).

A nucleotide binding fold in *CHS* sequences may represent the binding site for UDP-N-acetylglucosamine (the substrate). Interestingly, the region of the "Walker motif" which aligns with the *CHS* sequences is thought to form the binding site for the alpha and beta phosphates rather than the gamma phosphate of ATP (Fry *et al.*, 1986). This would be consistent with binding of the UDP portion of the substrate. Alternatively, the homologies may represent a genuine nucleotide binding site. This explanation is especially appealing because the nucleotide binding motif is less conserved in *CHS3* of *S. cerevisiae* (Bulawa, personal communication; Valdivieso *et al.*, 1991). Perhaps a subset of Chs enzymes are regulated by nucleotides.
The exact nature of the putative nucleotide binding fold in the *W. dermatitidis CHS* homologues most likely will be defined by constructions of chimeras and deletions, both of which require the cloning of full genes. To that end a *W. dermatitidis* genomic library has been constructed in Lambda gt11. The library has been screened with a degenerate probe representing all products

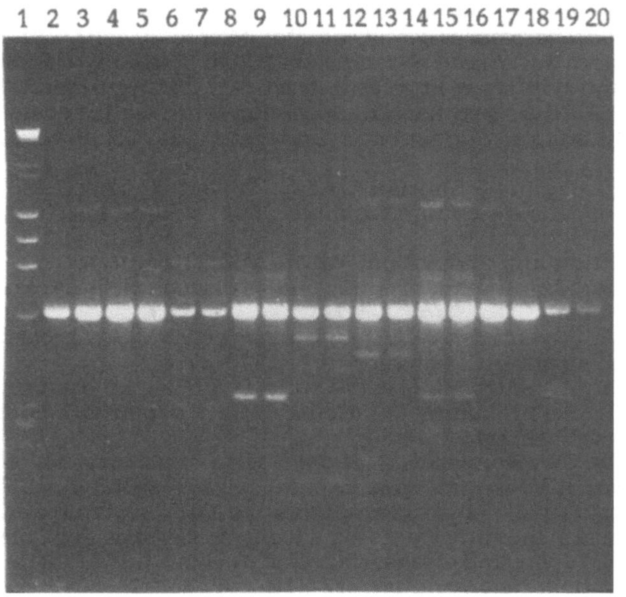

Figure 9. PCR products from reactions using using degenerate *CHS* primers to amplify genomic DNA as follows: (lane 1) Mol. wt markers, λ HindIII and øX174 HaeIII cut; (lanes 2 and 3), *W. dermatitidis* Mel3; (lanes 4 and 5) *W. dermatitidis* MC3W; (lanes 6 and 7) *Schizosaccharomyces pombe*; (lanes 8 and 9) *Exophiala jeanselmei*; (lanes 10 and 11) *Phaeococcus exophialae*; (lanes 12 and 13) *Phaeoannellomyces werneckii*; (lanes 14 and 15) *Rhinocladiella atrovirens*; (lanes 16 and 17) *W. dermatitidis* 8656; (lanes 18 and 19) *Xylohypha bantiana*; (lane 20) Negative control - no DNA (Momany and Szaniszlo, unpublished data).

amplified from *W. dermatitidis* genomic DNA with the *CHS* primers. Twenty-three positive clones have been identified by hybridization and are now being characterized.

It should be noted that any information gained by studying the *W. dermatitidis* *CHS* homologues will probably have applications to other fungi as well. The same set of degenerate primers that allowed us to amplify *CHS* homologues from *W. dermatitidis* have also been used to amplify homologues from a number of other fungal species (Bowen *et al.*, 1992) including a number of other melanized species thought to be related to *W. dermatitidis* (Figure 9). The *CHS* homologues detected in *W. dermatitidis* and these other fungi fall into three classes based on derived amino acid sequences, and share amazing levels of conservation within each class. Sequences within two classes have also been used to construct phylogenies. This phylogenetic information may prove quite valuable since *CHS* homologues have now been amplified and sequenced from a number of imperfect fungi, including a number of other pathogens whose teleomorphic classification at the genetic level is impossible even to guess.

Acknowledgements

The authors thank P. W. Robbins, C. E. Bulawa, J. L. Chen-Wu, A. R. Bowen, and J. Au-Young for generously sharing unpublished data, technical advise, and experimental ideas. This work was supported by grants from the Texas Applied Technology Program and the University of Texas Research Institute.

References

Au-Young J, Robbins PW (1990) Isolation of a chitin synthase gene from *Candida albicans* by expression in *Saccharomyces cerevisiae*. Mol. Microbiol. 4:197-207.

Bowen AR, Chen-Wu JL, Momany M, Young R, Szaniszlo PJ, Robbins PW (1992) Classification of fungal chitin synthases. Proc. Natl. Acad. Sci. USA 89: 519-523.

Bulawa CE, Osmond BC (1990) Chitin synthase I and chitin synthase II are not required for chitin synthesis *in vivo* in *Saccharomyces cerevisiae*. Proc. Natl. Acad. Sci. USA 87:7424-7428.

Bulawa CE, Slater M, Cabib E, Au-Young J, Sburlati A, Adair WL, Robbins PW (1986) The *S. cerevisiae* structural gene for chitin synthase is not required for chitin synthesis *in vivo*. Cell 46:213-225.

Bulawa CE, Wasco W (1991) Chitin and nodulation. Nature 253:710.

Cabib E (1987) The synthesis and degradation of chitin. Adv. Enzymol. Relat. Areas Mol. Biol. 59:59-101.

Cabib E, Bowers E, Sburlati A, Silverman SJ (1988) Fungal cell wall synthesis: construction of a biological structure. Microbiol. Sci. 4:370-375.

Cabib E, Silverman SJ, Sburlati A, Slater ML (1990) Chitin synthesis in yeast (*S. cerevisiae*). In: Kuhn PJ, Trinici APJ, Jung MJ, Goosey MW, Cooping KG (eds) Fungi. Springer-Verlag, New York, pp 31-44.

Cooper CR Jr, Harris JL, Jacobs CW, Szaniszlo PJ (1984) Effects of polyoxin AL on cellular development in *Wangiella dermatitidis*. Exp. Mycol. 8:349-363.

Dixon DM, Szaniszlo PJ, Polak A (1991) Dihydroxynaphthalene (DHN) melanin and its relationship with virulence in the early stages of phaeohyphomycosis. In: Cole GT, Hoch HC (eds) The fungal spore and disease initiation in plants and animals. Plenum Publishing Corporation, New York, pp 297-318.

Drubin DG (1991) Development of cell polarity in budding yeast. Cell 65:1093-1096.
Fry DC, Kuby SA, Mildan AS (1986) ATP-binding site of adenylate kinase: Mechanistic implications of its homology with *ras*-encoded p21, F_1-ATPase, and other nucleotide-binding proteins. Proc. Natl. Acad. Sci. USA 83:907-911.
Geis PA, CW Jacobs (1985) Polymorphism of *Wangiella dermatitidis*. In: Szaniszlo PJ (ed) Fungal dimorphism: with emphasis on fungi pathogenic for humans. Plenum Press, New York, pp 205-233.
Harold FM (1990) To shape a cell: an inquiry into the causes of morphogenesis of microorganisms. Microbiol. Rev. 54:381-431.
Harris JL, Szaniszlo PJ (1986) Localization of chitin in walls of *Wangiella dermatitidis* using colloidal gold-labeled chitinase. Mycologia 78:853-857.
Hartwell LH, Mortimer RK, Colutti J, Culotti M (1973) Genetic control of the cell division cycle in yeast: Genetic analysis of cdc mutants. Genetics 74:267-286.
Jacobs CW, Adams AEM, Szaniszlo PJ, Pringle JR (1988) Functions of microtubules in *Saccharomyces cerevisiae* cell cycle. J. Cell. Biol. 107:1409-1426.
Jacobs CW, Roberts RL, Szaniszlo PJ (1985) Reversal of multicellular-form development in a conditional morphological mutant of the fungus *Wangiella dermatitidis*. J. Gen. Microbiol. 131:1719-1728.
Jacobs CW, Szaniszlo PJ (1982) Microtubule function and its relation to cellular development and the yeast cell cycle in *Wangiella dermatitidis*. Arch. Microbiol. 133:155-161.
Kang MS, Narayansamy E, Mattia E, Au-Young J, Robbins PW, Cabib E (1984) Isolation of chitin synthetase from *Saccharomyces cerevisiae*; purification of an exzyme by entrapment in the reaction product. J. Biol. Chem. 259:14966-14972.
Mimura CS, Holbrook SR, Ames G (1991) Structural model of the nucleotide-binding conserved component of periplasmic permeases. Proc. Natl. Acad. Sci. USA 88:84-88.
Orlean P (1987) Two chitin synthases in *Saccharomyces cerevisiae*. J Biol Chem 202:5732-5739.
Oujezdsky KB, Grove SN, Szaniszlo PJ (1973) Morphological and structural changes during yeast-to-mold conversion of *Phialophora dermatitidis*. Mycologia 66:537-542.
Pringle JR, Hartwell LH (1981) The *Saccharomyces cerevisiae* cell cycle. In: Strathern JN, Jones EW, Broach JR (eds) The molecular biology of the yeast *Saccharomyces*: life cycle and inheritance. Cold Spring Harbor Monographs, Cold Spring Harbor, New York, pp 97-142.
Ride JP, Drysdale RB (1972) A rapid method for the chemical estimation of filamentous fungi in plant tissue. Physiol. Plant. Pathol. 2:7-15.
Roberts RL, Lo RJ, Szaniszlo PJ (1979) Nuclear division in temperature-sensitive multicellur mutants of *Wangiella dermatitidis*. J. Bacteriol. 137:1456-1458.
Roberts RL, Lo RJ, Szaniszlo PJ (1980) Induction of synchronous growth in the yeast phase of *Wangiella dermatitidis*. J. Bacteriol. 141:981-984.
Roberts RL, Szaniszlo PJ (1978) Temperature-sensitive multicellular mutants of *Wangiella dermatitidis*. J. Bacteriol. 135:622-632.
Roberts RL, Szaniszlo PJ (1980) Yeast-phase cell cycle of the polymorphic fungus *Wangilla dermatitidis*. J. Bacteriol. 144:721-731.
Rossman MG, Liljas A, Branden C, Banaszak LJ (1975) Evolutionary and structural relationships among dehydrogenases. In: Boyer D (ed) *The Enzymes* Vol XI Academic Press, New York, pp 61-102.
Sburlati A, Cabib E (1986) Chitin synthetase 2, a presumptive participant in septum formation in *Saccharomyces cerevisiae*. J. Biol. Chem. 261:15147-15152.
Silverman SJ (1989) Similar and different domains of chitin synthases 1 and 2 of *S. cerevisiae*: two domains with distinct functions. Yeast 5:459-467
Silverman SJ, Sburlati A, Slater ML, Cabib E (1988) Chitin synthase 2 is essential for septum formation and cell division in *Saccharomyces cerevesiae*. Proc. Natl. Acad. Sci. USA 85:4735-4739.
Slater ML, Sharrow SO, Gart JJ (1977) Cell cycle of *Saccharomyces cerevisiae* in populations growing at different rates. Proc. Natl. Acad. Sci. USA 74:3850-3854.
Szaniszlo PJ (1985) An introduction to dimorphism among zoopathogenic fungi. In:

Szaniszlo PJ (ed) Fungal dimorphism: with emphasis on fungi pathogenic for humans. Plenum Press, New York, pp 3-13.
Szaniszlo PJ, Geis PA, Jacobs CW, Cooper CR Jr, Harris JL (1983a) Cell-wall changes associated with yeast-to-multicellular-form conversion in *Wangiella dermatitidis*. In: Schlessinger D (ed) Microbiology-1983. American Society for Microbiology, Washington DC, pp 239-244.
Szaniszlo PJ, Hsieh PH, Marlowe JD (1976) Induction and ultrastructure of the multicellular (sclerotic) morphology in *Phialophora dermatitidis*. Mycologia 68:117-130.
Szaniszlo PJ, Jacobs CW, Geis PA (1983b) Dimorphism: morphological and biochemical aspects. In: Howard D (ed) Fungi pathogenic for humans and animals, part A (biology). Dekker, New York pp 323-436.
Valdivieso MH, Pieternella CM, Shaw JA, Cabib E, Duran A (1991) *CAL1*, A gene required for activity of chitin synthase 3 in *Saccharomyces cerevisiae*. J. Cell Biol. 114:101-109.
Walker JE, Saraste M, Kunswick MJ, Gay NJ (1982) Distantly related sequences in the α-and ß-subunits of ATP synthase, myosin, kinases and other ATP-requiring enzymes and a common nucleotide binding fold. EMBO J. 1:945-951.

Addition of unsaturated fatty acids down-modulates heat shock gene expression and produces attenuated strains in the fungus *Histoplasma capsulatum*

Bruno Maresca*°, Gustavo Di Lallo*, Celia L. Pardini*,
Gilda Petrella* and Luisella Carratù*
* International Institute of Genetics and Biophysics, CNR
Via Marconi, 12 - 80125 Naples
Italy

Introduction

Histoplasma capsulatum is the causative agent of histoplasmosis, a systemic fungal disease world-wide in occurrence and the most common respiratory mycotic infection affecting humans and animals. This organism, that consists of a pathogenic yeast phase present in human tissue and a saprobic mycelial phase found in soil, represents at a molecular level, the most extensively studied of the dimorphic pathogenic fungi (Maresca and Kobayashi, 1989). In culture, the transition from one phase to the other can be triggered reversibly by shifting the temperature of incubation between 25° (mycelia) and 37°C (yeast). This implies that each growth phase is an adaptation to two remarkably different environments. Therefore, it is likely that the temperature-induced phase transition and the events in the establishment of infection are intimately interrelated and, unlike the case in higher eukaryotes, the differentiation process in dimorphic fungi represents an adaptation to a new environment. In fact, the organism must face challenges that may not be strictly related to dimorphism to proceed towards phase transition (e.g., higher temperature, different redox potential and nutrients, presence of new degradative enzymes, etc.).
Heat shock genes (hs), among others, play a central role in the capacity of adapting to the new living conditions that dimorphic organisms face when they invade a mammalian host. The heat shock response has been described as a general homeostatic mechanism that protects cells and the entire organism from the deleterious effects of temporary environmental stresses, such as temperature increase, osmotic or pressure shock, etc. Recently, it has been shown that heat shock proteins (HSPs) play major roles in many cellular processes, and have a unique role in cell biology, from chronic degenerative diseases to immunology, from cancer research to interaction between host and parasites (Maresca and Lindquist, 1991). Heat shock

° Washington University, School of Medicine, Dept. of Internal Medicine, St. Louis, MO, 63110, USA

NATO ASI Series, Vol. H 69
Molecular Biology and its Application to Medical Mycology
Edited by B. Maresca, G. S. Kobayashi and H. Yamaguchi

proteins have been detected in many parasites (e.g., *Leishmania, Trypanosoma, Plasmodium, Giardia, Schistosoma*, etc.) and pathogenic fungi (*Histoplasma, Candida*, etc.) when temperature increases at the time of infection (Newport *et al.*, 1988; Polla BS, 1991; Maresca and Carratù, 1992) and have been proposed to play a fundamental role in the host/parasite relationship (Newport *et al.*, 1988; Polla BS, 1991; Maresca and Carratù, 1992). Furthermore, heat shock proteins (generally, HSP70) are among the dominant antigens recognized by the immune system to a large spectrum of eukaryotic pathogens (Polla BS, 1991; Maresca and Carratù, 1992). Thus, besides their role in host/pathogen interactions, these stress proteins assume immunological importance. However, even though the overall heat shock response in these organisms is very similar to that in better studied eucaryotic cells, specific differences exist. For example, in lower eucaryotes, such as yeast, after heat shock cells stop growing and then resume growth, possibly as a result of adaptation to the new conditions, in higher eucaryotic cells experimental conditions are such that the increase in temperature is transient and the temperature must be lowered to ensure survival. In pathogens, on the contrary, after host invasion, the temperature remains constant. In addition, while at the upper temperature range of higher eukaryotic cells and in *Saccharomyces cerevisiae* there is a block in *hsp83* mRNA maturation as well as in other intron-containing genes (Yost and Lindquist, 1988), in *H.capsulatum hsp82, tubulin* (Minchiotti *et al.*, 1991), *cdc2* and (Δ^9desaturase genes (Di Lallo *et al.*, 1992; Gargano *et al.*, 1992) are properly spliced under similar experimental conditions. In a dimorphic fungus adaptation to the host requires a response at the gene level and it is reasonable to postulate that several intron-containing DNA sequences play a vital role in adaptation to the new environment. Therefore, in dimorphic organisms such as *H.capsulatum* the heat shock response is not an artificial phenomenon and mRNA maturation machinery must remain functional after infection to allow the fungus to undergo morphogenesis and survive in the host (Maresca and Carratù, 1992). Furthermore, between a few to 24 hours after host invasion, parasites and dimorphic fungi do not induce the heat shock response at 37°C but at temperatures ranging between 39° and 41°C. In addition, it is reasonable to assume that during host invasion, when the parasite makes contact with host cells, the stress response is elicited independently of the temperature shock (Maresca and Carratù, 1992). However, it is not clear, so far, whether the expression of the heat shock gene family is part of the process of differentiation itself, or it is an epiphenomenon involved in adaptation to the new environmental conditions.

Heat shock and adaptation during phase transition

An example of this capacity to adapt to the new environment present in human tissue by *H.capsulatum* is the activation of heat shock genes during the induction of the pathogenic yeast phase (Minchiotti *et al.*, 1991; Caruso *et al.*, 1987; Minchiotti *et al.*, 1992). It has been demonstrated that a brief exposure of mycelial cells of the avirulent and temperature sensitive Downs strain to 37°C causes a rapid decline in intracellular ATP levels that parallels the uncoupling of oxidative phosphorylation (Lambowitz *et al.*, 1983). These observations led to the new idea that in *H.capsulatum* and other dimorphic

fungi, morphogenesis and adaptation to a temperature of 37°C is a consequence of a heat shock response. We have shown that induction of phase transition by shifting the temperature to 34°C in Downs or to 37°C in more virulent strains results in physiological and biochemical changes that are similar but less extreme. For example, at 34°C only partial uncoupling of respiration occurs in the Downs strain and electron transport efficiency is decreased by less than 50% of the initial values compared to the level measured at 37°C. However, Medoff *et al.* showed that temperatures (between 39° and 43°C) are required to mimic a similar effect in the more pathogenic strains such as G217B, G222B and G184B during mycelial to yeast phase transition (Medoff *et al.*, 1986). We have also analyzed the regulation of cloned *hsp70* and *hsp82* genes during heat shock at different temperatures in the thermosensitive Downs strain and in the temperature tolerant G222B and G217B strains. In the first 3 hours after the temperature shift, maximal transcription of *hsp70* and *hsp82* genes occurs at 34°C in Downs, whereas a temperature of 37°C is necessary to induce maximal transcription in G222B and G217B strains (Minchiotti *et al.*, 1991; Caruso *et al.*, 1987; Minchiotti *et al.*, 1992).
Work done in our and other laboratories has shown that in these isolates of *H.capsulatum* a close relationship exists between the level of heat shock gene expression, the level of thermosensitivity and the degree of pathogenicity. However, the normal level of expression can be rescued by incubating the cells for a short time at milder temperatures, even though heat shock genes are poorly expressed in temperature-sensitive and less pathogenic strains. This phenomenon, generally termed thermotolerance, occurs when organisms or cells are exposed to non lethal heat shock temperatures that induces a transient state of resistance to exposure to more elevated temperatures or other stresses. It has been demonstrated that thermotolerance is related to synthesis of heat shock proteins (Lindquist S, 1986) and that in *S.cerevisiae* HSP104 is required for tolerance to many forms of stress (Sanchez and Lindquist, 1990; Sanchez *et al.*, 1992). We have shown that the capacity of maintaining respiration coupled to ATP production depends on the strain analyzed and that the induction of heat shock proteins rescues ATP production at non-permissive temperatures in *H.capsulatum* (Patriarca *et al.*, 1992). A similar role of heat shock proteins was found in *S.cerevisiae* (Patriarca and Maresca, 1990).

Heat shock regulation

The cell response to abrupt increases in environmental temperatures is coordinated by the induction of the heat shock genes which occurs over a wide range of temperatures, from 4°C in the Antarctic fish *Nothotenia rossii* (Maresca *et al.*, 1988) to 88°C in the thermophilic bacterium *Sulfolobus* (Trent *et al.*, 1990). Despite extensive research concerning the mechanism(s) that regulates activation of heat shock gene expression, thus far no conclusion has been reached concerning the nature of a "primary sensor" capable of monitoring specific temperatures that induce heat shock gene transcription and how the signal(s) is/are transferred to the nucleus.
We have speculated that the heat injury is due to phase-transition of cellular membrane lipids similar to chilling injury described in cyanobacteria (Wada

et al., 1990) and plants (Murata *et al.*, 1992). In fact, it has been demonstrated that many of the physiological cellular responses to environmental changes are caused directly by modifications in membrane lipid structures, which affect either the overall fluidity or specific lipid domains (Quinn *et al.*, 1989). During seasonal or evolutionary temperature adaptation (homeoviscous adaptation), the conservation of a particular state of membrane fluidity following temperature changes is critical for a wide spectrum of membrane-associated protein functions (Cossins and Prosser, 1978; Hochachka and Somero, 1984). Since the phase transition temperature of cellular membranes depends on the degree of unsaturation of fatty acids of membrane lipids (Hochachka and Somero, 1984), we predicted that if a heat shock primary sensor (HSS) responsible for monitoring temperature variations were membrane-bound, modification of membrane fluidity (obtained, for example, by addition of fatty acids or membrane fluidifiers), would influence the level of expression of heat shock genes and thereby changing the temperature at which optimum heat shock response would occur.

The overall structural organization of heat shock genes and their regulation have been maintained throughout evolution with the characteristic heat shock element (HSE) consisting of a highly conserved nucleotide sequence (nnTCnnGAAnnTCnnnGAAnnTTCnn-GAAn; Amin *et al.*, 1988) and specific heat shock DNA binding protein(s) (heat shock factor, HSF). Thus far, specific HSFs have been identified in several organisms, such as *Drosophila* (Wu *et al.*, 1987), HeLa cells (Kingston *et al.*, 1987), *S.cerevisiae* (Sorger and Pelham, 1987) and *Tetrahymena* (Wiederrecht *et al.*, 1987). The heat shock response is characterized by rapid induction of this set of genes due to transcriptional activation (Sorger *et al.*, 1987). It has been proposed that in *S.cerevisiae* HSF binds DNA before and after heat shock and becomes phosphorylated after a temperature shift up, stimulating transcription (Sorger and Pelham, 1988). In *Drosophila* and human cells, HSF binds to DNA only after heat shock (Larson *et al.*, 1988). While this is likely to be the general mechanism of induction of heat shock genes, this model is not sufficient to explain the wide range of temperatures at which heat shock genes are induced in nature and how different stresses have distinct threshold levels in all organisms. In fact, it has yet to be established why each organism is capable of heat shock gene induction at its own specific physiological temperature. It has been suggested that specific binding of HSF could be mediated by direct temperature sensitivity of HSF itself (or that of additional transcription factors; Wu *et al.*, 1990) or that HSF could be activated by a temperature-regulated kinase (Sorger, 1991). Alternatively, HSP70 has been proposed as a thermometer that transduces the signal that activates gene transcription through its temperature-dependent interaction with HSF (Craig and Gross, 1991). However, it has yet to be established how HSP70, which is a highly conserved protein, detects the different temperature shifts necessary for heat activation (and stress activation) that exist in disparate organisms. According to this hypothesis, the HSP70/HSF structural conformation would be operative at a large range of temperatures and therefore a large number of temperature sensitive sensors would be required in animals and plants that experience seasonal acclimation.

The question we have addressed deals with the possibility that the specific temperature at which heat shock gene expression occurs might be altered in a cell by modifying its membrane fluidity, e.g., by changing its fatty acid constituents or by adding membrane fluidifiers. Homeoviscous adaptation is generally accomplished by incorporating different types of fatty acids during phospholipid synthesis or regulating the degree of unsaturation of the fatty acids itself (Maresca and Carratù, 1992). This is achieved as a result of gradual changes in the environmental temperature, atmospheric and osmotic pressures, salinity, etc. As a rule, the amount of saturated fatty acids (SFA) in membranes increases as cellular temperatures increase whereas lowering the temperature results in an increase of unsaturated fatty acids (UFAs). By modifying the transition temperature of membrane phospholipids, homeoviscous adaptation of membranes maintains a roughly constant fluid environment at any physiological temperature that allows enzymes to function properly.

In two strains of *H.capsulatum* that differ in virulence and in the optimum of temperature of heat shock mRNA induction we have demonstrated that it is possible to modulate heat shock gene transcription by addition of fatty acids. Furthermore, we have shown that benzyl alcohol, a fluidifying agent that decreases the phase transition temperature of membranes (Young *et al.*, 1991), had an effect similar to UFA on heat shock gene transcription. With the temperature sensitive Downs strain, in which the optimum of transcription occurs at 34°C, exposure of mycelia to palmitic or stearic acid (SFA) at 25°C induced a strong increase in heat shock transcription when cells were shifted to 37°C. Such increase in heat shock gene transcription was similar to that measurable when thermotolerance is induced (Yost and Lindquist, 1988). On the other hand, with the temperature-tolerant G217B strain, which has an optimum of 37°C for mRNA transcription, addition of oleic acid (UFA) to mycelia drastically reduced heat shock gene transcription when cells were shifted to 37°C. However, exposure to high concentration of oleic acid totally eliminated transcription of heat shock genes within the first 24 hours of exposure at this temperature. While the effect of fatty acids and benzyl alcohol on membrane fluidity was not measured directly, it has been demonstrated that these substances have a central role in all natural membranes and influence fluidity when added exogenously modulating membrane-bound enzymatic activities (Minchiotti *et al.*, 1991). We have also demonstrated that addition of SFA protects coupling of oxidative phosphorylation at temperatures that normally uncouple ATPase, viz., 37°C in the temperature susceptible Downs strain and 40°C in the temperature-tolerant G217B strain. On the other hand, addition of UFA uncoupled electron transport in Downs at 34°C and at 37°C in G217B strain.

The decline in the level of the heat shock response, along with other cellular functions that may have been influenced, affected the time required for mycelium-to-yeast morphologic transition. Mycelium to yeast transition with the Downs strain occurs in 7 to 8 days at 37°C, probably as a result of the uncoupling event and to the low level of heat shock gene transcription that take place in the first few minutes after the shift up in temperature. We previously showed that exposure of the Downs strain to an intermediate temperature of 34°C or by osmotic shock, induces a thermotolerant state that results in a restoration of normal heat shock transcription and ATP level, which shortens the phase transition to 3 days similar to the time seen with temperature tolerant strains (Young *et al.*, 1991).

We have determined the total fatty acid composition in the mycelial and yeast cells of Downs and G217B strains as well as in mycelia after 1 hr incubation at 37°C. The two mycelial strains grown at 25°C differed highly in the level of SFA, in particular stearic acid content is 41% lower in Downs compared to G217B. Major differences were also observed in fatty acid profile when mycelia were heat shocked 1 hour at 37°C. In fact, in the Downs strain there was a further drop in SFA (palmitic and stearic acids, respectively -16 and -28.8%), while in G217B they decreased only slightly. Consistent with the hypothesis that the Downs strain is a temperature sensitive variant with a defect in the regulation of the HSS, we have shown that incubation with SFA, by modifying membrane fluidity, restored normal heat shock transcription and ATPase activity shortening to 3 days the time necessary for the transformation to the yeast morphology. An opposite effect was observed at 37°C when UFA or benzyl alcohol were added to the G217B strain with total uncoupling of ATPase and drastic reduction of the heat shock response. These conditions also prolonged to 19 days the time required for the transition process to occur. Moreover, if the G217B strain was incubated with SFA and shifted from 25° to 40°C the time required for transition diminished from 8 to 3 days. The differences in the level of saturation of fatty acids can be explained by an alteration in the activity of one or more of the enzymes that regulates the degree of unsaturation of fatty acids. In particular, it is known that the ratio SFA/UFA is controlled by desaturase enzymes. For example D^9-desaturase catalyzes double bond formation within fatty acyl chains into saturated palmitic (16:0) or stearic acids (18:0) precursors. Difference in enzymatic activity or in the regulation of expression may explain in the two strains the difference in composition in unsaturated fatty acid of membrane and its fluidity. We have now cloned and sequenced *Ole1* (D^9-desaturase) gene from the temperature tolerant G217B strain (Gargano *et al.*, 1992) and are cloning the homologous gene from the Downs strain to compare their nucleotide sequences and possible difference of expression. Experiments in a mouse model of histoplasmosis with cells treated with UFA have shown a significant decrease in the degree of virulence (from ca. 70% to 20%) probably as a consequence of the reduced capacity to adapt to the new environment.

It seems likely that membrane fluidity plays a central role not only in regulating the activity of membrane-bound enzymes but also directly influences the expression of specific genes. The possibility of genetically interfere with membrane fluidity by transforming virulent strains with cloned desaturase genes may produce cells incapable of properly express the adaptation process (heat shock response) and be less virulent. This approach represents an entirely new strategy to establish an attenuated form of a pathogen and, to our knowledge, a new (and general) procedure for the production of a new class of vaccines for dimorphic pathogens.

Acknowledgements

This work was supported by a Contract from Science and Technology for Development programme of the Commission of the European Community, TS2-M132-I, a Contract from Ministero della Sanità - Istituto Superiore di Sanità, 5° Progetto AIDS 1992, NIH Grant AI 29609 and a NATO travel award grant # 5-205/RG890437

References

Amin, J, Anathan, J, Voelmy, R (1988) Key features of heat shock regulatory elements. Mol. Cell. Biol. 8: 3761-3769.

Caruso, M, Sacco, M, Medoff, G, Maresca, B (1987) Heat shock 70 gene is differentially expressed in *H.capsulatum* strains with different level of thermotolerance and pathogenicity. Mol. Microbiol. 1:151-158.

Cossins, AR, Prosser, CL (1978) Evolutionary adaptations of membranes to temperaturee. Proc. Natl. Acad. Sci. USA, 75: 2040-2043.

Craig, EA, Gross, CA (1991) Is *hsp70* the cellular thermometer? Trends Biochem. 16: 135-140.

Di Lallo, G, Gargano, S, Maresca, B (1992) manuscript in preparation.

Gargano, S, Di Lallo, G, Maresca, B (1992) manuscript in preparation.

Hochachka, PW, Somero, GN (1984) Biochemical Adaptation, pp. 537, Princeton Univ. Press, Princeton, N.J.

Kingston, RE, Schuetz, TJ, Larin, Z (1987) Heat inducible human factor that binds to a human *hsp70* promoter. Mol. Cell. Biol. 7: 1530-1534.

Lambowitz, AM, Kobayashi, GS, Painter, A, Medoff, G (1983) Possible relationship of morphogenesis in pathogenic fungus *Histoplasma capsulatum* to heat shock response. Nature 303: 806-809.

Larson, JS, Schuetz, TJ, Kingston, RE (1988) Activation *in vitro* of sequence-specific DNA binding by a human regulatory factor. Nature 335: 372-375. Erratum (1988) 336: 184.

Lindquist, S (1986) The heat shock response. Ann. Rev. Biochem. 55: 1151-1191.

Maresca, B, Carratù, L (1992) The biology of the heat shock resaponse in parasites. Parasitol. Today 8: 260-266.

Maresca, B, Kobayashi, GS (1989) Dimorphism in *Histoplasma capsulatum*: a model for the study of cell differentiation in pathogenic fungi. Microbiol. Review 53: 186-209.

Maresca, B, Lindquist, S, Eds. (1991) Heat Shock, Springer-Verlag, New York.

Maresca, B, Patriarca, EJ, Goldenberg, C, Sacco, M (1988) Heat shock and cold adaptation in Antarctic fishes: a molecular approach. Comp. Biochem. Physiol. 90B: 623-630.

Medoff, G, Maresca, B, Lambowitz, AM, Kobayashi, GS, Painter, A, Sacco, M, Carratù, L (1986) Correlation between pathogenicity and temperature sensitivity in different strains of *Histoplasma capsulatum*. J. Clin. Invest. 78: 1638-1647.

Minchiotti, G, Gargano, S, Maresca, B (1991) The intron-containing hsp82 gene of the dimorphic pathogenic fungus Histoplasma capsulatum is properly spliced in severe heat shock conditions. Mol. Cell. Biol. 11: 5624-5630.

Minchiotti, G, Gargano, S, Maresca, B (1992) Cloning and expression *hsp82* gene of the dimorphic pathogenic fungus *H.capsulatum*. Biochem. Biophys. Acta 1131: 103-107.

Murata, N, Ishizaki-Nishizawa, O, Higashi, S, Hayashi, H, Tasaka, Y, Nishida, I (1992) Genetically engineered alteration in the chilling sensitivity of plants. Nature 356: 710-713

Newport, G, Culpepper, J, Agabian ,N (1988) Parasite Heat-shock Proteins. Parasitol. Today 4: 306-312.

Patriarca, EJ, Kobayashi, GS, Maresca, (1992) Mitochondrial activity and heat shocck response during morphogenesis in the dimorphic pathogenic fungus *Histoplasma capsulatum*. Biochem. Cell Biol. 70: 207-214.

Patriarca, EJ, Maresca, B (1990) Acquired thermotolerance following heat shock protein synthesis prevents impairment of mitochondrial ATPase activity at elevated temperatures in *Saccharomyces cerevisiae*. Exp. Cell Res. 190: 57-64.

Polla, BS (1991) Combined March issues of Parasitol. Today and Immunol. Today A38-A41.

Quinn, PJ, Joo, F, Vigh, L (1989) The role of unsaturated lipids in membrane structure and stability. Prog. Biophys Mol. Biol. 53: 71-103.

Sanchez, Y, Lindquist, S (1990) HSP104 required for induced thermotolerance. Science 248: 1112-1115.

Sanchez, Y, Taulien, J, Borkovic, KA, Lindquist, S (1992) Hsp104 is required for tolerance to many forms of stress. EMBO J. 11: 2357-2364.

Sorger, PK (1991) Heat shock factor and the heat shock response. Cell 65: 363-366.

Sorger, PK, Lewis, MJ, Pelham, HRB (1987) Heat shock factor is regulated differently in yeast and HeLa cells. Nature 329: 81-84.

Sorger, PK, Pelham, HRB (1987) Purification and characterization of a heat-shock element binding protein from yeast. EMBO J. 6: 3035-3041.

Sorger, PK, Pelham, HRB (1988) Yeast heat shock factor is an essential DNA-binding protein that exhibits temperature-dependent phosporylation. Cell 54: 855-864.

Trent, JD, Osipiuk, J, Pinkau, T (1990) Acquired thermotolerance and heat shock in the extremely thermophilic archeabacterium *Sulfolobus sp.* strain B12. J. Bacteriol. 172:1478-1484.

Wada, H, Gombos, Z, Murata, N (1990) Enhancement of chilling tolerance of a cyanobacterium by genetic manipulation of fatty acid desaturation. Nature 347: 200-203.

Wiederrecht, G, Shuey, DJ, Kibbe, WA, Parker, CS (1987) The *Saccharomyces* and *Drosophila* heat shock transcription factors are identical in size and DNA binding properties. Cell 48: 507-515.

Wu, C, Wilson, S, Walker, B, Dawid, I, Paisley, T, Zimarino, V, Ueda H (1987) Purification and properties of *Drosophila* heat shock activator protein. Science 238: 1247-1253.

Wu, C, Zimarino, V, Tsai, C, Walker, B, Wilson, S (1990) Trancriptional regulation of heat shock genes. In Stress Proteins in Biology and Medicine (Morimoto, R.I., Tissieres, A. and Georgopoulos, C., Eds.) pp. 429-442, Cold Spring Harbor Press.

Yost, HJ, Lindquist, S (1988) Translation of unspliced transcripts after heat shock. Science 242: 1544-1548.

Young, D, Garbe, T, Lathigra, R (1991) Heat shock proteins and mycobacterial infection.In *Heat Shock* (Maresca, B. and Lindquist, S., Eds) pp. 203-208, Springer Verlag, New York.

Subject index

NATO ASI Series H

NATO ASI Series H

Vol. 22: **Neural Development and Regeneration. Cellular and Molecular Aspects.** Edited by A. Gorio, J. R. Perez-Polo, J. de Vellis, and B. Haber. 711 pages. 1988.

Vol. 23: **The Semiotics of Cellular Communication in the Immune System.** Edited by E.E. Sercarz, F. Celada, N.A. Mitchison, and T. Tada. 326 pages. 1988.

Vol. 24: **Bacteria, Complement and the Phagocytic Cell.** Edited by F. C. Cabello und C. Pruzzo. 372 pages. 1988.

Vol. 25: **Nicotinic Acetylcholine Receptors in the Nervous System.** Edited by F. Clementi, C. Gotti, and E. Sher. 424 pages. 1988.

Vol. 26: **Cell to Cell Signals in Mammalian Development.** Edited by S.W de Laat, J.G. Bluemink, and C.L. Mummery. 322 pages. 1989.

Vol. 27: **Phytotoxins and Plant Pathogenesis.** Edited by A. Graniti, R. D. Durbin, and A. Ballio. 508 pages. 1989.

Vol. 28: **Vascular Wilt Diseases of Plants. Basic Studies and Control.** Edited by E. C. Tjamos and C. H. Beckman. 590 pages. 1989.

Vol. 29: **Receptors, Membrane Transport and Signal Transduction.** Edited by A. E. Evangelopoulos, J. P. Changeux, L. Packer, T. G. Sotiroudis, and K.W.A. Wirtz. 387 pages. 1989.

Vol. 30: **Effects of Mineral Dusts on Cells.** Edited by B.T. Mossman and R.O. Begin. 470 pages. 1989.

Vol. 31: **Neurobiology of the Inner Retina.** Edited by R. Weiler and N.N. Osborne. 529 pages. 1989.

Vol. 32: **Molecular Biology of Neuroreceptors and Ion Channels.** Edited by A. Maelicke. 675 pages. 1989.

Vol. 33: **Regulatory Mechanisms of Neuron to Vessel Communication in Brain.** Edited by F. Battaini, S. Govoni, M.S. Magnoni, and M. Trabucchi. 416 pages. 1989.

Vol. 34: **VectorsasTools for the Study of Normal and Abnormal Growth and Differentiation.** Edited by H. Lother, R. Dernick, and W. Ostertag. 477 pages. 1989.

Vol. 35: **Cell Separation in Plants: Physiology, Biochemistry and Molecular Biology.** Edited by D. J. Osborne and M. B. Jackson. 449 pages. 1989.

Vol. 36: **Signal Molecules in Plants and Plant-Microbe Interactions.** Edited by B.J.J. Lugtenberg. 425 pages. 1989.

Vol. 37: **Tin-Based Antitumour Drugs.** Edited by M. Gielen. 226 pages. 1990.

Vol. 38: **The Molecular Biology of Autoimmune Disease.** Edited by A.G. Demaine, J-P. Banga, and A.M. McGregor. 404 pages. 1990.

Vol. 39: **Chemosensory Information Processing.** Edited by D. Schild. 403 pages. 1990.

Vol. 40: **Dynamics and Biogenesis of Membranes.** Edited by J. A. F. Op den Kamp. 367 pages. 1990.

Vol. 41: **Recognition and Response in Plant-Virus Interactions.** Edited by R. S. S. Fraser. 467 pages. 1990.

NATO ASI Series H

Vol. 42: **Biomechanics of Active Movement and Deformation of Cells.**
Edited by N. Akkas. 524 pages. 1990.

Vol. 43: **Cellularand MolecularBiologyof Myelination.**
Edited by G. Jeserich, H. H. Althaus, and T. V. Waehneldt. 565 pages. 1990.

Vol. 44: **Activation and Desensitization of Transducing Pathways.**
Edited by T. M. Konijn, M. D. Houslay, and P. J. M. Van Haastert. 336 pages. 1990.

Vol. 45: **Mechanism of Fertilization: Plants to Humans.**
Edited by B. Dale. 710 pages. 1990.

Vol .46: **Parallels in Cell to Cell Junctions in Plants and Animals.**
Edited by A. W Robards, W. J . Lucas, J . D. Pitts, H . J . Jongsma, and D. C. Spray. 296 pages. 1990.

Vol. 47: **Signal Perception andTransduction in Higher Plants.**
Edited by R. Ranjeva and A. M. Boudet. 357 pages. 1990.

Vol. 48: **Calcium Transport and Intracellular Calcium Homeostasis.**
Edited by D. Pansu and F. Bronner. 456 pages. 1990.

Vol. 49: **Post-Transcriptional Control of Gene Expression.**
Edited by J. E. G. McCarthy and M. F. Tuite. 671 pages. 1990.

Vol. 50: **Phytochrome Properties and Biological Action.**
Edited by B. Thomas and C. B. Johnson. 337 pages. 1991.

Vol. 51: **Cell to Cell Signals in Plants and Animals.**
Edited by V. Neuhoff and J. Friend. 404 pages. 1991.

Vol. 52: **Biological Signal Transduction.**
Edited by E. M . Ross and K . W. A. Wirtz. 560 pages. 1991.

Vol. 53: **Fungal Cell Wall and Immune Response.**
Edited by J. P. Latge and D. Boucias. 472 pages. 1991.

Vol. 54: **The Early Effects of Radiation on DNA.**
Edited by E. M. Fielden and P. O'Neill. 448 pages. 1991.

Vol. 55: **The Translational Apparatus of Photosynthetic Organelles.**
Edited by R. Mache, E. Stutz, and A. R. Subramanian. 260 pages. 1991.

Vol. 56: **Cellular Regulation by Prote in Phosphorylation.**
Edited by L. M. G. Heilmeyer, Jr. 520 pages. 1991.

Vol. 57: **MolecularTechniques in Taxonomy.**
Edited by G . M . Hewitt, A. W. B. Johnston, and J. P. W. Young . 420 pages. 1991.

Vol. 58: **Neurocytochemical Methods.**
Edited by A. Calas and D. Eugene. 352 pages. 1991.

Vol. 59: **Molecular Evolution of the Major Histocompatibility Complex.**
Edited by J. Klein and D. Klein. 522 pages. 1991.

Vol. 60: **Intracellular Regulation of Ion Channels.**
Edited by M. Morad and Z. Agus. 261 pages. 1992.

Vol. 61: **Prader-Willi Syndrome and Other Chromosome 15q Deletion Disorders.**
Edited by S. B. Cassidy. 277 pages. 1992.

NATO ASI Series H

Vol. 62: **Endocytosis. From Cell Biology to Health, Disease and Therapie.**
Edited by P. J. Courtoy. 547 pages. 1992.

Vol. 63: **Dynamics of Membrane Assembly.**
Edited by J. A. F. Op den Kamp. 402 pages. 1992.

Vol. 64: **Mechanics of Swelling. From Clays to Living Cells and Tissues.**
Edited by T. K. Karalis. 802 pages. 1992.

Vol. 65: **Bacteriocins, Microcins and Lantibiotics.**
Edited by R. James, C. Lazdunski, and F. Pattus. 530 pages. 1992.

Vol. 66: **Theoretical and Experimental Insights into Immunology.**
Edited by A. S. Perelson and G. Weisbuch. 497 pages. 1992.

Vol. 67: **Flow Cytometry. New Developments.**
Edited by A. Jacquemin-Sablon. 1993.

Vol. 68: **Biomarkers. Research and Application in the Assessment of Environmental Health.**
Edited by D. B. Peakall and L. R. Shugart. 138 pages. 1993.

Vol. 69: **Molecular Biology and its Application to Medical Mycology.** Edited by B. Maresca, G. S. Kobayashi, and H. Yamaguchi. 271 pages. 1993.

MIX
Papier aus verantwortungsvollen Quellen
Paper from responsible sources
FSC® C105338

FSC
www.fsc.org

If you have any concerns about our products,
you can contact us on
ProductSafety@springernature.com

In case Publisher is established outside the EU,
the EU authorized representative is:
Springer Nature Customer Service Center GmbH
Europaplatz 3, 69115 Heidelberg, Germany

Printed by Libri Plureos GmbH
in Hamburg, Germany